Graduate Texts in Mathematics 286

Graduate Texts in Mathematics

Graduate Texts in Mathematics bridge the gap between passive study and creative understanding, offering graduate-level introductions to advanced topics in mathematics. The volumes are carefully written as teaching aids and highlight characteristic features of the theory. Although these books are frequently used as textbooks in graduate courses, they are also suitable for individual study.

More information about this series at http://www.springer.com/series/136

Daniel Hug • Wolfgang Weil

Lectures on Convex Geometry

 Springer

Daniel Hug
Institute of Stochastics
Karlsruhe Institute of Technology (KIT)
Karlsruhe, Germany

Wolfgang Weil
Department of Mathematics
Karlsruhe Institute of Technology (KIT)
Karlsruhe, Germany

ISSN 0072-5285   ISSN 2197-5612   (electronic)
Graduate Texts in Mathematics
ISBN 978-3-030-50182-2   ISBN 978-3-030-50180-8   (eBook)
https://doi.org/10.1007/978-3-030-50180-8

This Springer imprint is published by the registered company Springer Nature Switzerland AG.
The registered company address is: Gewerbestrasse 11, 6330 Cham, Switzerland

To the memory of Wolfgang Weil
To Rolf Schneider on his 80th birthday

*I am very grateful for many years of
friendship and close scientific interaction.*

Freiburg and Karlsruhe, 17 March 2020
Daniel Hug

Preface

Convexity is an elementary property of sets and functions. A subset A of an affine space is convex if it contains all the segments joining any two points of A. In other words, A is convex if for all $x, y \in A$ and $\alpha \in [0, 1]$ the point $\alpha x + (1 - \alpha)y$ also lies in A. This simple algebraic property has surprisingly many and far-reaching consequences of a geometric nature, but it also has basic topological consequences as well as deep analytical implications. The notion of convexity can be transferred from sets to real-valued functions via their epigraphs. The epigraph of a real-valued function f on $\mathbb{R}^n$ is the set of all points in $\mathbb{R}^{n+1}$ lying on or above its graph. Then a function f is convex if the epigraph of f is a convex subset of $\mathbb{R}^{n+1}$. On the other hand, a convex set can be naturally described by various convex functions such as its convex indicator function, its distance function or its support function. The interplay between convex sets and functions turns out to be particularly fruitful and goes well beyond a formal correspondence. The results on convex sets and functions play a central role in many mathematical fields, in particular in functional analysis, optimization theory, probability theory and stochastic geometry.

In this book, we concentrate on convex sets in $\mathbb{R}^n$, which is the prototype of a finite-dimensional real affine space. In infinite-dimensional spaces, other methods often have to be used and different types of problems occur. In the first part of the book, we focus on the classical aspects of convexity. Starting with convex sets and their basic algebraic, combinatorial, metric and extremal properties in Chap. 1, we consider convex functions in Chap. 2. In particular, regularity properties of convex functions and support functions of convex sets are introduced and studied to some extent.

We then provide an introduction to the Brunn–Minkowski theory in Chap. 3. We set the foundations by describing the space of convex bodies (compact convex sets). Our main goal in this chapter is to introduce and study important functionals on the space of convex bodies. First, we provide an elementary approach to the volume and surface area of convex bodies via approximation with polytopes. Then we introduce the mixed volume as a multilinear functional on n-tuples of convex bodies and establish its main properties. As special cases of mixed volumes, we obtain the Minkowski functionals, which are also known as intrinsic volumes or quermassintegrals. A major source of deep relationships between various functionals on convex bodies is the famous Brunn–Minkowski inequality, which provides a lower bound for the volume of the sum of two sets in terms of the volumes of the individual sets. Among the consequences of the Brunn–Minkowski inequality are Minkowski's inequalities for special mixed volumes and, in particular, the isoperimetric inequality. Finally, we present a proof of the famous Alexandrov–Fenchel inequality for mixed volumes. Along the way, we derive general versions of the Brunn–Minkowski inequality and of Minkowski's inequality for mixed volumes. Since symmetrization methods are important throughout mathematics, we briefly discuss Steiner symmetrization and demonstrate how it can be used in the proof of geometric inequalities.

In Chap. 4, we treat further central topics in the Brunn–Minkowski theory that have proved to be highly relevant, including surface area measures, projection functions and zonoids (or projection bodies). Furthermore, we take the opportunity to provide a short introduction to geometric valuation theory, which deals with finitely additive functionals on the space of convex bodies. Finally, in Chap. 5 we introduce invariant measures on topological groups and homogeneous spaces, tailored to the present purpose of deriving basic formulas of integral geometry in Euclidean space. Geometric valuation theory offers a convenient route to a variety of such formulas.

The first two chapters of this book should be easily accessible for an interested reader. Apart from a sound knowledge of linear algebra and elementary real analysis (in Chap. 2), no deeper technical prerequisites from other fields of mathematics are initially required. Later, we shall occasionally use results from functional analysis. In some parts, we require basic familiarity with notions of set-theoretic topology and measure theory.

Although convex geometry is a classical topic, there have been many fascinating developments in the field in recent years. This is particularly true for the analytic aspects of convexity and for its various connections to probability theory, geometric functional analysis, geometric tomography and other branches of mathematics. Some excellent monographs (see, e.g., [4, 22, 38, 41, 55, 56, 81]) describe recent developments in convex geometry or convex geometric analysis at the frontier of current research and its applications (see, e.g., [35, 84]). In contrast to these volumes, the purpose of the present text is to provide a reasonably self-contained introduction to the topic which paves the way to the more advanced and specialized literature.

This book evolved from courses and seminars which have been given repeatedly by the authors. The numerous exercises and the supplementary material at the end of each section constitute an essential part of this textbook. Some of the exercises are routine or cover a short argument omitted in the text, but there are also more demanding problems, and there is plenty of supplementary information, stated in the form of problems, which is useful for readers willing to delve deeper into the subject. For some of the problems (marked with an ∗), complete solutions are provided in Chap. 6. For the solutions of all other problems, an Instructor's Solution

Manual will be available to faculty who are teaching a course using this textbook as the official textbook.

Karlsruhe, Germany   Daniel Hug
Karlsruhe, Germany   Wolfgang Weil

Contents

List of Symbols

$\langle \cdot , \cdot \rangle$	Scalar product
$\| \cdot \|$	Euclidean norm derived from $\langle \cdot , \cdot \rangle$
B^n	Euclidean unit ball with centre at the origin 0
$\mathbb{S}^{n-1}$	Euclidean unit sphere
lin A	Linear hull of A (smallest linear subspace containing A)
aff A	Affine hull of A (smallest affine subspace containing A)
dim A	Dimension of A ($=$ dimension of aff A)
int A	Interior of A (relative to the ambient space)
relint A	Relative interior of A (interior with respect to aff A)
cl A	Closure of A
bd A	Boundary of A (relative to the ambient space)
relbd A	Relative boundary of A (boundary with respect to aff A as the ambient space)
$[x, y]$	Closed segment connecting the points $x, y \in \mathbb{R}^n$
conv A	Convex hull of a set $A \subset \mathbb{R}^n$
ext A	Set of extreme points of a closed convex set A
exp A	Set of exposed points of a closed convex set A
epi f	Epigraph of a function f
dom f	Effective domain of a function f
$\mathcal{K}^n$	Set of all nonempty compact convex sets (convex bodies) in $\mathbb{R}^n$
$\mathcal{K}^n_0$	Set of convex bodies in $\mathbb{R}^n$ with nonempty interiors
$\mathcal{P}^n$	Set of all convex polytopes in $\mathbb{R}^n$
$\mathcal{P}^n_0$	Set of convex polytopes in $\mathbb{R}^n$ with nonempty interiors
$p(A, \cdot)$	Metric projection map onto a closed convex set A
$h_A = h(A, \cdot)$	Support function of a nonempty convex set A
f^*	Conjugate function of a convex function f
$\|f\|$	Supremum of $|f|$ on its domain
$f'(x, u)$	Directional derivative of a real-valued function f at x in direction u
$\partial f(x)$	Subdifferential of a convex function f at a point x

grad $f(x)$	Gradient of a function f at a point x
$\nabla f(x)$	Alternative notation for grad $f(x)$
$\partial^2 f(x)$	Hessian matrix of a real-valued function f at a point x
$df(x)$	Differential of a function $f : \mathbb{R}^n \to \mathbb{R}^m$ at a point x, considered as a linear map
$V(K)$	Volume of a convex body K
$F(K)$	Surface area of a convex body K
$\overline{B}(K)$	Mean width of a convex body K
$V_j(K)$	jth intrinsic volume of a convex body K
$W_i(K)$	ith quermassintegral of a convex body K
$V(K_1, \ldots, K_n)$	Mixed volume of $K_1, \ldots, K_n \in \mathcal{K}^n$
$S_j(K, \cdot)$	jth area measure of a convex body K
$S(K_1, \ldots, K_{n-1}, \cdot)$	Mixed area measure of $K_1, \ldots, K_{n-1} \in \mathcal{K}^n$
$G(n, k)$	Set of all k-dimensional linear subspaces of $\mathbb{R}^n$ (linear Grassmannian)
$A(n, k)$	Set of all k-dimensional affine subspaces of $\mathbb{R}^n$ (affine Grassmannian)
$O(n)$	Set of all orthogonal transformations (rotations) of $\mathbb{R}^n$
$SO(n)$	Set of proper (orientation preserving) rotations of $\mathbb{R}^n$
$G(n)$	Set of proper rigid motions of $\mathbb{R}^n$
$U^\perp$	Orthogonal complement of $U \in G(n, k)$
$K\vert U$	Orthogonal projection of $K \in \mathcal{K}^n$ to U
λ_n	Lebesgue measure on $\mathbb{R}^n$
λ_F	Lebesgue measure on $F \in A(n, k)$
σ	Spherical Lebesgue measure on $\mathbb{S}^{n-1}$
$\mathcal{H}^s$	s-dimensional Hausdorff measure
κ_n	Lebesgue measure of B^n
ω_n	Spherical Lebesgue measure of $\mathbb{S}^{n-1}$

Preliminaries and Notation

Throughout the book, we work in n-dimensional Euclidean space $\mathbb{R}^n$. Elements of $\mathbb{R}^n$ are denoted by lowercase letters like $x, y, \ldots, a, b, \ldots$, scalars by Greek letters $\alpha, \beta, \ldots$ and (real) functions by $f, g, \ldots$ We identify the vector space structure and the affine structure of $\mathbb{R}^n$, i.e., we do not distinguish between vectors and points. The coordinates of a point $x \in \mathbb{R}^n$ are used only occasionally; therefore, we indicate them as $x = (x^{(1)}, \ldots, x^{(n)})^\top$ or, more traditionally, as $x = (x_1, \ldots, x_n)^\top$. If we consider sequences of points in $\mathbb{R}^n$, then the former notation with an additional lower index is used. We equip $\mathbb{R}^n$ with its usual topology, generated by the standard scalar product

$$\langle x, y \rangle := x_1 y_1 + \cdots + x_n y_n, \quad x, y \in \mathbb{R}^n,$$

and the corresponding Euclidean norm

$$\|x\| := ((x_1)^2 + \cdots + (x_n)^2)^{1/2}, \quad x \in \mathbb{R}^n.$$

If $x \in \mathbb{R}^n \setminus \{0\}$, then $x^\perp := \{z \in \mathbb{R}^n : \langle z, x \rangle = 0\}$ denotes the linear subspace consisting of all vectors which are orthogonal to x. More generally, for $U \subset \mathbb{R}^n$ we define $U^\perp := \{z \in \mathbb{R}^n : \langle z, u \rangle = 0 \text{ for all } u \in U\}$. In particular, if U is a k-dimensional linear subspace of $\mathbb{R}^n$, then $U^\perp$ is an $(n-k)$-dimensional linear subspace, the linear subspace orthogonal to U. By B^n we denote the (Euclidean) unit ball

$$B^n := \{x \in \mathbb{R}^n : \|x\| \leq 1\},$$

and by

$$\mathbb{S}^{n-1} := \{x \in \mathbb{R}^n : \|x\| = 1\}$$

the (Euclidean) unit sphere. We also write $B^n(x, r)$ for the (Euclidean) ball with centre $x \in \mathbb{R}^n$ and radius $r \geq 0$, and we simply write $B^n(r)$ instead of $B^n(0, r)$.

Sometimes we make (explicit) use of the Euclidean metric

$$d(x, y) := \|x - y\|, \quad x, y \in \mathbb{R}^n.$$

Occasionally, it is convenient to write $\frac{x}{\alpha}$ instead of $\frac{1}{\alpha}x$, for $x \in \mathbb{R}^n$ and $\alpha \in \mathbb{R}$.

Convex sets in $\mathbb{R}^1$ are often not very exciting (they are open, closed or half-open, bounded or unbounded intervals), usually results on convex sets are only interesting for $n \geq 2$. Nevertheless, the case $n = 1$, for instance, can be important as the starting point for an induction argument. In some situations, results only make sense if $n \geq 2$, although we shall not emphasize this in all cases. As a rule, $A, B, \ldots$ denote general (convex or nonconvex) sets, $K, L, \ldots$ will be used for compact convex sets (convex bodies), and we prefer to write $P, Q, \ldots$ for (convex) polytopes.

In a vector space, linear subspaces are subsets which are again vector spaces, and, equivalently, they are closed with respect to addition of vectors and scalar multiplication of vectors. In an affine space $\mathbb{A}$ with underlying vector space V, the corresponding substructure is an *affine subspace* (or affine flat). Affine subspaces are the subsets of $\mathbb{A}$ which are obtained by attaching to a point $a \in \mathbb{A}$ the vectors of a linear subspace $U \subset V$, for which we write $F = a + U$. While the point a is not uniquely determined by F, the linear subspace U is determined by F; in fact, for any point $b \in a + U$, we have $a + U = b + U$. The dimension of F is defined as the dimension of U. It is easy to see that intersections of affine subspaces are affine subspaces or the empty set. For a given set $\emptyset \neq A \subset \mathbb{A}$, the *affine hull* of A is the intersection of all affine subspaces of $\mathbb{A}$ which contain A. The affine hull of the empty set is the empty set and taking affine hulls preserves inclusions. Finally, we point out that the notions of linear independence of vectors in a vector space and of *affine independence* of points in an affine space are closely related concepts (see Exercise 1.1.1). As usual, we say that points $x_1, \ldots, x_k \in \mathbb{A}$ are affinely independent if for $\lambda_1, \ldots, \lambda_k \in \mathbb{R}$ with $\lambda_1 + \cdots + \lambda_k = 0$ the condition $\lambda_1 x_1 + \cdots + \lambda_k x_k = 0$ implies that $\lambda_1 = \cdots = \lambda_k = 0$. Here, the condition $\lambda_1 x_1 + \cdots + \lambda_k x_k = 0$ together with $\lambda_1 + \cdots + \lambda_k = 0$ means that $\lambda_2(x_2 - x_1) + \cdots + \lambda_k(x_k - x_1) = 0$ for $k \geq 2$.

A particular (canonical) example of an affine subspace is a vector space considered as an affine space over itself. For points $a, b \in \mathbb{A}$, we write $b - a$ for the vector from a to b (the unique vector which yields b if attached to a). If $a_1, \ldots, a_k \in \mathbb{A}$ and $\lambda_1, \ldots, \lambda_k \in \mathbb{R}$ with $\sum_{i=1}^{k} \lambda_i = 1$, then we write $\lambda_1 a_1 + \cdots + \lambda_k a_k$ for the point in $\mathbb{A}$ such that

$$\sum_{i=1}^{k} \lambda_i a_i = a_1 + \sum_{i=2}^{k} \lambda_i (a_i - a_1).$$

It is easy to check that the role of a_1 can be taken by any of the points $a_1, \ldots, a_k$ without changing the right-hand side. The point $\lambda_1 a_1 + \cdots + \lambda_k a_k$ is called an affine combination of $a_1, \ldots, a_k$. We say that $A \subset \mathbb{A}$ is affine if *affine combinations* of any two points of A are again in A, that is, if $\lambda a + (1 - \lambda)b \in A$ for $a, b \in A$ and $\lambda \in \mathbb{R}$.

A linear map $h : V \to W$ from a vector space V to a vector space W is a map which satisfies $h(\lambda v + \mu w) = \lambda h(v) + \mu h(w)$ for $v, w \in V$ and $\lambda, \mu \in \mathbb{R}$. An *affine map* $f : \mathbb{A} \to \mathbb{B}$ from an affine space $\mathbb{A}$ to an affine space $\mathbb{B}$ is a map which satisfies $f(\lambda a + (1 - \lambda)c) = \lambda f(a) + (1 - \lambda)f(c)$ for $a, c \in \mathbb{A}$ and $\lambda \in \mathbb{R}$.

The following notation will be used throughout this text:

lin A	Linear hull of A (smallest linear subspace containing A)
aff A	Affine hull of A (smallest affine subspace containing A)
dim A	Dimension of A ($=$ dimension of aff A)
int A	Interior of A (relative to the ambient space)
relint A	Relative interior of A (interior with respect to aff A)
cl A	Closure of A
bd A	Boundary of A (relative to the ambient space)
relbd A	Relative boundary of A (boundary with respect to aff A)

If f is a function on $\mathbb{R}^n$ with values in $\mathbb{R}$ or in the extended real line $[-\infty, \infty]$ and if A is a subset of the latter, we frequently abbreviate the set $\{x \in \mathbb{R}^n : f(x) \in A\}$ by $\{f \in A\}$. By a linear form f on $\mathbb{R}^n$ we mean a linear map $f : \mathbb{R}^n \to \mathbb{R}$. Hyperplanes $E \subset \mathbb{R}^n$ can be shortly written as $E = \{f = \alpha\}$, where f is a linear form, $f \neq 0$, and $\alpha \in \mathbb{R}$; note that this representation of E in terms of f and α is not unique. Since hyperplanes are sets of points, writing $E = \{f = \alpha\}$ we mean $E = \{x \in \mathbb{R}^n : f(x - 0) = \alpha\}$, where $x - 0$ is the vector pointing from the origin $0 \in \mathbb{R}^n$ to the point x. In the following, as usual we identify the point x and the position vector $x - 0$ of x with respect to the origin. If $x_0 \in \mathbb{R}^n$ is chosen such that $f(x_0 - 0) = \alpha$, then $E = \{x \in \mathbb{R}^n : f(x - x_0) = 0\}$. The closed half-spaces bounded by E then are $\{f \geq \alpha\}$ and $\{f \leq \alpha\}$, and the open half-spaces bounded by E are $\{f > \alpha\}$ and $\{f < \alpha\}$. Using the scalar product, it is often convenient to write hyperplanes and (closed) supporting half-spaces in the form

$$H(u, t) := \{x \in \mathbb{R}^n : \langle x, u \rangle = t\},$$

$$H^-(u, t) := \{x \in \mathbb{R}^n : \langle x, u \rangle \leq t\},$$

$$H^+(u, t) := \{x \in \mathbb{R}^n : \langle x, u \rangle \geq t\},$$

where $u \in \mathbb{R}^n \setminus \{0\}$ and $t \in \mathbb{R}$.

The Euclidean metric on $\mathbb{R}^n$ induces a topology, which is independent of the specific metric (or norm) that is considered. The smallest σ-algebra containing the open sets is the Borel σ-algebra, which is denoted by $\mathcal{B}(\mathbb{R}^n)$. More generally, for a topological space $(T, \mathcal{T})$ with the system $\mathcal{T}$ of open sets, the induced Borel σ-algebra is denoted by $\mathcal{B}(T)$. In particular, this applies to the unit sphere $\mathbb{S}^{n-1}$, which inherits the topology of $\mathbb{R}^n$. We write λ_n for the Lebesgue measure on $\mathbb{R}^n$. If F is a k-dimensional affine subspace of $\mathbb{R}^n$, we write λ_F for Lebesgue measure on F, with the Euclidean metric induced from the ambient $\mathbb{R}^n$ (the resulting measure is independent of the choice of an origin in F). If the underlying affine subspace F is clear from the context, we simply write λ_k if dim $F = k$. For spherical Lebesgue measure on $\mathbb{S}^{n-1}$, we write σ (without indicating the dimension, which will always be clear from the context). Finally, we write $\mathcal{H}^s$, $s \geq 0$, for the s-dimensional Hausdorff measure on $\mathbb{R}^n$. The normalization of the Hausdorff measures is chosen in such a way that on $\mathcal{B}(\mathbb{R}^n)$, the Lebesgue measure λ_n and the n-dimensional Hausdorff measure $\mathcal{H}^n$ are equal. Moreover, spherical Lebesgue measure and the $(n-1)$-dimensional Hausdorff measure $\mathcal{H}^{n-1}$ coincide on $\mathbb{S}^{n-1}$. Integrating with respect to Lebesgue measure in $\mathbb{R}^n$, we simply write 'dx' instead of '$\lambda_n(dx)$' if this is convenient. Also note that for $A \in \mathcal{B}(\mathbb{S}^{n-1})$, we have

$$\sigma(A) = n\lambda_n(\{tu : t \in [0, 1], u \in A\}).$$

Hence, $\omega_n := \mathcal{H}^{n-1}(\mathbb{S}^{n-1})$ and $\kappa_n := \lambda_n(B^n)$ are related by $\omega_n = n\kappa_n$.

For sets A, B, we write $A \subset B$ if $a \in A$ implies that $a \in B$, in particular, the sets may be equal. The abbreviation w.l.o.g. means 'without loss of generality' and is used sometimes to reduce the argument to a special case. The logical symbols $\forall$ (for all) and $\exists$ (exists) are occasionally (rarely) used in formulas. As usual, $\square$ denotes the end of a proof. Finally, we write $|A|$ for the cardinality of a set A.

Chapter 1
Convex Sets

Convexity is a basic but fundamental notion in mathematics. A subset of $\mathbb{R}^n$ is called convex if for any two of its points, the whole segment connecting these points is contained in the set. This first chapter provides a brief introduction to some of the basic ideas and results related to convex sets. We shall see that convexity is stable under natural operations and transformations. It is remarkable at first sight that convexity is a key concept underlying several classical combinatorial results. Moreover, convex sets have characteristic metric properties related to the notions of support and separation. To derive these properties, we shall use the metric projection onto convex sets. Finally, in this chapter extreme points will be shown to provide an efficient description of convex sets.

Throughout this book we restrict ourselves to a Euclidean framework, although the definition of convexity and some of the results are of a purely affine nature.

1.1   Algebraic Properties

We usually identify an affine space with its associated vector space. However, we shall deliberately use the words "vector" and "point". The definition of a convex set requires just the structure of $\mathbb{R}^n$ as an affine space. In particular, it is related to (but different from) the notions of a linear and an affine subspace.

Definition 1.1   A set $A \subset \mathbb{R}^n$ is *convex* if $(1 - \alpha)x + \alpha y \in A$ for $x, y \in A$ and $\alpha \in [0, 1]$.

We illustrate Definition 1.1 with a couple of examples. The proofs of the corresponding claims are left as easy exercises for the reader.

© Springer Nature Switzerland AG 2020
D. Hug, W. Weil, *Lectures on Convex Geometry*, Graduate Texts
in Mathematics 286, https://doi.org/10.1007/978-3-030-50180-8_1

Example 1.1 Apart from the empty set, the whole space $\mathbb{R}^n$ or one-pointed sets, the simplest convex sets are segments. For $x, y \in \mathbb{R}^n$ we denote by

$$[x, y] := \{(1 - \alpha)x + \alpha y : \alpha \in [0, 1]\}$$

the *closed segment* between x and y. Similarly,

$$(x, y) := \{(1 - \alpha)x + \alpha y : \alpha \in (0, 1)\}$$

is the *open segment*, and we define half-open segments $(x, y]$ and $[x, y)$ in an analogous way. Note that if $x = y$, then all these segments are equal to $\{x\}$. Clearly, a set is convex if and only if with any two points from the set, the segment between these points is also contained in the set. To verify that $[x, y]$ is indeed convex, it is sufficient to observe that for $\alpha, \beta, \tau \in [0, 1]$ we have

$$(1 - \tau)\left[(1 - \alpha)x + \alpha y\right] + \tau\left[(1 - \beta)x + \beta y\right]$$
$$= \left[(1 - \tau)(1 - \alpha) + \tau(1 - \beta)\right] x + \left[(1 - \tau)\alpha + \tau\beta\right] y,$$

where the coefficients on the right-hand side are in $[0, 1]$ and sum up to 1.

Example 1.2 Other trivial examples are the affine flats in $\mathbb{R}^n$, that is, sets of the form $a + U$, where $a \in \mathbb{R}^n$ is a fixed point and $U \subset \mathbb{R}^n$ is a linear subspace.

Example 1.3 Fix a linear form $f \neq 0$ and some $\alpha \in \mathbb{R}$. If $\{f = \alpha\}$ is the representation of a hyperplane, then the *open halfspaces* $\{f < \alpha\}, \{f > \alpha\}$ and the *closed halfspaces* $\{f \leq \alpha\}, \{f \geq \alpha\}$ are convex.

Example 1.4 Further convex sets are the (not necessarily Euclidean) *balls*

$$B(r) := \{x \in \mathbb{R}^n : \|x\|_\circ \leq r\}, \quad r \geq 0,$$

and their translates. Here, $\| \cdot \|_\circ$ can be any norm (not just the Euclidean norm). In fact, if $x, y \in B(r)$ and $\alpha \in [0, 1]$, using the properties of a norm, we obtain

$$\|(1 - \alpha)x + \alpha y\|_\circ \leq (1 - \alpha)\|x\|_\circ + \alpha\|y\|_\circ \leq (1 - \alpha)r + \alpha r = r.$$

Example 1.5 Remove an arbitrary subset from the boundary of a Euclidean ball $B^n(x, r) = \{z \in \mathbb{R}^n : \|z - x\| \leq r\}$ with centre $x \in \mathbb{R}^n$ and radius $r \geq 0$. The resulting set is still convex. Note that this statement is no longer true if the Euclidean norm $\|\cdot\|$ is replaced for instance by the maximum norm (when the norm balls would then be cubes).

Example 1.6 Another convex set as well as a nonconvex set are shown in Fig. 1.1.

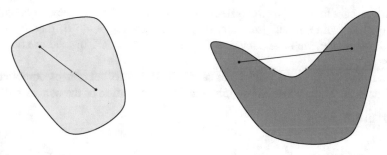

Fig. 1.1 A convex set (left) and a nonconvex set (right)

Definition 1.2 Let $k \in \mathbb{N}$, let $x_1, \ldots, x_k \in \mathbb{R}^n$, and let $\alpha_1, \ldots, \alpha_k \in [0, 1]$ be such that $\alpha_1 + \cdots + \alpha_k = 1$. Then $\alpha_1 x_1 + \cdots + \alpha_k x_k$ is called a *convex combination* of the points $x_1, \ldots, x_k$.

Theorem 1.1 *A set $A \subset \mathbb{R}^n$ is convex if and only if all convex combinations of points in A lie in A.*

Proof First, assume that all convex combinations of points in A lie in A. Taking $k = 2$ in the definition of a convex combination, we see that A is convex.

For the other direction, suppose that A is convex and $k \in \mathbb{N}$. We use induction on k.

For $k = 1$, the assertion is trivially true.

For the step from $k - 1$ to k, $k \geq 2$, assume $x_1, \ldots, x_k \in A$ and $\alpha_1, \ldots, \alpha_k \in [0, 1]$ with $\alpha_1 + \cdots + \alpha_k = 1$. We may assume that $\alpha_k \neq 1$. Then we define

$$\beta_i := \frac{\alpha_i}{1 - \alpha_k}, \quad i = 1, \ldots, k - 1,$$

hence $\beta_i \in [0, 1]$ and $\beta_1 + \cdots + \beta_{k-1} = 1$. By the induction hypothesis, we have $\beta_1 x_1 + \cdots + \beta_{k-1} x_{k-1} \in A$. Since A is convex, we conclude that

$$\sum_{i=1}^{k} \alpha_i x_i = (1 - \alpha_k) \left(\sum_{i=1}^{k-1} \beta_i x_i \right) + \alpha_k x_k \in A,$$

which completes the argument. $\square$

If $\{A_i : i \in I\}$ is an arbitrary family of convex sets (in $\mathbb{R}^n$), then the (possibly empty) intersection $\bigcap_{i \in I} A_i$ is convex. In particular, for a given set $A \subset \mathbb{R}^n$, the intersection of all convex sets containing A is convex.

Definition 1.3 For $A \subset \mathbb{R}^n$, the set

$$\operatorname{conv} A := \bigcap \{ C \subset \mathbb{R}^n : A \subset C, C \text{ is convex} \}$$

is called the *convex hull* of A.

With respect to the inclusion order, conv A is the smallest convex set containing A. It is clear from the definition that if $A, B \subset \mathbb{R}^n$ and $A \subset B$, then conv $A \subset$ conv B. Moreover, we have conv $\emptyset = \emptyset$ and conv $\mathbb{R}^n = \mathbb{R}^n$. Instead of conv A we also write $\text{conv}(A)$.

The following theorem shows that conv A is the set of all convex combinations of points in A. Thus, we obtain an intrinsic characterization of the convex hull.

Theorem 1.2 *For $A \subset \mathbb{R}^n$,*

$$
\text{conv } A = \left\{ \sum_{i=1}^{k} \alpha_i x_i : k \in \mathbb{N}, x_1, \ldots, x_k \in A, \alpha_1, \ldots, \alpha_k \in [0, 1], \sum_{i=1}^{k} \alpha_i = 1 \right\}.
$$

Proof Let B denote the set on the right-hand side. If C is a convex set containing A, then any convex combination of points from A is also a convex combination of points from C. By Theorem 1.1 we conclude that $B \subset C$. Hence, we get $B \subset$ conv A.

On the other hand, the set B is convex, since

$$
\beta(\alpha_1 x_1 + \cdots + \alpha_k x_k) + (1 - \beta)(\gamma_1 y_1 + \cdots + \gamma_m y_m)
$$
$$
= \beta\alpha_1 x_1 + \cdots + \beta\alpha_k x_k + (1 - \beta)\gamma_1 y_1 + \cdots + (1 - \beta)\gamma_m y_m,
$$

for $x_i, y_j \in A$ and coefficients $\beta, \alpha_i, \gamma_j \in [0, 1]$ with $\alpha_1 + \cdots + \alpha_k = 1$ and $\gamma_1 + \cdots + \gamma_m = 1$, and

$$
\beta\alpha_1 + \cdots + \beta\alpha_k + (1 - \beta)\gamma_1 + \cdots + (1 - \beta)\gamma_m = \beta + (1 - \beta) = 1.
$$

Since B contains A, we also get conv $A \subset B$. $\square$

Remark 1.1 Trivially, A is convex if and only if $A = \text{conv } A$.

Remark 1.2 Later, in Sect. 1.2, we shall give an improved version of Theorem 1.2 (Carathéodory's theorem), where the number k of points used in the representation of conv A is bounded by $n + 1$.

Remark 1.3 For $x_1, \ldots, x_k \in \mathbb{R}^n$, we have

$$
\text{conv}\{x_1, \ldots, x_k\} = \left\{ \sum_{i=1}^{k} \alpha_i x_i : \alpha_1, \ldots, \alpha_k \in [0, 1], \sum_{i=1}^{k} \alpha_i = 1 \right\}.
$$

Definition 1.4 For sets $A, B \subset \mathbb{R}^n$ and $\alpha, \beta \in \mathbb{R}$, we put

$$
\alpha A + \beta B := \{\alpha x + \beta y : x \in A, y \in B\}.
$$

The set $\alpha A + \beta B$ is called a *linear combination* or *Minkowski combination* of the sets A, B, the operation $+$ is called *vector addition* or *Minkowski addition*.

Special cases get special names:

$A + B$	the *sum set*
$A + x$ (the case $B = \{x\}$)	a *translate* of A by the vector $x \in \mathbb{R}^n$
αA	a *multiple* of A for $\alpha \in \mathbb{R}$
$\alpha A + x$ (for $\alpha \geq 0$)	a *homothetic image* of A
$-A := (-1)A$	the *reflection* of A (in the origin)
$A - B := A + (-B)$	the *difference* of A and B

Example 1.7 To get an idea of what the sum $A + B$ of two sets $A, B \subset \mathbb{R}^n$ looks like, it may be helpful to observe that

$$A + B = \bigcup_{a \in A}(a + B) = \bigcup_{b \in B}(A + b).$$

Specifically,

- the Minkowski sum of a (regular) triangle and its reflection in the origin is a (regular) hexagon (see Fig. 1.2);
- the boundary of a unit square + the boundary of a unit square is a square with edge length 2. This shows that the Minkowski sum of two nonconvex sets can be a convex set.

Remark 1.4 If $A, B \subset \mathbb{R}^n$ are convex and $\alpha, \beta \in \mathbb{R}$, then $\alpha A + \beta B$ is convex. To see this, let $a_1, a_2 \in A$, $b_1, b_2 \in B$, $\gamma \in [0, 1]$. Then

$$(1 - \gamma)[\alpha a_1 + \beta b_1] + \gamma[\alpha a_2 + \beta b_2]$$
$$= \alpha[(1 - \gamma)a_1 + \gamma a_2] + \beta[(1 - \gamma)b_1 + \gamma b_2] \in \alpha A + \beta B,$$

which yields the assertion.

Remark 1.5 In general, we have $A + A \neq 2A$ and $A - A \neq \{0\}$. However, for a convex set A and $\alpha, \beta \geq 0$, we have $\alpha A + \beta A = (\alpha + \beta)A$. The latter property characterizes convexity of a set A. In fact, the inclusion $(\alpha + \beta)A \subset \alpha A + \beta A$ is

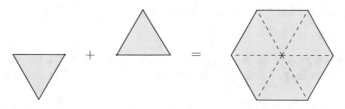

Fig. 1.2 The Minkowski sum $\triangledown + \triangle = (-\triangle) + \triangle$ is a hexagon ○

always true. For the converse, for $a_1, a_2 \in A$ and $\alpha + \beta > 0$ we use that

$$\alpha a_1 + \beta a_2 = (\alpha + \beta) \left[\frac{\alpha}{\alpha + \beta} a_1 + \frac{\beta}{\alpha + \beta} a_2 \right] \in (\alpha + \beta) A,$$

if A is convex. Conversely, if $(1 - \lambda) A + \lambda A = A$ for $\lambda \in [0, 1]$, it follows that A is convex.

We next show that affine transformations preserve convexity.

Theorem 1.3 *Let $A \subset \mathbb{R}^n$ and $B \subset \mathbb{R}^m$ be convex, and let $f : \mathbb{R}^n \to \mathbb{R}^m$ be an affine map. Then*

$$f(A) := \{ f(x) : x \in A \}$$

and

$$f^{-1}(B) := \{ x \in \mathbb{R}^n : f(x) \in B \}$$

are convex.

Proof We show that $f(A)$ is convex. For this, let $y_1, y_2 \in f(A)$ and $\lambda \in [0, 1]$. Then there are $a_1, a_2 \in A$ such that $y_i = f(a_i)$ for $i = 1, 2$ and

$$(1 - \lambda) y_1 + \lambda y_2 = (1 - \lambda) f(a_1) + \lambda f(a_2) = f((1 - \lambda) a_1 + \lambda a_2) \in f(A),$$

since f is affine and the convexity of A yields $(1 - \lambda) a_1 + \lambda a_2 \in A$.

To show that $f^{-1}(B)$ is convex, let $x_1, x_2 \in f^{-1}(B)$ and $\lambda \in [0, 1]$. Then $f(x_1), f(x_2) \in B$. Since f is affine and B is convex, we get

$$f((1 - \lambda) x_1 + \lambda x_2) = (1 - \lambda) f(x_1) + \lambda f(x_2) \in B,$$

which shows that $(1 - \lambda) x_1 + \lambda x_2 \in f^{-1}(B)$. $\square$

Corollary 1.1 *The projection of a convex set onto an affine subspace is convex.*

The converse is obviously false, since (e.g.) a shell bounded by two concentric balls is not convex but has convex projections.

Definition 1.5

(a) The intersection of finitely many closed halfspaces is called a *polyhedral set*.
(b) The convex hull of finitely many points $x_1, \ldots, x_k \in \mathbb{R}^n$ is called a (convex) *polytope*.
(c) The convex hull of affinely independent points is called a *simplex*. An *r-simplex* is the convex hull of $r + 1$ affinely independent points.

Note that the empty set and the whole space are polyhedral sets. While polytopes are always compact sets, polyhedral sets are closed but may be unbounded.

Intuitively speaking, the vertices of a polytope P form a minimal subset of points of P the convex hull of which is equal to P. A vertex of a polytope P can also be characterized as a point x of P for which $P \setminus \{x\}$ is still convex. We take the latter property as a definition of a vertex.

Definition 1.6 A point x of a polytope P is called a *vertex* of P if $P \setminus \{x\}$ is convex. The set of all vertices of P is denoted by vert P.

Theorem 1.4 *Let P be a polytope in $\mathbb{R}^n$, and let $x_1, \ldots, x_k \in \mathbb{R}^n$ be distinct points.*

(a) *Suppose that $P = \operatorname{conv}\{x_1, \ldots, x_k\}$. Then x_1 is a vertex of P if and only if $x_1 \notin \operatorname{conv}\{x_2, \ldots, x_k\}$.*
(b) *P is the convex hull of its vertices.*

Proof (a) Let x_1 be a vertex of P. Then $P \setminus \{x_1\}$ is convex and $x_1 \notin P \setminus \{x_1\}$. Since $\{x_2, \ldots, x_k\} \subset P \setminus \{x_1\}$, we get $\operatorname{conv}\{x_2, \ldots, x_k\} \subset P \setminus \{x_1\}$, and thus $x_1 \notin \operatorname{conv}\{x_2, \ldots, x_k\}$.

If x_1 is not a vertex of P, then there exist distinct points $a, b \in P \setminus \{x_1\}$ and $\lambda \in (0, 1)$ such that $x_1 = (1 - \lambda)a + \lambda b$. Hence there exist $\mu_1, \ldots, \mu_k \in [0, 1]$ and $\tau_1, \ldots, \tau_k \in [0, 1]$ with $\mu_1 + \cdots + \mu_k = 1$ and $\tau_1 + \cdots + \tau_k = 1$ such that $\mu_1, \tau_1 \neq 1$ and

$$a = \sum_{i=1}^{k} \mu_i x_i, \qquad b = \sum_{i=1}^{k} \tau_i x_i.$$

Thus we get

$$x_1 = \sum_{i=1}^{k} ((1 - \lambda)\mu_i + \lambda\tau_i) x_i,$$

from which it follows that

$$x_1 = \sum_{i=2}^{k} \frac{(1 - \lambda)\mu_i + \lambda\tau_i}{1 - (1 - \lambda)\mu_1 - \lambda\tau_1} x_i, \tag{1.1}$$

where $(1 - \lambda)\mu_1 + \lambda\tau_1 \neq 1$. The right-hand side of (1.1) is a convex combination of $x_2, \ldots, x_k$, since

$$\sum_{i=2}^{k} \frac{(1 - \lambda)\mu_i + \lambda\tau_i}{1 - (1 - \lambda)\mu_1 - \lambda\tau_1} = \frac{(1 - \lambda)(1 - \mu_1) + \lambda(1 - \tau_1)}{1 - (1 - \lambda)\mu_1 - \lambda\tau_1} = 1,$$

hence $x_1 \in \operatorname{conv}\{x_2, \ldots, x_k\}$.

(b) Using (a), we can successively remove points from $\{x_1, \ldots, x_k\}$ which are not vertices without changing the convex hull (as follows from Exercise 1.1.6). Moreover, if $x \in \text{vert}(P) \setminus \{x_1, \ldots, x_k\}$, then $P = \text{conv}\{x, x_1, \ldots, x_k\}$ and hence $x \notin \text{conv}\{x_1, \ldots, x_k\} = P$ by (a), a contradiction.   □

Remark 1.6 A polyhedral set is closed and convex. Polytopes, as convex hulls of finite sets, are closed and bounded, hence compact.

Remark 1.7 For a polytope P, Theorem 1.4 shows that $P = \text{conv}(\text{vert } P)$. This is a special case of Minkowski's theorem, which is proved in Sect. 1.5.

Remark 1.8 Polyhedral sets and polytopes are somehow dual notions. We shall see later in Sect. 1.4 that the set of polytopes coincides with the set of bounded polyhedral sets.

Remark 1.9 The polytope property is preserved by basic operations for sets. In particular, if P, Q are polytopes in $\mathbb{R}^n$, then the following sets are polytopes as well:

(i) $\text{conv}(P \cup Q)$,

(ii) $P \cap Q$,

(iii) $\alpha P + \beta Q$ for $\alpha, \beta \in \mathbb{R}$,

(iv) $f(P)$ for an affine map $f : \mathbb{R}^n \to \mathbb{R}^m$.

The assertions (i), (iii), (iv) are the subject of Exercises 1.1.7, 1.1.8 and 1.1.9. The assertion (ii) is not as straightforward. The proof that $P \cap Q$ is a polytope will follow later quite easily from the mentioned connection between polytopes and bounded polyhedral sets.

Remark 1.10 A more general form of Theorem 1.4 (with almost the same proof) is the subject of Exercise 1.1.20.

Remark 1.11 If P is the convex hull of affinely independent points $x_0, \ldots, x_r$, then P is an r-simplex and each of the points x_i is a vertex of P. An r-simplex P has dimension $\dim P = r$.

Simplices are characterized by the property that their points are unique convex combinations of the vertices.

Theorem 1.5 *A convex set $A \subset \mathbb{R}^n$ is a simplex if and only if there exist $x_0, \ldots, x_k \in A$ such that each $x \in A$ has a unique representation as a convex combination of $x_0, \ldots, x_k$.*

Proof Let A be a k-simplex. Then $A = \text{conv}\{x_0, \ldots, x_k\}$ with affinely independent $x_0, \ldots, x_k \in \mathbb{R}^n$. The existence of a representation as requested follows from Theorem 1.2. The uniqueness assertion is implied by Exercise 1.1.1, (a) $\Rightarrow$ (e).

For the converse, we again use Theorem 1.2, but now Exercise 1.1.3.   □

Exercises and Supplements for Sect. 1.1

1. Points $x_1, \ldots, x_m \in \mathbb{R}^n$ are called *affinely independent* if for $\lambda_1, \ldots, \lambda_m \in \mathbb{R}$ with $\sum_{i=1}^m \lambda_i = 0$ it follows from $\sum_{i=1}^m \lambda_i x_i = 0$ that $\lambda_1 = \cdots - \lambda_m = 0$. For $x \in \mathbb{R}^n$, we put $\tau(x) := (x, 1) \in \mathbb{R}^{n+1}$. Then the following conditions are equivalent for given points $x_1, \ldots, x_m \in \mathbb{R}^n$:

   (a) $x_1, \ldots, x_m$ are affinely independent.
   (b) For $m \geq 2$: $x_2 - x_1, \ldots, x_m - x_1$ are linearly independent.
   (c) $\tau(x_1), \ldots, \tau(x_m)$ are linearly independent.
   (d) Whenever $\lambda_1, \ldots, \lambda_m, \mu_1, \ldots, \mu_m \in \mathbb{R}$ are such that $\sum_{i=1}^m \lambda_i = \sum_{i=1}^m \mu_i$ and $\sum_{i=1}^m \lambda_i x_i = \sum_{i=1}^m \mu_i x_i$, then $\lambda_1 = \mu_1, \ldots, \lambda_m = \mu_m$.
   (e) Whenever $\lambda_1, \ldots, \lambda_m, \mu_1, \ldots, \mu_m \geq 0$ are such that $\sum_{i=1}^m \lambda_i = \sum_{i=1}^m \mu_i$ and $\sum_{i=1}^m \lambda_i x_i = \sum_{i=1}^m \mu_i x_i$, then $\lambda_1 = \mu_1, \ldots, \lambda_m = \mu_m$.

2. Show that a set $\emptyset \neq A \subset \mathbb{R}^n$ is affine if and only if it is an affine subspace. State and prove analogues of Theorems 1.1, 1.2 and 1.3 for affine combinations, affine hulls and images/preimages of affine sets under affine transformations.

3.* Assume that $x_1, \ldots, x_k \in \mathbb{R}^n$ are such that each $x \in \text{conv}\{x_1, \ldots, x_k\}$ is a unique convex combination of $x_1, \ldots, x_k$. Show that $x_1, \ldots, x_k$ are affinely independent.

4. (a) Show that $A \subset \mathbb{R}^n$ is convex if and only if $\alpha A + \beta A = (\alpha + \beta)A$ holds for $\alpha, \beta \geq 0$.
   (b) Which nonempty sets $A \subset \mathbb{R}^n$ are characterized by $\alpha A + \beta A = (\alpha + \beta)A$, for all $\alpha, \beta \in \mathbb{R}$?

5. Let $A \subset \mathbb{R}^n$ be closed. Show that A is convex if and only if $A + A = 2A$ holds.

6. Let $A, B \subset \mathbb{R}^n$ and assume that $A \subset \text{conv}(B)$. Show that $\text{conv}(A \cup B) = \text{conv}(B)$.

7.* Let $A, B \subset \mathbb{R}^n$. Show that $\text{conv}(\text{conv}(A) \cup \text{conv}(B)) = \text{conv}(A \cup B)$.

8.* Let $A, B \subset \mathbb{R}^n$. Show that $\text{conv}(A + B) = \text{conv}(A) + \text{conv}(B)$.

9. Let $A \subset \mathbb{R}^n$ and let $f : \mathbb{R}^n \to \mathbb{R}^m$ be an affine map. Show that $\text{conv}(f(A)) = f(\text{conv}(A))$.

10. Let $A, B \subset \mathbb{R}^n$ be nonempty convex sets, and let $x \in \mathbb{R}^n$. Show that

    (a) $\text{conv}(\{x\} \cup A) = \{(1 - \lambda)x + \lambda a : \lambda \in [0, 1], a \in A\}$.
    (b) If $A \cap B = \emptyset$, then

$$\text{conv}(\{x\} \cup A) \cap B = \emptyset \qquad \text{or} \qquad \text{conv}(\{x\} \cup B) \cap A = \emptyset.$$

11. Let $K, L \subset \mathbb{R}^n$ be nonempty, closed convex sets. Assume that $K \cup L$ is convex. Then $K \cap L \neq \emptyset$ and $(K \cap L) + (K \cup L) = K + L$.

12. For a set $A \subset \mathbb{R}^n$ let

$$\ker A := \{x \in A : [x, y] \subset A \text{ for } y \in A\}$$

be the *kernel* of A. Show that ker A is convex. Show by an example that $A \subset B$ does not imply that ker $A \subset$ ker B.

13.* A set

$$R := \{x + \alpha y : \alpha \geq 0\}, \quad x \in \mathbb{R}^n, y \in \mathbb{S}^{n-1},$$

is called a *ray* (starting at x with direction y).

Let $A \subset \mathbb{R}^n$ be convex and unbounded. Show that A contains a ray.

Hint: Start with the case of a closed set A. For the general case, Theorem 1.10 is useful.

14. For a set $A \subset \mathbb{R}^n$, the *polar* A° is defined as

$$A^\circ := \{x \in \mathbb{R}^n : \langle x, y \rangle \leq 1 \text{ for } y \in A\}.$$

Prove the following assertions.

(a) A° is closed, convex and contains 0.
(b) If $A \subset B$, then $A^\circ \supset B^\circ$.
(c) $(A \cup B)^\circ = A^\circ \cap B^\circ$.
(d) If P is a polytope, then P° is a polyhedral set.
(e) If $r > 0$, then $(r \cdot A)^\circ = r^{-1} \cdot A^\circ$.
(f) If $r > 0$, then $B^n(0, r)^\circ = B^n(0, 1/r)$.

15. (a) If $\| \cdot \|' : \mathbb{R}^n \to [0, \infty)$ is a norm, show that the corresponding unit ball $B' := \{x \in \mathbb{R}^n : \|x\|' \leq 1\}$ is convex and symmetric (that is, $B' = -B'$).

(b) Show that

$$\| \cdot \|_1 : \mathbb{R}^n \to [0, \infty), \quad x = (x_1, \ldots, x_n) \mapsto \sum_{i=1}^n |x_i|,$$

and

$$\| \cdot \|_\infty : \mathbb{R}^n \to [0, \infty), \quad x = (x_1, \ldots, x_n) \mapsto \max_{i=1,\ldots,n} |x_i|,$$

are norms. Describe the corresponding unit balls B_1 and B_∞.

(c) Show that for an arbitrary norm $\| \cdot \|' : \mathbb{R}^n \to [0, \infty)$ there are constants $\alpha, \beta > 0$ such that

$$n^{-1}\alpha \| \cdot \|_1 \leq \alpha \| \cdot \|_\infty \leq \| \cdot \|' \leq \beta \| \cdot \|_1 \leq n\beta \| \cdot \|_\infty.$$

Describe these inequalities in terms of the corresponding unit balls B_1, B_∞, B'.

Hint: First show the inequalities between $\|\cdot\|_\infty$ and $\|\cdot\|_1$ and then the upper bound for $\|\cdot\|'$. Then prove that

$$\inf\{\|x\|' : x \in \mathbb{R}^n, \|x\|_\infty = 1\} > 0,$$

and deduce the lower bound for $\|\cdot\|'$ from that.

(d) Use (c) to show that all norms on $\mathbb{R}^n$ are equivalent.

16. Let $A \subset \mathbb{R}^n$ be a *locally finite* set (this means that $A \cap B^n(r)$ is a finite set, for $r \geq 0$). For each $x \in A$, we define the *Voronoi cell*

$$C(x, A) := \{z \in \mathbb{R}^n : \|z - x\| \leq \|z - y\| \text{ for } y \in A\},$$

consisting of all points $z \in \mathbb{R}^n$ which have x as their nearest point (or one of their nearest points) in A.

(a) Show that the Voronoi cells $C(x, A), x \in A$, are closed and convex.

(b) If conv $A = \mathbb{R}^n$, show that the Voronoi cells $C(x, A), x \in A$, are bounded and polyhedral, hence they are convex polytopes.

(c) Show by an example that the condition conv $A = \mathbb{R}^n$ is not necessary for the boundedness of the Voronoi cells $C(x, A), x \in A$.

17. Show that the set $\mathcal{A}$ of all convex subsets of $\mathbb{R}^n$ is a complete lattice with respect to the inclusion order.

Hint: Define the "join" (sup, $\vee$) and the "meet" (inf, $\wedge$) by

$$A \vee B := A \cap B,$$

$$A \wedge B := \text{conv}(A \cup B),$$

$$\sup \mathcal{M} := \bigcap_{A \in \mathcal{M}} A, \quad \mathcal{M} \subset \mathcal{A},$$

$$\inf \mathcal{M} := \text{conv}\left(\bigcup_{A \in \mathcal{M}} A\right), \quad \mathcal{M} \subset \mathcal{A}.$$

18. Let $P = \text{conv}\{x_0, \ldots, x_n\}$ be an n-simplex in $\mathbb{R}^n$. Denote by E_i the affine hull of $\{x_0, \ldots, x_n\} \setminus \{x_i\}$ and by H_i the closed halfspace bounded by E_i and with $x_i \in H_i, i = 0, \ldots, n$.

(a) Show that $x_i \in \text{int } H_i, i = 0, \ldots, n$.

(b) Show that $P = \bigcap_{i=0}^{n} H_i$.

(c) Show that $P \cap E_i$ is an $(n-1)$-simplex, $i = 0, \ldots, n$.

19. Let $K, L \subset \mathbb{R}^n$ be nonempty and closed sets. Suppose that $K \cup L$ is convex. Show that $K \cap L \neq \emptyset$.

20. Let $M \subset \mathbb{R}^n$ be a set and $x \in \mathbb{R}^n \setminus M$. Define $A := \text{conv}(\{x\} \cup M)$. Show that $A \setminus \{x\}$ is convex if and only if $x \notin \text{conv}(M)$.

1.2   Combinatorial Properties

Combinatorial problems arise naturally in connection with polytopes. In the following, however, we discuss problems of general convex sets which are of a combinatorial nature, since they involve the cardinality of points or sets. The most important results in this part of convex geometry (which is called *Combinatorial Geometry*) are the theorems of Carathéodory, Helly and Radon.

Theorem 1.6 (Radon's Theorem) *Let* $x_1, \ldots, x_m \in \mathbb{R}^n$ *be affinely dependent points. Then there exists a partition* $\{1, \ldots, m\} = I \cup J$, $I \cap J = \emptyset$, *such that*

$$\mathrm{conv}\{x_i : i \in I\} \cap \mathrm{conv}\{x_j : j \in J\} \neq \emptyset.$$

Proof Let $x_1, \ldots, x_m \in \mathbb{R}^n$ be affinely dependent. Then there exist numbers $\alpha_1, \ldots, \alpha_m \in \mathbb{R}$, not all zero, such that

$$\sum_{i=1}^{m} \alpha_i x_i = 0 \quad \text{and} \quad \sum_{i=1}^{m} \alpha_i = 0.$$

Define $I := \{i \in \{1, \ldots, m\} : \alpha_i \geq 0\}$ and $J := \{1, \ldots, m\} \setminus I$. Then

$$\alpha := \sum_{i \in I} \alpha_i = \sum_{j \in J} (-\alpha_j) > 0.$$

Hence

$$y := \sum_{i \in I} \frac{\alpha_i}{\alpha} x_i = \sum_{j \in J} \frac{-\alpha_j}{\alpha} x_j \in \mathrm{conv}\{x_i : i \in I\} \cap \mathrm{conv}\{x_j : j \in J\},$$

which proves the theorem.   □

Any sequence of $n + 2$ points in $\mathbb{R}^n$ is affinely dependent. As a consequence of Radon's Theorem, we next derive Helly's Theorem (in a particular version). It provides an answer to a question of the following type. Let $A_1, \ldots, A_m$ be a sequence of sets such that any s of these sets enjoy a certain property (for instance, having nonempty intersection). Is it true that then all sets of the sequence enjoy this property as well?

Theorem 1.7 (Helly's Theorem) *Let* $A_1, \ldots, A_m$ *be convex sets in* $\mathbb{R}^n$, $m \geq n+1$. *If each* $n + 1$ *of the sets* $A_1, \ldots, A_m$ *have nonempty intersection, then*

$$\bigcap_{i=1}^{m} A_i \neq \emptyset.$$

Proof We proceed by induction with respect to $m \geq n + 1$. For $m = n + 1$ there is nothing to show. Let $m \geq n + 2$, and assume that the assertion is true for $m - 1$ sets. Hence there are

$$x_i \in A_1 \cap \cdots \cap \check{A}_i \cap \cdots \cap A_m$$

(A_i is omitted) for $i = 1, \ldots, m$. The sequence $x_1, \ldots, x_m$ of $m \geq n + 2$ points is affinely dependent. By Radon's theorem there is a partition $\{1, \ldots, m\} = I \cup J$, $I \cap J = \emptyset$, such that

$$x \in \text{conv}\{x_i : i \in I\} \cap \text{conv}\{x_j : j \in J\} \neq \emptyset.$$

Since $x_i \in A_j$ for $i \in I$ and $j \in J$, we have $x_i \in \cap_{j \in J} A_j$, and therefore

$$x \in \text{conv}\{x_i : i \in I\} \subset \bigcap_{j \in J} A_j. \tag{1.2}$$

Furthermore, since $x_i \in A_j$ for $i \in J$ and $j \in I$, we have $x_i \in \cap_{j \in I} A_j$, and therefore

$$x \in \text{conv}\{x_i : i \in J\} \subset \bigcap_{j \in I} A_j. \tag{1.3}$$

Thus (1.2) and (1.3) yield $x \in A_1 \cap \cdots \cap A_m$. □

Helly's Theorem has interesting and sometimes surprising applications. For some of them, we refer to the exercises. In general, the theorem cannot be extended to infinite families of convex sets (see Exercise 1.1.1). An exception is the case of compact sets.

Theorem 1.8 (Helly's Theorem) *Let $\mathcal{A}$ be a family of at least $n+1$ compact convex sets in $\mathbb{R}^n$ ($\mathcal{A}$ may be infinite) and assume that any $n + 1$ sets in $\mathcal{A}$ have a nonempty intersection. Then, there is a point $x \in \mathbb{R}^n$ which is contained in all sets of $\mathcal{A}$.*

Proof By Theorem 1.7, every finite subfamily of $\mathcal{A}$ has a nonempty intersection. For compact sets, this implies that

$$\bigcap_{A \in \mathcal{A}} A \neq \emptyset.$$

In fact, if $\bigcap_{A \in \mathcal{A}} A = \emptyset$, then

$$\bigcup_{A \in \mathcal{A}} (\mathbb{R}^n \setminus A) = \mathbb{R}^n.$$

By the covering property, any compact set $A_0 \in \mathcal{A}$ is covered by finitely many open sets $\mathbb{R}^n \setminus A_1, \ldots, \mathbb{R}^n \setminus A_k$ with $A_i \in \mathcal{A}$ for $i = 1, \ldots, k$. This yields

$$\bigcap_{i=0}^{k} A_i = \emptyset,$$

a contradiction.   □

The following result will be frequently used later on.

Theorem 1.9 (Carathéodory's Theorem) *For a set $A \subset \mathbb{R}^n$ and $x \in \mathbb{R}^n$ the following two assertions are equivalent:*

(a) $x \in \operatorname{conv} A$,
(b) *there is an r-simplex P, for some $r \in \{0, \ldots, n\}$, with vertices in A and such that $x \in P$.*

Proof (b) $\Rightarrow$ (a): Since $\operatorname{vert} P \subset A$, we have $x \in P = \operatorname{conv} \operatorname{vert} P \subset \operatorname{conv} A$.
(a) $\Rightarrow$ (b): By Theorem 1.2, $x = \alpha_1 x_1 + \cdots + \alpha_k x_k$ with $k \in \mathbb{N}$, $x_1, \ldots, x_k \in A$, $\alpha_1, \ldots, \alpha_k \in (0, 1]$ and $\alpha_1 + \cdots + \alpha_k = 1$. Let k be the minimal number for which such a representation is possible, that is, x is not in the convex hull of any $k - 1$ points of A. We now show that $x_1, \ldots, x_k$ are affinely independent. In fact, assume that there are numbers $\beta_1, \ldots, \beta_k \in \mathbb{R}$, not all zero, such that

$$\sum_{i=1}^{k} \beta_i x_i = 0 \quad \text{and} \quad \sum_{i=1}^{k} \beta_i = 0.$$

Let $J \neq \emptyset$ be the set of indices $i \in \{1, \ldots, k\}$ for which $\beta_i > 0$. Choose $i_0 \in J$ such that

$$\frac{\alpha_{i_0}}{\beta_{i_0}} = \min_{i \in J} \frac{\alpha_i}{\beta_i}.$$

Then, we have

$$x = \sum_{i=1}^{k} \left(\alpha_i - \frac{\alpha_{i_0}}{\beta_{i_0}} \beta_i \right) x_i$$

with

$$\alpha_i - \frac{\alpha_{i_0}}{\beta_{i_0}} \beta_i \geq 0, \quad \sum_{i=1}^{k} \left(\alpha_i - \frac{\alpha_{i_0}}{\beta_{i_0}} \beta_i \right) = 1 \quad \text{and} \quad \alpha_{i_0} - \frac{\alpha_{i_0}}{\beta_{i_0}} \beta_{i_0} = 0.$$

This is a contradiction to the minimality of k.   □

Assume that x is a convex combination of affinely dependent points $x_1, \ldots, x_k$. Then the preceding argument shows that one of these points is redundant and x is a convex combination of at most $k - 1$ of these points. Repeating this procedure, we arrive at a representation of x as a convex combination of an affinely independent subset of $\{x_1, \ldots, x_k\}$. In particular, any point of the convex hull of A is the convex combination of at most $n + 1$ points of A. Clearly, the choice of these points from A will depend on the point x.

For the results of this section, linear analogues can be stated and proved which are concerned with vectors, linear hulls, convex cones, positive hulls instead of points, affine hulls, convex sets, and convex hulls. Often results can either be proved in a similar way or deduced from the corresponding counterpart.

Exercises and Supplements for Sect. 1.2

1. (a) Show by an example that Theorem 1.8 is wrong in general if the sets in $\mathcal{A}$ are only assumed to be closed (and not necessarily compact).
   (b) Show by an example that the result is also wrong in general if the sets are bounded but not closed.
   (c) Construct an example of four sets in the plane, three of which are compact and convex (one can even choose rectangles), such that any three of the sets have a nonempty intersection, but such that the intersection of all sets is the empty set.
   (d) Show that Theorem 1.8 remains true if all sets in $\mathcal{A}$ are closed and convex and one of the sets is compact and convex.

2. (a) Let $\mathcal{R}$ be a finite set of paraxial rectangles. For any two rectangles $R, R' \in \mathcal{R}$ let $R \cap R' \neq \emptyset$. Show that all rectangles in $\mathcal{R}$ have a common point.
   (b) Let S be a finite family of arcs in $\mathbb{S}^1$, each of which is contained in an open semi-circular arc of the circle. Any three arcs of S have a point in common. Show that all arcs have a point in common.
   Is it sufficient to assume that any two arcs of S have a common point?

3. In an old German fairy tale, a brave little tailor claimed the fame to have 'killed seven at one blow'. A closer examination showed that the victims were in fact flies which had landed on a toast covered with jam. The tailor had used a fly-catcher of convex shape for his sensational victory. As the remains of the flies on the toast showed, it was possible to kill any three of them with one stroke of the (suitably) shifted fly-catcher without even turning the direction of the handle.
   Is it possible that the tailor told the truth (if it is assumed that the flies are points)?

4. Let $k \in \mathbb{N}$ and $k \geq n + 1$. Let $A, A_1, \ldots, A_k \subset \mathbb{R}^n$ be nonempty and convex. Assume that for any set $I \subset \{1, \ldots, k\}$ with $|I| = n + 1$ there is a vector $t_I \in \mathbb{R}^n$ such that

$$A_i \subset A + t_I \text{ for } i \in I.$$

Show that there is a vector $t \in \mathbb{R}^n$ such that $t \in A_i + (-A)$ for $i \in \{1, \ldots, k\}$. If the sets $A_1, \ldots, A_k$ are singletons, then $A_i \subset A + t$ for $i \in \{1, \ldots, k\}$.

5. Let $\mathcal{F}$ be a family of parallel closed segments in $\mathbb{R}^2$, $|\mathcal{F}| \geq 3$. Suppose that for any three segments in $\mathcal{F}$ there is a line intersecting all three segments. Show that there is a line in $\mathbb{R}^2$ intersecting all segments in $\mathcal{F}$. (The problem is slightly easier if it is assumed that $\mathcal{F}$ is a finite family of segments.)

6. Prove the following version of Carathéodory's theorem:
   Let $A \subset \mathbb{R}^n$ and $x_0 \in A$ be fixed. Then conv A is the union of all simplices with vertices in A and such that x_0 is one of the vertices.

7.* Establish the following refined form of Carathéodory's theorem (due to Fenchel, Stoelinga, Bunt, see also [7] for a discussion):
   Let $A \subset \mathbb{R}^n$ be a set with at most n connected components. Then conv A is the union of all simplices with vertices in A and dimension at most $n - 1$. In other words, any point of conv A is in the convex hull of at most n points of A.

8. Suggestions for further reading: The combinatorial results of this section have been extended and applied in various directions. For instance there exist colourful, fractional, dimension-free and topological versions and generalizations of the theorems of Radon, Helly and Carathéodory (see [1, 6, 8, 9]). For a colourful version of Carathéodory's theorem, see also Exercise 1.4.7.

9. Applications of combinatorial results to containment problems are discussed in [52, 63]. Here are two examples from these works. The first is considered in [63] by E. Lutwak:
   Let $K, L \subset \mathbb{R}^n$ be compact convex sets. Suppose that for every simplex Δ such that $L \subset \Delta$, there exists a $v \in \mathbb{R}^n$ such that $K + v \subset \Delta$. Then there exists a $v_0 \in \mathbb{R}^n$ such that $K + v_0 \subset L$.
   An inscribed counterpart is discussed in [52]:
   Suppose that $K, L \subset \mathbb{R}^n$ have nonempty interiors. If every simplex contained in K can be translated inside L, then K can be translated inside L.

10. Let $K \subset \mathbb{R}^n$ be an n-dimensional compact convex set. Show that there exists a point $c \in K$ such that whenever $a \in K$, $b \in \operatorname{bd} K$ with $c \in [a, b]$, then

$$\|a - c\| \leq \frac{n}{n+1} \|a - b\|. \tag{$*$}$$

Hint: Consider the sets

$$K_x := x + \frac{n}{n+1}(K - x), \qquad x \in K.$$

Verify that for any points $x_0, \ldots, x_n \in K$ we have

$$\frac{1}{n+1}(x_0 + \cdots + x_n) \in \bigcap K_{x_i}.$$

Now Helly's theorem can be applied.

11.* Let $K \subset \mathbb{R}^n$ be an n-dimensional compact convex set. Show that $c \in K$ has the property $(*)$ stated in Exercise 1.2.10 if and only if $-(K - c) \subset n(K - c)$.

12. In $\mathbb{R}^2$ the points

$$x_1 = \begin{pmatrix} 1 \\ 0 \end{pmatrix}, \quad x_2 = \begin{pmatrix} 1 \\ 3 \end{pmatrix}, \quad x_3 = \begin{pmatrix} 4 \\ 3 \end{pmatrix}, \quad x_4 = \begin{pmatrix} 4 \\ 0 \end{pmatrix} \quad \text{and} \quad x = \begin{pmatrix} 7/4 \\ 5/4 \end{pmatrix}$$

are given. Confirm that

$$x = \frac{1}{2}x_1 + \frac{1}{4}x_2 + \frac{1}{6}x_3 + \frac{1}{12}x_4.$$

Use the method of proof for Carathéodory's theorem to express x as a convex combination of x_1, x_2, x_3.

13. Is the decomposition in Radon's theorem uniquely determined for $m = n + 2$ points in $\mathbb{R}^n$?

   Hint: See [31].

14. Let $u_1, \ldots, u_m \in \mathbb{R}^n \setminus \{0\}$. Show that

$$0 \in \text{conv}\{u_1, \ldots, u_m\} \iff \mathbb{R}^n = \bigcup_{i=1}^{m} H^{|}(u_i, 0).$$

   Let $\mathbb{R}^n = \bigcup_{i=1}^{m} H^+(u_i, 0)$. Show that there is a set $I \subset \{1, \ldots, m\}$ with at most $n + 1$ elements such that

$$\mathbb{R}^n = \bigcup_{i \in I} H^+(u_i, 0).$$

   In words: If N closed halfspaces containing the origin in their boundaries cover $\mathbb{R}^n$, then at most $n + 1$ of these halfspaces are needed to cover $\mathbb{R}^n$.

1.3 Topological Properties

Although convexity is a purely algebraic property, it has some useful topological consequences. For instance, we shall see that a nonempty convex set always has a nonempty relative interior. In order to prove this seemingly obvious fact, we first need an auxiliary result. We recall the following definitions and basic observations.

- Intersections of affine subspaces are affine subspaces (or the empty set).
- The affine hull, aff A, of a nonempty set $A \subset \mathbb{R}^n$ is the intersection of all affine subspaces containing A.

- $\dim(A) = \dim(\operatorname{aff} A)$, where $\emptyset \neq A \subset \mathbb{R}^n$, and this is the maximal number $k \in \mathbb{N}$ for which there are affinely independent points $x_0, \dots, x_k \in A$.
- The relative interior, relint A, of a set A is the interior with respect to aff A as the ambient space.

These statements should be checked and illustrated with various examples.

Proposition 1.1 *If* $P = \operatorname{conv}\{x_0, \dots, x_k\}$ *is a k-simplex in $\mathbb{R}^n$, for an integer $k \in \{1, \dots, n\}$, then*

$$\operatorname{relint} P = \left\{ \alpha_0 x_0 + \dots + \alpha_k x_k \in \mathbb{R}^n : \alpha_i \in (0, 1), \alpha_0 + \dots + \alpha_k = 1 \right\}.$$

In particular, $(x_0 + \dots + x_k)/(k+1) \in \operatorname{relint} P$.

Proof First, we have

$$P = \left\{ \sum_{i=0}^{k} \alpha_i x_i \in \mathbb{R}^n : \alpha_i \in [0, 1], \sum_{i=0}^{k} \alpha_i = 1 \right\}$$

$$= x_0 + \left\{ \sum_{i=1}^{k} \alpha_i (x_i - x_0) \in \mathbb{R}^n : \alpha_i \in [0, 1], \sum_{i=1}^{k} \alpha_i \leq 1 \right\}.$$

Since $x_1 - x_0, \dots, x_k - x_0$ are linearly independent, the linear isomorphism between k-dimensional vector spaces given by

$$F : \mathbb{R}^k \to L := \operatorname{lin}\{x_1 - x_0, \dots, x_k - x_0\}, \quad (\beta_1, \dots, \beta_k)^\top \mapsto \sum_{i=1}^{k} \beta_i (x_i - x_0),$$

is a homeomorphism. Hence we deduce

$$\operatorname{relint} P = x_0 + F \circ F^{-1}(\operatorname{relint}(P - x_0)) = x_0 + F(\operatorname{relint} F^{-1}(P - x_0))$$

$$= x_0 + F \left(\operatorname{relint} \left\{ (\alpha_1 \dots, \alpha_k)^\top : \alpha_i \in [0, 1], \sum_{i=1}^{k} \alpha_i \leq 1 \right\} \right)$$

$$= x_0 + F \left(\left\{ (\alpha_1 \dots, \alpha_k)^\top : \alpha_i \in (0, 1), \sum_{i=1}^{k} \alpha_i < 1 \right\} \right)$$

$$= x_0 + \left\{ \sum_{i=1}^{k} \alpha_i (x_i - x_0) : \alpha_i \in (0, 1), \sum_{i=1}^{k} \alpha_i < 1 \right\}$$

$$= \left\{ \alpha_0 x_0 + \dots + \alpha_k x_k \in \mathbb{R}^n : \alpha_i \in (0, 1), \alpha_0 + \dots + \alpha_k = 1 \right\},$$

which proves the assertion. $\qquad\qquad\qquad\qquad\qquad\qquad\qquad\qquad\qquad\qquad\qquad\quad\square$

Theorem 1.10 *If $A \subset \mathbb{R}^n$, $A \neq \emptyset$, is convex, then* relint $A \neq \emptyset$.

Proof If dim $A = k$, then A contains $k + 1$ affinely independent points and hence a k-simplex P. If $k = 0$, there is nothing to show. Hence, suppose that $k \geq 1$. Then, by Proposition 1.1, there is some $x \in$ relint P. For each such x we have $x \in$ relint A. $\qquad \square$

Theorem 1.10 shows that for the investigation of a fixed convex set A, it is useful to consider the affine hull of A as the basic space, since then A has interior points. We will often take advantage of this fact by assuming that the affine hull of A is the whole space $\mathbb{R}^n$. Therefore, proofs in the following frequently start with a sentence claiming that we may assume (w.l.o.g.) that the convex set under consideration has dimension n. (Of course, the reader should check these assertions carefully in each case.)

A further consequence of convexity is that topological notions like interior or closure of a (convex) set can be expressed in purely geometric terms.

Theorem 1.11 *If $A \subset \mathbb{R}^n$ is convex, then*

$$\text{cl } A = \{x \in \mathbb{R}^n : \exists y \in A \text{ with } [y, x) \subset A\}, \tag{a}$$

$$\text{relint } A = \{x \in \mathbb{R}^n : \forall y \in \text{aff}(A) \setminus \{x\} \exists z \in (x, y) \text{ with } [x, z] \subset A\}, \tag{b}$$

$$\text{relint } A = \{x \in \mathbb{R}^n : \forall y \in \text{aff}(A) \setminus \{x\} \exists z \in A \text{ with } x \in (z, y)\}. \tag{c}$$

Again, we first provide an auxiliary result which is of independent interest and will be applied several times throughout the text.

Proposition 1.2 *If $A \subset \mathbb{R}^n$ is convex, $x \in$ cl A, $y \in$ relint A, then $[y, x) \subset$ relint A.*

Proof As explained above, we may assume dim $A = n$. Let $x \in$ cl A, $y \in$ relint A and $z \in (y, x)$, that is, $z = \alpha y + (1 - \alpha)x$, $\alpha \in (0, 1)$. We have to show that $z \in$ int A. Since $x \in$ cl A, there exists a sequence $x_k \in A$, $k \in \mathbb{N}$, with $x_k \to x$ as $k \to \infty$. Since $y \in$ int A, there exists an open ball $V \subset A$ centered at y. The points $y_k := \frac{1}{\alpha}(z - (1 - \alpha)x_k)$ converge to y, as $k \to \infty$. Hence, $y_k \in V$ if k is large enough. The convexity of A implies that $z \in \alpha V + (1 - \alpha)x_k \subset A$. Since $\alpha V + (1 - \alpha)x_k$ is open, we have $z \in$ int A (see Fig. 1.3 for an illustration). $\qquad \square$

Fig. 1.3 Illustration for the proof of Proposition 1.2

Proof (of Theorem 1.11) The case $A = \emptyset$ is trivial, hence we assume now that $A \neq \emptyset$.

(a) Let B denote the set on the right-hand side of (a). Then we obviously have $B \subset \operatorname{cl} A$. To show the converse inclusion, let $x \in \operatorname{cl} A$. By Theorem 1.10 there is a point $y \in \operatorname{relint} A$, hence by Proposition 1.2 we have $[y, x) \subset \operatorname{relint} A \subset A$. Therefore, $x \in B$.

(b) The set on the right-hand side of (b) is denoted by C. If $x \in \operatorname{relint} A$ and $y \in \operatorname{aff}(A) \setminus \{x\}$, then $z := (1 - \varepsilon)x + \varepsilon y \in A$ if $\varepsilon \in (0, 1)$ is sufficiently small and hence $z \in [x, z] \subset A$. This yields $\operatorname{relint} A \subset C$. For the converse, let $x \in C$. By Theorem 1.10, we can choose $y \in \operatorname{relint} A$. If $y = x$, we get $x \in \operatorname{relint} A$. Hence, suppose that $y \neq x$. Then $2x - y = x + (x - y) \in \operatorname{aff}(A) \setminus \{x\}$. The definition of C implies that there exists a $z \in (x, 2x - y)$ with $z \in [x, z] \subset A$. Then $x \in (y, z)$ and Proposition 1.2 shows that $x \in \operatorname{relint} A$.

(c) The set on the right-hand side of (c) is denoted by D. If $x \in \operatorname{relint} A$ and $y \in \operatorname{aff}(A) \setminus \{x\}$, then $z_\varepsilon := (1 - \varepsilon)x + y \in \operatorname{aff}(A)$ if $\varepsilon \in \mathbb{R}$. We choose $\varepsilon < 0$ and $|\varepsilon|$ so small that $z_\varepsilon \in A$. Then

$$x = \frac{1}{1 - \varepsilon} z_\varepsilon + \frac{-\varepsilon}{1 - \varepsilon} y \in (z_\varepsilon, y).$$

Now let $x \in D$. We can choose $y \in \operatorname{relint} A$. There is nothing to show if $x = y$. Suppose that $x \neq y$. By assumption, there is some $z \in A$ such that $x \in (y, z) \subset [y, z) \subset \operatorname{relint} A$. $\qquad\square$

Remark 1.12 Theorem 1.11 shows that (and how) topological notions like the interior and the closure of a set can be defined for convex sets A on a purely algebraic basis, without the need to specify a topology on the underlying space. This fact can be used in arbitrary real vector spaces V (without a given topology) to introduce and study "topological properties" of convex sets.

In view of Remark 1.12, we deduce the following two corollaries from Theorem 1.11 and Proposition 1.2, instead of giving a direct proof based on the topological notions 'relint' and 'cl'.

Corollary 1.2 *For a convex set $A \subset \mathbb{R}^n$, the sets $\operatorname{relint} A$ and $\operatorname{cl} A$ are convex.*

Proof The convexity of $\operatorname{relint} A$ follows immediately from Proposition 1.2.

For the convexity of $\operatorname{cl} A$, suppose that $x_1, x_2 \in \operatorname{cl} A$ and $\alpha \in (0, 1)$. From Theorem 1.11 (a), we get points $y_1, y_2 \in A$ with $[y_1, x_1) \subset A$, $[y_2, x_2) \subset A$. Hence

$$\alpha[y_1, x_1) + (1 - \alpha)[y_2, x_2) \subset A.$$

Since $\alpha y_1 + (1 - \alpha)y_2 \in A$ and

$$[\alpha y_1 + (1 - \alpha)y_2, \alpha x_1 + (1 - \alpha)x_2) \subset \alpha[y_1, x_1) + (1 - \alpha)[y_2, x_2) \subset A,$$

we obtain $\alpha x_1 + (1 - \alpha)x_2 \in \operatorname{cl} A$, again from Theorem 1.11 (a). $\qquad\square$

Corollary 1.3 *For a convex set $A \subset \mathbb{R}^n$,*

$$\operatorname{cl} A = \operatorname{cl} \operatorname{relint} A$$

and

$$\operatorname{relint} A = \operatorname{relint} \operatorname{cl} A.$$

Proof The inclusion $\operatorname{cl} \operatorname{relint} A \subset \operatorname{cl} A$ is obvious. Let $x \in \operatorname{cl} A$. By Theorem 1.10 there is a $y \in \operatorname{relint} A$ and by Proposition 1.2 we have $[y, x) \subset \operatorname{relint} A$. But then clearly $x \in \operatorname{cl} \operatorname{relint} A$.

The inclusion $\operatorname{relint} A \subset \operatorname{relint} \operatorname{cl} A$ is again obvious. Let $x \in \operatorname{relint} \operatorname{cl} A$. Since $\operatorname{cl} A$ is convex by Corollary 1.2, we can apply Theorem 1.11 in aff $A = $ aff $\operatorname{cl} A$ to $\operatorname{cl} A$. By Theorem 1.10 there exists some $y \in \operatorname{relint} A$. If $y = x$, then $x \in \operatorname{relint} A$. If $y \neq x$, we obtain $z \in \operatorname{cl} A$ such that $x \in (z, y) \subset \operatorname{relint} A$, by Proposition 1.2. $\square$

We finally study the topological properties of the convex hull operator. For a closed set $A \subset \mathbb{R}^n$, the convex hull conv A need not be closed. A simple example is given by the set

$$A := \{(t, t^{-1}) : t > 0\} \cup \{(0, 0)\} \subset \mathbb{R}^2.$$

However, the convex hull operator behaves well with respect to open and compact sets.

Theorem 1.12 *If $A \subset \mathbb{R}^n$ is (relatively) open, then conv A is (relatively) open. If $A \subset \mathbb{R}^n$ is compact, then conv A is compact.*

Proof Let A be open and $x \in \operatorname{conv} A$. Then there exist $x_i \in A$ and $\alpha_i \in (0, 1]$, $i \in \{1, \ldots, k\}$, such that $x = \alpha_1 x_1 + \cdots + \alpha_k x_k$ and $\alpha_1 + \cdots + \alpha_k = 1$. We can choose a ball U around the origin such that $U + x_i \subset A \subset \operatorname{conv} A$, $i = 1, \ldots, k$. Since

$$U + x = \alpha_1(U + x_1) + \cdots + \alpha_k(U + x_k) \subset \operatorname{conv} A,$$

we have $x \in \operatorname{int} \operatorname{conv} A$, hence conv A is open. For a relatively open set, we argue in the same way, in the affine hull of the set.

Now let A be compact. Since A is contained in a ball $B^n(r)$, we have conv $A \subset B^n(r)$, that is, conv A is bounded. In order to show that conv A is closed, let $x_k \to x$, $x_k \in \operatorname{conv} A$, for $k \in \mathbb{N}$. By Theorem 1.9, each x_k has a representation

$$x_k = \alpha_{k0} x_{k0} + \cdots + \alpha_{kn} x_{kn}$$

with

$$\alpha_{ki} \in [0, 1], \quad \sum_{i=0}^{n} \alpha_{ki} = 1, \quad \text{and} \quad x_{ki} \in A.$$

Since A and $[0, 1]$ are compact, we find a subsequence $(k_r)_{r \in \mathbb{N}}$ in $\mathbb{N}$ such that the $2n + 2$ sequences $(x_{k_r j})_{r \in \mathbb{N}}$, $j = 0, \ldots, n$, and $(\alpha_{k_r j})_{r \in \mathbb{N}}$, $j = 0, \ldots, n$, all converge. We denote the limits by y_j and β_j, $j = 0, \ldots, n$. Then, we have $y_j \in A$, $\beta_j \in [0, 1]$, $\beta_0 + \cdots + \beta_n = 1$, and $x = \beta_0 y_0 + \cdots + \beta_n y_n$. Hence, $x \in \text{conv } A$. □

Remark 1.13 Theorem 1.12 in particular implies again that a convex polytope P is compact.

Remark 1.14 We give an alternative argument for the first part of Theorem 1.12 (following a suggestion of Mathew Penrose). Let A be relatively open and $x \in \text{conv } A$. Then there exist $x_i \in A$ and $\alpha_i \in (0, 1]$, $i \in \{1, \ldots, k\}$, such that $x = \alpha_1 x_1 + \cdots + \alpha_k x_k$ and $\alpha_1 + \cdots + \alpha_k = 1$. If $k = 1$, the assertion is clear. If $k \geq 2$ (and hence $\alpha_1 \neq 1$), then we have

$$x = \alpha_1 x_1 + (1 - \alpha_1) \underbrace{\sum_{j=2}^{k} \frac{\alpha_j}{1 - \alpha_1} x_j}_{=:y}.$$

Since $x_1 \in A = \text{relint } A \subset \text{relint conv } A$ and $y \in \text{conv } A$, Proposition 1.2 yields that $x \in (x_1, y) \subset \text{relint conv } A$.

Remark 1.15 For an alternative argument for the second part of Theorem 1.12, define

$$C := \{(\alpha_0, \ldots, \alpha_n, x_0, \ldots, x_n) \in [0, 1]^{n+1} \times A^{n+1} : \alpha_0 + \cdots + \alpha_n = 1\}$$

and

$$f : C \to \text{conv } A, \quad f(\alpha_0, \ldots, \alpha_n, x_0, \ldots, x_n) := \sum_{i=0}^{n} \alpha_i x_i.$$

Clearly, f is continuous and C is compact. Hence $f(C)$ is compact. By Carathéodory's theorem, $f(C) = \text{conv } A$, which shows that $\text{conv } A$ is compact.

Exercises and Supplements for Sect. 1.3

1. Let $P = \text{conv}\{a_0, \ldots, a_n\}$ be an n-simplex in $\mathbb{R}^n$ and $x \in \text{int } P$.
   Show that the polytopes

$$P_i := \text{conv}\{a_0, \ldots, a_{i-1}, x, a_{i+1}, \ldots, a_n\}, \quad i = 0, \ldots, n,$$

   are n-simplices with pairwise disjoint interiors and that

$$P = \bigcup_{i=0}^{n} P_i.$$

2. Show that, for $A \subset \mathbb{R}^n$,

$$\text{cl conv } A = \bigcap \{B \subset \mathbb{R}^n : B \supset A, \ B \text{ closed and convex}\}.$$

3.*Let $A, B \subset \mathbb{R}^n$ be convex.

   (a) Show that $\text{relint}(A + B) = \text{relint } A + \text{relint } B$.
   (b) If A (or B) is bounded, show that $\text{cl}(A + B) = \text{cl } A + \text{cl } B$.
   (c) Show by an example that (b) may be wrong if neither A nor B is assumed to be bounded.

4. Let $A, B \subset \mathbb{R}^n$ be convex, A closed, B compact. Show that $A + B$ is closed (and convex). Give an example which shows the need of the assumption of compactness of one of the sets A, B for this statement.

5. Let $K, L \subset \mathbb{R}^n$ be nonempty closed sets. Suppose that $K \cup L$ is convex and $\dim(L) < \dim(K)$. Show that then $L \subset K$.

6. The diameter $\text{diam}(A)$ of a nonempty bounded set $A \subset \mathbb{R}^d$ is defined by

$$\text{diam}(A) := \sup\{\|x - y\| : x, y \in A\}.$$

   Let $A \subset \mathbb{R}^d$ be nonempty and bounded. Show that $\text{diam}(A) = \text{diam}(\text{conv}(A))$ and $\text{diam}(A) = \text{diam}(\text{cl}(A))$.

7. Let $A \subset \mathbb{R}^n$ be a convex set which meets each line in a closed set. Show that then A is a closed set. Is this conclusion still correct if the assumption of convexity is dropped?

1.4 Support and Separation

Convex sets are sets which contain with their elements also all convex combinations of these elements. In this section, we consider a description of convex sets which is of a dual nature, in the sense that convex sets A are obtained as intersections of halfspaces. For such a result, we have to assume that A is a closed set.

We start with results on the metric projection which are of independent interest.

Theorem 1.13 *Let $A \subset \mathbb{R}^n$ be nonempty, closed and convex. Then for each $x \in \mathbb{R}^n$, there is a unique point $p(A, x) \in A$ satisfying*

$$\|p(A, x) - x\| = \inf_{y \in A} \|y - x\|.$$

Definition 1.7 The mapping $p(A, \cdot) : \mathbb{R}^n \to A$ is called the *metric projection* onto the nonempty, closed and convex set $A \subset \mathbb{R}^n$.

See Fig. 1.4 for an illustration of the metric projection $p(A, x)$ of a point x onto a set A.

Proof (of Theorem 1.13) For $x \in A$, we obviously have $p(A, x) = x$. For $x \notin A$, there is some $r > 0$ such that $A \cap B^n(x, r) \neq \emptyset$, and hence

$$\inf_{y \in A} \|y - x\| = \inf_{y \in A \cap B^n(x,r)} \|y - x\|.$$

Since $A \cap B^n(x, r)$ is compact and $f(y) := \|y - x\|$, $y \in \mathbb{R}^n$, defines a continuous map, there is a point $y_0 \in A \cap B^n(x, r)$ realizing the minimum of f on A.

If $y_1 \in A$ is a second point realizing this minimum, then $y_2 := \frac{1}{2}(y_0 + y_1) \in A$ and $\|y_2 - x\| < \|y_0 - x\|$, by Pythagoras' theorem (see Fig. 1.5).

This is a contradiction and hence the metric projection $p(A, x)$ is unique.

Fig. 1.4 The metric projection $p(A, x)$ of the point x onto the convex set A

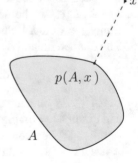

Fig. 1.5 Uniqueness of the nearest point in a closed convex set

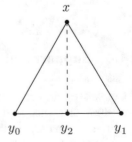

In the last step, we can also argue analytically. Put $a := \|y_0 - x\| = \|y_1 - x\|$. Then

$$a^2 \leq \left\| \frac{1}{2}(y_0 + y_1) - x \right\|^2 = \left\| \frac{y_0 - x}{2} + \frac{y_1 - x}{2} \right\|^2$$

$$= \frac{1}{4}a^2 + \frac{1}{4}a^2 + \frac{1}{2}\langle y_0 - x, y_1 - x \rangle$$

$$\leq \frac{1}{2}a^2 + \frac{1}{2}\|y_0 - x\| \|y_1 - x\| \leq \frac{1}{2}a^2 + \frac{1}{2}a^2 = a^2,$$

where we used the Cauchy–Schwarz inequality. By the equality condition of this inequality, we deduce that $y_0 - x = \lambda(y_1 - x)$ for some $\lambda > 0$. But then $\lambda = 1$ and thus $y_0 - x = y_1 - x$, hence $y_0 = y_1$. $\qquad\square$

Remark 1.16 As the above proof shows, the existence of a nearest point $p(A, x)$ is guaranteed for all closed sets A. The convexity of A implies that $p(A, x)$ is uniquely determined. A more general class of sets consists of closed sets A for which the uniqueness of $p(A, x)$ holds at least in an ε-neighbourhood of A, that is, for $x \in A + \varepsilon B^n$, with $\varepsilon > 0$. Such sets are called sets of *positive reach*, and the largest ε for which uniqueness of the metric projection holds is called the *reach* of A. Convex sets thus have reach ∞.

Definition 1.8 Let $A \subset \mathbb{R}^n$ be closed and convex, and let $E = \{f = \alpha\}$ be a hyperplane. Then E is called a *supporting hyperplane* of A if $A \cap E \neq \emptyset$ and A is contained in one of the two closed halfspaces $\{f \leq \alpha\}$ or $\{f \geq \alpha\}$. A halfspace containing A and bounded by a supporting hyperplane of A is called a *supporting halfspace* of A, the set $A \cap E$ is called a *support set* and any $x \in A \cap E$ is called a *supporting point*. If E is a supporting hyperplane of A, we shortly say that the hyperplane E supports A.

Example 1.8 The set

$$A := \{(a, b) \in \mathbb{R}^2 : b \geq a^{-1}, a > 0\}$$

is closed and convex. The line $g := \{a + b = 2\}$ is a supporting line, since $(1, 1) \in A \cap g$ and $A \subset \{a + b \geq 2\}$. The lines $h := \{a = 0\}$ and $k := \{b = 0\}$ bound the set A, but are not supporting lines, since they do not have a point in common with A.

Theorem 1.14 *Let $A \subset \mathbb{R}^n$ be nonempty, closed and convex. Let $x \in \mathbb{R}^n \setminus A$. Then the hyperplane E through $p(A, x)$, orthogonal to $x - p(A, x)$, supports A. Moreover, the halfspace H bounded by E and not containing x is a supporting halfspace.*

Proof Obviously, $x \notin E$. Since $p(A, x) \in E \cap A$, it remains to show that $A \subset H$. Assume that there is a $y \in A$, $y \notin H$. Then $\langle y - p(A, x), x - p(A, x) \rangle > 0$. We consider the orthogonal projection $\bar{y}$ of x onto the line through $p(A, x)$ and y. By

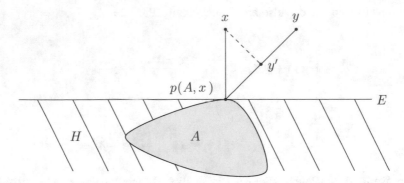

Fig. 1.6 Supporting hyperplanes through image points of the metric projection map. Illustration for the first proof of Theorem 1.14

Pythagoras' theorem, $\|\bar{y}-x\| < \|p(A,x)-x\|$. If $\bar{y} \in (p(A,x), y]$, we put $y' := \bar{y}$ (see Fig. 1.6). Otherwise, we have $y \in (p(A,x), \bar{y}]$ and put $y' := y$.

In both cases we obtain a point $y' \in (p(A,x), y] \subset A$ with $\|y' - x\| < \|p(A,x) - x\|$. This is a contradiction, hence we conclude $A \subset H$. $\Box$

Proof (Second Proof) Let $a \in A$. Then $F : t \mapsto \|(1-t)p(A,x) + ta - x\|^2$, $t \in [0,1]$, has a global minimum at $t = 0$. Hence $F'(0) \geq 0$, that is,

$$\langle a - p(A,x), x - p(A,x) \rangle \leq 0.$$

Since $a \in A$ is arbitrary, it follows that

$$A \subset \{z \in \mathbb{R}^n : \langle z - p(A,x), x - p(A,x) \rangle \leq 0\} =: H,$$

$p(A,x) \in E$, and $x \in \mathbb{R}^n \setminus H$. $\Box$

Corollary 1.4 *Every nonempty, closed convex set $A \subset \mathbb{R}^n$, $A \neq \mathbb{R}^n$, is the intersection of all closed halfspaces which contain A. More specifically, A is the intersection of all of its supporting halfspaces.*

Proof Obviously, A lies in the intersection B of its supporting halfspaces. For $x \notin A$, Theorem 1.14 yields the existence of a supporting halfspace H of A with $x \notin H$. Hence $x \notin B$. $\Box$

Theorem 1.14 and Corollary 1.4 do not imply that every boundary point of A is a support point. In order to show such a result, we approximate $x \in \text{bd}\, A$ by points x_k from $\mathbb{R}^n \setminus A$ and consider the corresponding supporting hyperplanes E_k which exist by Theorem 1.14. For $x_k \to x$, we want to define a supporting hyperplane in x as the limit of the E_k. A first step in this direction is to prove that $p(A, x_k) \to p(A, x)$ (where $p(A, x) = x$), that is, $p(A, \cdot)$ is continuous. We even show now that $p(A, \cdot)$ is Lipschitz continuous with Lipschitz constant 1.

Theorem 1.15 *Let $A \subset \mathbb{R}^n$ be nonempty, closed and convex. Then*

$$\|p(A, x) - p(A, y)\| \leq \|x - y\|$$

for $x, y \in \mathbb{R}^n$.

Proof In the proof, we abbreviate $p(A, \cdot)$ by p. Let $x, y \in \mathbb{R}^n$. By the second proof for Theorem 1.14 we then obtain, by choosing $a = p(y)$ for the first inequality and arguing by symmetry for the second,

$$\langle x - p(x), p(y) - p(x) \rangle \leq 0 \quad \text{and} \quad \langle y - p(y), p(x) - p(y) \rangle \leq 0.$$

Note that this holds for $x \notin A$ and $y \notin A$, respectively, but the inequalities remain true also for $x \in A$ and $y \in A$, since then both sides are zero. Addition of these two inequalities yields

$$\langle p(y) - p(x), p(y) - y + x - p(x) \rangle \leq 0,$$

and therefore

$$\|p(y) - p(x)\|^2 \leq \langle p(y) - p(x), y - x \rangle \leq \|p(y) - p(x)\| \cdot \|y - x\|,$$

where the Cauchy–Schwarz inequality was used for the last estimate. If $p(x) \neq p(y)$, this yields the required inequality. The case $p(x) = p(y)$ is trivial. $\quad\sqcup$

Theorem 1.16 (Support Theorem) *Let $A \subset \mathbb{R}^n$ be closed and convex. Then through each boundary point of A there exists a supporting hyperplane.*

Proof For given $x \in \text{bd}\, A$ and $k \in \mathbb{N}$, we choose $x_k \in B^n(x, 1)$, $x_k \notin A$, and such that $\|x - x_k\| < \frac{1}{k}$. Then

$$\|x - p(A, x_k)\| = \|p(A, x) - p(A, x_k)\| \leq \|x - x_k\| < \frac{1}{k}$$

by Theorem 1.15. Since x_k, $p(A, x_k)$ are interior points of $B^n(x, 1)$, there is a (unique) boundary point y_k of $B^n(x, 1)$ such that $x_k \in (p(A, x_k), y_k)$. Theorem 1.14 then implies that $p(A, y_k) = p(A, x_k)$. In view of the compactness of $B^n(x, 1)$, we may choose a converging subsequence $y_{k_r} \to y$. By Theorem 1.15, $p(A, y_{k_r}) \to p(A, y)$ and $p(A, y_{k_r}) = p(A, x_{k_r}) \to p(A, x) = x$, hence $p(A, y) = x$. Since $y \in \text{bd}\, B^n(x, 1)$, we also know that $x \neq y$. The assertion now follows from Theorem 1.14. $\quad\square$

Remark 1.17 Supporting hyperplanes, halfspaces and points can be defined for nonconvex sets A as well. However, in general they only exist if $\text{conv}\, A$ is closed and not all of $\mathbb{R}^n$. Then $\text{conv}\, A$ is the intersection of all supporting halfspaces of A.

Some of the previous results can be interpreted as separation theorems, in the sense of the following definition.

Definition 1.9 For two sets $A, B \subset \mathbb{R}^n$ and a hyperplane $E = \{f = \alpha\}$, we say that E *separates* A and B if either $A \subset \{f \leq \alpha\}, B \subset \{f \geq \alpha\}$ or $A \subset \{f \geq \alpha\}, B \subset \{f \leq \alpha\}$. We say that convex sets A and B can be *properly separated* if there exists a separating hyperplane which does not contain both A and B.

Theorem 1.14 then says that a closed convex set A and a point $x \notin A$ can be separated by a hyperplane (there is even a separating hyperplane which has positive distance to both, A and x). This result can be extended to compact convex sets B (instead of the point x). Moreover, Theorem 1.16 states that each boundary point of A can be separated from A by a hyperplane. The following result provides a necessary and sufficient condition for the proper separation of two nonempty convex sets.

Theorem 1.17 (Separation Theorem) *Let $A, B \subset \mathbb{R}^n$ be nonempty and convex sets in $\mathbb{R}^n$. Then*

$$\text{relint } A \cap \text{relint } B = \emptyset$$

if and only if A and B can be properly separated.

Proof Suppose that $\text{relint } A \cap \text{relint } B = \emptyset$. Then, clearly $0 \notin \text{relint } A - \text{relint } B$, and hence Exercise 1.3.3 (a) implies that $0 \notin \text{relint}(A - B)$.

First, observe that $\text{cl}(A - B)$ is nonempty, closed and convex. We distinguish two cases. If $0 \notin \text{cl}(A - B)$, we apply Theorem 1.14. If $0 \in \text{cl}(A - B)$, we apply Theorem 1.16. In both cases, we obtain a hyperplane $E = \{f = 0\}$ through 0 with $A - B \subset \{f \leq 0\}$. Then $f(a) \leq f(b)$ for $a \in A$ and $b \in B$. Put $\alpha := \sup_{a \in A} f(a) \leq f(b) < \infty$, for any $b \in B$, so that $A \subset \{f \leq \alpha\}$. Since $f(b) \geq \alpha$, for $b \in B$, we conclude that $B \subset \{f \geq \alpha\}$. In order to obtain a properly separating hyperplane, we apply this argument in the affine hull of $A \cup B$ and then extend the hyperplane with respect to the affine hull to a hyperplane of $\mathbb{R}^n$ by adding the linear subspace orthogonal to the affine hull.

For the converse, simply observe that if E is a separating hyperplane for A and B and $x_0 \in \text{relint } A \cap \text{relint } B$, then we first get $x_0 \in E$ and then $A, B \subset E$. Hence E cannot be a properly separating hyperplane. $\square$

Remark 1.18 Let $A, B \subset \mathbb{R}^n$ be convex, A closed, B compact, and assume that $A \cap B = \emptyset$. Then there is a hyperplane $\{f = \gamma\}$ and an $\varepsilon > 0$ such that $A \subset \{f \geq \gamma + \varepsilon\}$ and $B \subset \{f \leq \gamma - \varepsilon\}$. In this case, we say that A and B are *strongly separated* by the hyperplane $\{f = \gamma\}$. This strong separation result is the subject of Exercise 1.4.4.

Remark 1.19 Some of the properties which we derived are characteristic for convexity. For example, a closed set $A \subset \mathbb{R}^n$ such that each $x \notin A$ has a unique metric projection onto A, must be convex (Motzkin's Theorem). Also the Support Theorem has a converse. A closed set $A \subset \mathbb{R}^n$, $\text{int } A \neq \emptyset$, such that each boundary point is a support point, must also be convex. For proofs of these results, see for instance [81, Theorem 1.2.4 and Theorem 1.3.3], [93, Theorem 7.5.5] or Exercises 1.4.1 and 1.4.2.

Remark 1.20 Let $A \subset \mathbb{R}^n$ be nonempty, closed and convex. Then, for each direction $u \in \mathbb{S}^{n-1}$, there is a supporting hyperplane $E(u)$ of A in direction u (i.e., with outer normal u) if and only if A is compact (see Exercise 1.4.3).

Remark 1.21 In infinite-dimensional topological vector spaces V, similar support and separation theorems hold true. However, there are some important differences, mainly due to the fact that convex sets A in V need not have relative interior points. Therefore a common assumption is that int $A \neq \emptyset$. Otherwise it is possible that A is closed but does not have any support points, or, in the other direction, that every point of A is a support point (although A does not lie in a hyperplane).

For the rest of this section, we consider convex polytopes and show that every polytope P can be obtained as the intersection of finitely many supporting halfspaces of P. In other words, we show that polytopes are polyhedral sets. In the following definition, we distinguish support sets according to their dimension.

Definition 1.10 A support set F of a closed convex set $A \subset \mathbb{R}^n$ is called a *k-support set* if dim $F = k$, where $k \in \{0, \ldots, n-1\}$. The 1-support sets of A are called *edges*, and the $(n-1)$-support sets of A are called *facets*. Sometimes we also call the support sets of A of dimension dim$(A) - 1$ facets of A.

Theorem 1.18 *The 0-support sets of a polytope $P \subset \mathbb{R}^n$ are precisely the sets of the form $\{x\}$, $x \in$ vert P.*

Proof Let $\{x\}$ be a 0-support set of P. Hence there is a supporting hyperplane $\{f = \beta\}$ such that $P \subset \{f \leq \beta\}$ and $P \cap \{f = \beta\} = \{x\}$. But then $P \setminus \{x\} = P \cap \{f < \beta\}$ is convex, hence $x \in$ vert P.

Conversely, let $x \in$ vert P and let vert $(P) \setminus \{x\} = \{x_1, \ldots, x_k\}$, $k \geq 1$ (the case vert $(P) = \{x\}$ is obvious). Then, $x \notin Q := \mathrm{conv}\{x_1, \ldots, x_k\}$. By Theorem 1.14 there exists a supporting hyperplane $\{f = \alpha\}$ of Q through $p(Q, x)$ with supporting halfspace $\{f \leq \alpha\}$ and such that $\beta := f(x) > \alpha$. Let $y \in P$, that is,

$$y = \sum_{i=1}^{k} \alpha_i x_i + \alpha_{k+1} x, \quad \alpha_i \geq 0, \quad \sum_{i=1}^{k+1} \alpha_i = 1.$$

Then

$$f(y) = \sum_{i=1}^{k} \alpha_i \underbrace{f(x_i)}_{\leq \alpha < \beta} + \alpha_{k+1} f(x) \leq \beta$$

and equality holds if and only if $\alpha_1 = \cdots = \alpha_k = 0$ and $\alpha_{k+1} = 1$, that is, $y = x$. Hence $\{f \leq \beta\}$ is a supporting halfspace and $P \cap \{f = \beta\} = \{x\}$, thus x is a 0-support set of P. $\qquad\square$

Remark 1.22 In the following, we will no longer distinguish strictly between 0-support sets and vertices, although the former are sets consisting of one point and the latter are points.

Theorem 1.19 *Let $P \subset \mathbb{R}^n$ be a polytope with* vert $P = \{x_1, \ldots, x_k\}$ *and let F be a support set of P. Then $F = \text{conv}\{x_i : x_i \in F\}$.*

Proof Assume that $F = P \cap \{f = \alpha\}$ and $P \subset \{f \leq \alpha\}$. Suppose (w.l.o.g.) that $x_1, \ldots x_m \in F$, for some $m \in \{1, \ldots, k\}$, and $x_{m+1}, \ldots, x_k \notin F$. Then, $x_{m+1}, \ldots, x_k \in \{f < \alpha\}$, that is, $f(x_j) = \alpha - \delta_j, \delta_j > 0, j = m + 1, \ldots, k$.

Let $x \in P$ with $x = \alpha_1 x_1 + \cdots + \alpha_k x_k, \alpha_i \geq 0$, and $\alpha_1 + \cdots + \alpha_k = 1$. Then

$$f(x) = \alpha_1 f(x_1) + \cdots + \alpha_k f(x_k) = \alpha - \alpha_{m+1}\delta_{m+1} - \cdots - \alpha_k \delta_k.$$

Hence, $x \in F$ if and only if $\alpha_{m+1} = \cdots = \alpha_k = 0$. $\qquad\square$

Remark 1.23 Theorem 1.19 implies, in particular, that a support set of a polytope is a polytope and that there are only finitely many support sets.

Theorem 1.20 *Every polytope $P \subset \mathbb{R}^n$ is a polyhedral set.*

Proof If $\dim P = k < n$ and $E := \text{aff } P$, we first observe that E can be written as an intersection of $r = 2(d - k)$ halfspaces $\tilde{H}_1, \ldots, \tilde{H}_r$ in $\mathbb{R}^n$, $E = \bigcap_{j=1}^{r} \tilde{H}_j$. Suppose we already know that P is polyhedral in E, that is,

$$P = \bigcap_{i=1}^{m} H_i,$$

where $H_i \subset E$ are k-dimensional halfspaces. If $E^{\perp}$ denotes the linear subspace orthogonal to the linear subspace which spans E, then

$$P = \bigcap_{i=1}^{m}(H_i \oplus E^{\perp}) \cap \bigcap_{j=1}^{r} \tilde{H}_j.$$

Hence, it follows that P is polyhedral in $\mathbb{R}^n$.

It remains to treat the case $\dim P = n$. For this, let $F_1, \ldots, F_m$ be the support sets of P and let $H_1, \ldots, H_m$ be the corresponding supporting halfspaces, that is, $P \subset H_i$ and $F_i = P \cap \text{bd } H_i$ for $i = 1, \ldots, m$. Then we have

$$P \subset H_1 \cap \cdots \cap H_m =: P'.$$

Assume there is an $x \in P' \setminus P$. We choose $y \in \operatorname{int} P$ and consider $[y, x] \cap P$. Since P is compact and convex (and $x \notin P$), there is a $z \in (y, x)$ with $\{z\} = [y, x] \cap \operatorname{bd} P$. By the support theorem there is a supporting hyperplane of P through z, and hence there is a support set F_i of P with $z \in F_i \subset \operatorname{bd} H_i$. On the other hand, since $y \in \operatorname{int} H_i$, $x \in P' \subset H_i$, and $z \in (y, x)$, we have $z \in \operatorname{int} H_i$, a contradiction. □

Exercises and Supplements for Sect. 1.4

1. Let $A \subset \mathbb{R}^n$ be closed and $\operatorname{int} A \neq \emptyset$. Show that A is convex if and only if every boundary point of A is a support point. Does the assertion remain true without the assumption $\operatorname{int} A \neq \emptyset$?

2. Let $A \subset \mathbb{R}^n$ be closed. Suppose that for each $x \in \mathbb{R}^n$ there is a unique point $p(A, x) \in A$ such that $\|x - p(A, x)\| = \min\{\|x - y\| : y \in A\}$. Show that A is convex (Motzkin's theorem).

3.* Let $A \subset \mathbb{R}^n$ be nonempty, closed and convex. Show that A is compact if and only if for each $u \in \mathbb{S}^{n-1}$ there is some $\alpha \in \mathbb{R}$ such that $A \subset H^-(u, \alpha)$.

4. Let $A, B \subset \mathbb{R}^n$ be convex, A closed, B compact, and assume that $A \cap B = \emptyset$. Show that there is a hyperplane $\{f = \gamma\}$ and there is an $\varepsilon > 0$ such that $A \subset \{f \geq \gamma + \varepsilon\}$ and $B \subset \{f \leq \gamma - \varepsilon\}$.

5. A Bavarian farmer is the happy owner of a large herd of happy cows, consisting of totally black and totally white animals. One day he finds them sleeping in the sun in his largest meadow. Watching them, he notices that for any four cows it would be possible to build a straight fence separating the black cows from the white ones.

   Show that the farmer could build a straight fence, separating the whole herd into black and white animals.

   Hint: Cows are lazy. When they sleep, they sleep—even if you build a fence across the meadow. Warning: Cows are not convex and certainly they are not points.

6. Let $F_1, \ldots, F_m$ be the facets of an n-dimensional polytope $P \subset \mathbb{R}^n$, and let $H_1, \ldots, H_m$ be the corresponding supporting halfspaces containing P. Show that

$$P = \bigcap_{i=1}^{m} H_i. \tag{$*$}$$

(This is a generalization of the representation shown in the proof of Theorem 1.20.) Show further that the representation $(*)$ is minimal in the sense that, for each representation

$$P = \bigcap_{i \in I} \tilde{H}_i,$$

with a family of halfspaces $\{\tilde{H}_i : i \in I\}$, we have $\{H_1, \ldots, H_m\} \subset \{\tilde{H}_i : i \in I\}$.

7. Let $A_1, \ldots, A_{n+1} \subset \mathbb{R}^n$ be sets. Suppose that $x \in \bigcap_{i=1}^{n+1} \operatorname{conv}(A_i)$. Then there are points $a_i \in A_i$ for $i = 1, \ldots, n+1$ such that $x \in \operatorname{conv}\{a_1, \ldots, a_{n+1}\}$.

   Interpretation The points of the set A_i are assigned the colour i. Then the result says that if a point lies in the convex hull of the points having colour i, for $i = 1, \ldots, n+1$, then the point is in the convex hull of a colourful simplex the vertices of which show the $n+1$ different colours. In the special case of equal sets $A_1 = \ldots = A_{n+1} =: A$ (all the points of A exhibit all $n+1$ colours), the assertion follows from Carathéodory's theorem. For this reason, the assertion is a colourful version of Carathéodory's theorem.

8.* Let $A \subset \mathbb{R}^n$ be closed and convex. Show that the intersection of a family of support sets of A is a support set of A or the empty set.

9. Let $K_1, \ldots, K_m \subset \mathbb{R}^n$ be compact convex sets with $\bigcap_{i=1}^{m} K_i = \emptyset$. Show that there are closed halfspaces $H_1^+, \ldots, H_m^+ \subset \mathbb{R}^n$ such that $K_i \subset H_i^+$ for $i = 1, \ldots, m$ and such that $\bigcap_{i=1}^{m} H_i^+ = \emptyset$.

10. Let $A := \{(0, y, 1)^\top \in \mathbb{R}^3 : y \in \mathbb{R}\}$ and $B := \{(x, y, z)^\top \in \mathbb{R}^3 : x, y, z \geq 0, xy \geq z^2\}$. Show that A, B are disjoint, closed, convex sets. Determine the distance of the sets, that is,

$$d(A, B) := \inf\{\|a - b\| : a \in A, b \in B\}.$$

   Determine all hyperplanes separating A and B.

11.* For $A \subset \mathbb{R}^n$ and $a \in A$, let

$$N(A, a) := \{u \in \mathbb{R}^n : \langle u, x - a \rangle \leq 0 \text{ for } x \in A\}.$$

   Prove the following assertions.

   (a) $N(A, a)$ is a closed convex cone and $N(A, a) = N(\operatorname{cl} A, a)$. (A set $C \subset \mathbb{R}^n$ is called a cone if $\lambda C \subset C$ for $\lambda \geq 0$.)
   (b) If A is convex, then $N(A, a) = N(A \cap B^n(a, \varepsilon), a)$, for $\varepsilon > 0$.
   (c) If A is convex and $a \in \operatorname{bd} A$, then $\dim N(A, a) \geq 1$, whereas $N(A, a) = \{0\}$ if $a \in \operatorname{int} A$.

   The set $N(A, a)$ is called the *normal cone* of A at a.

12.* Let $A, B \subset \mathbb{R}^n$, $a \in A$ and $b \in B$. Show that

$$N(A + B, a + b) = N(A, a) \cap N(B, b).$$

13.* Let $A, B \subset \mathbb{R}^n$ be convex sets and $c \in A \cap B$. Suppose that $\operatorname{relint}(A) \cap \operatorname{relint}(B) \neq \emptyset$. Show that

$$N(A \cap B, c) = N(A, c) + N(B, c).$$

   In particular, the sum $N(A, c) + N(B, c)$ is closed.

1.5   Extremal Representations

In the previous section, we have seen that the trivial representation of a closed convex set $A \subset \mathbb{R}^n$ as the intersection of all closed convex sets containing A can be improved to a nontrivial one, where A is represented as the intersection of the supporting halfspaces of A. On the other hand, we have the trivial representation of A as the set of all convex combinations of points of A. Therefore, it is natural to discuss the corresponding nontrivial problem of finding a subset $B \subset A$, as small as possible, for which $A = \text{conv } B$ holds. Although there are some general results for closed convex sets A (see, e.g., Exercises 1.5.10 and 1.5.11), we shall concentrate on the compact case, where we can give a simple solution for this problem which is easy to state.

Definition 1.11 Let $A \subset \mathbb{R}^n$ be closed and convex. A point $x \in A$ is called an *extreme point* of A if x cannot be represented as a nontrivial convex combination of points of A, that is, if $x = \alpha y + (1 - \alpha)z$ with $y, z \in A$ and $\alpha \in (0, 1)$, implies that $x = y = z$. The set of all extreme points of A is denoted by $\text{ext } A$.

Remark 1.24 If A is a closed halfspace in $\mathbb{R}^n$ and $n \geq 2$, then $\text{ext } A = \emptyset$. In general, $\text{ext } A \neq \emptyset$ if and only if A does not contain any lines (see Exercise 1.5.1).

Remark 1.25 For $x \in A$, we have $x \in \text{ext } A$ if and only if $A \setminus \{x\}$ is convex. In fact, assume that $x \in \text{ext } A$. Let $y, z \in A \setminus \{x\}$. Then $[y, z] \subset A$. If $[y, z] \not\subset A \setminus \{x\}$, then $x = \alpha y + (1 - \alpha)z$ for some $\alpha \in (0, 1)$. Since $x \in \text{ext } A$, it follows that $x = y = z$, a contradiction. Hence $[y, z] \subset A \setminus \{x\}$, i.e., $A \setminus \{x\}$ is convex. Conversely, assume that $A \setminus \{x\}$ is convex. Let $y, z \in A$ and let $\alpha \in (0, 1)$ be such that $x = \alpha y + (1 - \alpha)z$. If $y \neq x$ and $z \neq x$, then $y, z \in A \setminus \{x\}$ and therefore $x \in [y, z] \subset A \setminus \{x\}$, a contradiction. Therefore, $y = x$ or $z = x$, which implies that $x = y = z$.

Remark 1.26 For a polytope P, Remark 1.25 yields that $\text{ext } P = \text{vert } P$.

Remark 1.27 If $\{x\}$ is a support set of A, then $x \in \text{ext } A$. The converse is false, as the simple example of a planar set A shows, where A is the sum of a circle and a segment. Each of the points x_i is extreme, but $\{x_i\}$ is not a support set (see Fig. 1.7).

Remark 1.27 explains why the following definition is relevant.

Fig. 1.7 A convex set A and the extreme points $x_1, \ldots, x_4$

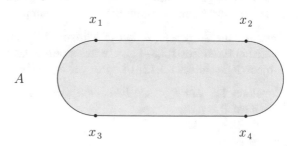

Definition 1.12 Let $A \subset \mathbb{R}^n$ be closed and convex. A point $x \in A$ is called an *exposed point* if $\{x\}$ is a support set (that is, a 0-support set) of A. The set of all exposed points of A is denoted by $\exp A$.

Remark 1.28 In view of Remark 1.27 we have $\exp A \subset \text{ext} A$.

Theorem 1.21 (Minkowski's Theorem) *Let $K \subset \mathbb{R}^n$ be compact and convex. Let $A \subset K$. Then $K = \text{conv} A$ if and only if $\text{ext} K \subset A$. In particular, $K = \text{conv}(\text{ext} K)$.*

Proof Suppose $K = \text{conv} A$ and $x \in \text{ext} K$. Assume $x \notin A$. Then $A \subset K \setminus \{x\}$. Since $K \setminus \{x\}$ is convex, $K = \text{conv} A \subset K \setminus \{x\}$, a contradiction.

In the other direction, we need only show that $K = \text{conv} \, \text{ext} K$. We prove this by induction on n. For $n = 1$, a compact convex subset of $\mathbb{R}^1$ is a segment $[a, b]$ and $\text{ext}[a, b] = \{a, b\}$.

Let $n \geq 2$ and suppose the result holds in dimension $n - 1$. Since $\text{ext} K \subset K$, we obviously have $\text{conv} \, \text{ext} K \subset K$. To prove the reverse inclusion, let $x \in K$ and let g be an arbitrary line through x. Then $g \cap K = [y, z]$ with $x \in [y, z]$ and $y, z \in \text{bd} K$. By the support theorem, y, z are support points, that is, there are supporting hyperplanes E_y, E_z of K with $y \in K_1 := E_y \cap K$ and $z \in K_2 := E_z \cap K$. By the induction hypothesis, applied in E_y, E_z, we get

$$K_1 = \text{conv}(\text{ext} K_1), \quad K_2 = \text{conv}(\text{ext} K_2).$$

We have $\text{ext} K_1 \subset \text{ext} K$. Namely, consider $u \in \text{ext} K_1$ and suppose that $u = \alpha v + (1 - \alpha)w$ for $v, w \in K$ and $\alpha \in (0, 1)$. Since u lies in the supporting hyperplane E_y, the same must hold for v and w. Hence $v, w \in K_1$ and since $u \in \text{ext} K_1$, we obtain $u = v = w$. Therefore, $u \in \text{ext} K$.

In the same way, we get $\text{ext} K_2 \subset \text{ext} K$ and thus

$$x \in [y, z] \subset \text{conv}\,(\text{conv}(\text{ext} K_1) \cup \text{conv}(\text{ext} K_2))$$

$$\subset \text{conv}(\text{ext} K),$$

which completes the proof.   □

Remark 1.29 If $K \subset \mathbb{R}^n$ is compact and convex then $\text{ext} K$ is a closed set for $n = 2$, but in general this is not true for $n \geq 3$ as examples show. In view of this, it is perhaps surprising that $\text{conv} \, \text{ext} K$ is still compact, as Theorem 1.21 shows.

Remark 1.30 A generalization of Minkowski's theorem to closed convex sets is treated in Exercises 1.5.8–11. A representation theorem for polyhedral cones is the subject of Exercises 1.5.12–14.

Corollary 1.5 *Let $P \subset \mathbb{R}^n$ be compact and convex. Then P is a polytope if and only if $\text{ext} P$ is finite.*

Proof If P is a polytope, then Theorem 1.4 and Remark 1.26 show that ext P is finite. For the converse, assume that ext P is finite, hence ext $P = \{x_1, \ldots, x_k\}$. Theorem 1.21 then shows $P = \mathrm{conv}\{x_1, \ldots, x_k\}$, hence P is a polytope. □

Now we are able to prove a converse of Theorem 1.20.

Theorem 1.22 *Let $P \subset \mathbb{R}^n$ be a bounded polyhedral set. Then P is a polytope.*

Proof Clearly, P is compact and convex. We show that ext P is finite.

Let $x \in$ ext P and assume $P = \bigcap_{i=1}^{k} H_i$ with halfspaces H_i bounded by the hyperplanes $E_i, i = 1, \ldots, k$. We consider the convex set

$$D := \bigcap_{i=1}^{k} A_i,$$

where

$$A_i = \begin{cases} E_i, & x \in E_i, \\ \mathrm{int}\, H_i, & x \notin E_i. \end{cases}$$

Then $x \in D \subset P$. Since x is an extreme point and D is relatively open as the intersection of an affine subspace and an open set, we get dim $D = 0$, hence $D = \{x\}$. Since there are only finitely many different sets D, ext P must be finite. The result now follows from Corollary 1.5. □

Remark 1.31 This result together with Theorem 1.20 now shows that the intersection of finitely many polytopes is again a polytope.

If in Theorem 1.21 we replace the set ext K by exp K, the corresponding result will be wrong in general, as simple examples show (compare Theorem 1.21 and Remark 1.27). In particular, even in the plane the set of exposed points of a compact convex set need not be compact. There is, however, a modified version of Theorem 1.21 which holds for exposed points.

Theorem 1.23 *Let $K \subset \mathbb{R}^n$ be compact and convex. Then*

$$K = \mathrm{cl}\,\mathrm{conv}(\exp K).$$

Proof For the proof (see Fig. 1.8 for an illustration), we can assume that K consists of more than one point. Since K is compact, for each $x \in \mathbb{R}^n$ there exists a point $y_x \in K$ farthest away from x, that is, a point with

$$\|y_x - x\| = \max_{y \in K} \|y - x\|.$$

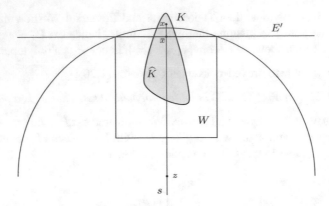

Fig. 1.8 Illustration for the proof of Theorem 1.23

The hyperplane E through y_x orthogonal to $y_x - x \neq 0$ is then a supporting hyperplane of K and we have $E \cap K = \{y_x\}$, hence $y_x \in \exp K$. Let

$$\widehat{K} := \operatorname{cl} \operatorname{conv} \{y_x : x \in \mathbb{R}^n\}.$$

Then $\widehat{K} \subset K$, thus $\widehat{K}$ is compact and convex.

Assume that there exists an $x \in K \setminus \widehat{K}$. Then, by Theorem 1.14 the hyperplane E' through $\bar{x} := p(\widehat{K}, x)$ with normal $x - \bar{x}$ is a supporting hyperplane of $\widehat{K}$. Consider the ray $s := x + [0, \infty)(\bar{x} - x)$. On s we can find a point z with

$$\|x - z\| > \max_{y \in \widehat{K}} \|y - z\|. \tag{1.4}$$

To see this, we first choose an orthogonal spherical cylinder

$$W \subset \{z \in \mathbb{R}^n : \langle z - \bar{x}, x - \bar{x} \rangle \leq 0\},$$

with radius r and height $2r$, where r is large enough so that $\widehat{K} \subset W$, and such that $\bar{x}$ is the center of one of the two bases of W. Then we choose $z \in s$ such that $\|z - \bar{x}\| \geq \max\{r, (r^2 - \delta^2)/(2\delta)\}$ with $\delta := \|x - \bar{x}\|$. This implies that

$$\|z - a\|^2 \leq \|z - \bar{x}\|^2 + r^2 < \|z - x\|^2 = (\|z - \bar{x}\| + \delta)^2$$

for $a \in W$, and hence there is a ball B with center $z \in s$ and such that $\widehat{K} \subset W \subset B$, but $x \notin B$, which yields (1.4).

By definition of $\widehat{K}$, there exists a $y_z \in \widehat{K}$ with

$$\|y_z - z\| = \max_{y \in K} \|y - z\| \geq \|x - z\| > \max_{y \in \widehat{K}} \|y - z\|,$$

by (1.4), a contradiction. Therefore, $K = \widehat{K}$. Since $y_x \in \exp K$, for $x \in \mathbb{R}^n$, we obtain

$$K = \widehat{K} \subset \mathrm{cl\ conv\ exp}\ K \subset K,$$

hence $K = \mathrm{cl\ conv\ exp}\ K$. □

Corollary 1.6 (Straszewicz's Theorem) *Let $K \subset \mathbb{R}^n$ be compact and convex. Then*

$$\mathrm{ext}\ K \subset \mathrm{cl\ exp}\ K.$$

Proof By Theorems 1.23 and 1.12, we have

$$K = \mathrm{cl\ conv\ exp}\ K \subset \mathrm{cl\ conv\ cl\ exp}\ K = \mathrm{conv\ cl\ exp}\ K \subset K,$$

hence

$$K = \mathrm{conv\ cl\ exp}\ K.$$

By Theorem 1.21, this implies that $\mathrm{ext}\ K \subset \mathrm{cl\ exp}\ K$. □

Exercises and Supplements for Sect. 1.5

1. Let $A \subset \mathbb{R}^n$ be closed and convex. Show that $\mathrm{ext}\ A \neq \emptyset$ if and only if A does not contain any line.
2. Let $K \subset \mathbb{R}^n$ be compact and convex.

   (a) If $n = 2$, show that $\mathrm{ext}\ K$ is closed.
   (b) If $n \geq 3$, show by an example that $\mathrm{ext}\ K$ need not be closed.

3.* Let $A \subset \mathbb{R}^n$ be closed and convex. A subset $M \subset A$ is called an *extreme set* in A (or a *face* of A) if M is convex and if $x, y \in A$, $(x, y) \cap M \neq \emptyset$ implies that $[x, y] \subset M$. The set A and $\emptyset$ are faces of A, all other faces are called proper. Prove the following assertions.

   (a) Extreme sets M are closed.
   (b) Each support set of A is extreme.
   (c) If $M, N \subset A$ are extreme, then $M \cap N$ is extreme. (This extends to arbitrary families of extreme sets.)
   (d) If M is extreme in A and $N \subset M$ is extreme in M, then N is extreme in A.
   (e) If $M, N \subset A$ are extreme and $M \neq N$, then $\mathrm{relint}\ M \cap \mathrm{relint}\ N = \emptyset$.
   (f) Let B be a nonempty and relatively open subset of A. Then there is a unique face F of A such that $B \subset \mathrm{relint}(F)$.

(g) Let $\mathcal{E}(A) := \{M \subset A : M \text{ extreme}\}$. Then $A = \bigcup\limits_{M \in \mathcal{E}(A)} \text{relint } M$ is a
disjoint union.

(h) Extreme sets of A of dimension $\dim(A) - 1$ are always support sets.

4. A real (n, n)-matrix $A = ((\alpha_{ij}))$ is called *doubly stochastic* if $\alpha_{ij} \geq 0$ and

$$\sum_{k=1}^{n} \alpha_{kj} = \sum_{k=1}^{n} \alpha_{ik} = 1$$

for $i, j \in \{1, \ldots, n\}$. A doubly stochastic matrix with components in $\{0, 1\}$ is called a *permutation matrix*.
Prove the following statements.

(a) The set $K \subset \mathbb{R}^{n^2}$ of doubly stochastic matrices is compact and convex.
(b) The extreme points of K are precisely the permutation matrices.

5. Let $P \subset \mathbb{R}^n$ be a polyhedral set, but not an affine subspace (a flat). Let

$$P = \text{aff}(P) \cap \bigcap_{i=1}^{m} H^-(u_i, \alpha_i)$$

with $\alpha_1, \ldots, \alpha_m \in \mathbb{R}$ and $u_1, \ldots, u_m \in \mathbb{R}^n \setminus \{0\}$ be a representation of P in which none of the halfspaces $H^-(u_i, \alpha_i)$ can be omitted. Put $F_i := P \cap H(u_i, \alpha_i)$. Then

(a) $\text{relint}(P) = \bigcap_{i=1}^{m} \{x \in P : \langle x, u_i \rangle < \alpha_i\}$.
(b) $\text{relbd}(P) = \bigcup_{i=1}^{m} F_i$.
(c) $F_1, \ldots, F_m$ are precisely the facets of P.
(d) Each proper face F of P is equal to the intersection of all facets of P which contain F.
(e) The number of faces of P is finite. Each face of P is a support set and polyhedral (or empty).

6. Let $P \subset \mathbb{R}^n$ be a polyhedral set, but not an affine subspace (a flat). Prove the following statements.

(a) If $0 \leq j \leq k$, $F^j \in \mathcal{F}_j(P)$, $F^k \in \mathcal{F}_k(P)$ and $F^j \subset F^k$, then there are $F^i \in \mathcal{F}_i(P)$ for $i = j+1, \ldots, k-1$ with $F^j \subset F^{j+1} \subset \cdots \subset F^{k-1} \subset F^k$.
(b) If $0 \leq j \leq k < \dim(P)$ and $F^j \in \mathcal{F}_j(P)$, then $F^j = \bigcap\{F \in \mathcal{F}_k(P) : F^j \subset F\}$.
(c) If $\dim(P) = n$, then each $(n-2)$-dimensional face of P is contained in precisely two facets of P.

7. Let $\emptyset \neq P \subset \mathbb{R}^n$ be closed and convex. If the number of different support sets of P is finite, then P is a polyhedral set.

8. Let $A \subset \mathbb{R}^n$ be a closed convex set with $A \neq \operatorname{conv}(\operatorname{relbd}(A))$. Then A is a flat or a semi-flat.
9. Let $\emptyset \neq A \subset \mathbb{R}^n$ be closed and convex. The *recession cone* of A is defined by

$$\operatorname{rec}(A) := \{u \in \mathbb{R}^n : A + u \subset A\}.$$

Show that the recession cone of A is a closed convex cone.
For each $x \in \mathbb{R}^n$ and $u \in \mathbb{R}^n$, let $G^+(x, u) := \{x + \lambda u : \lambda \geq 0\}$. Then, for an arbitrary $x \in A$,

$$\operatorname{rec}(A) = \{u \in \mathbb{R}^n : G^+(x, u) \subset A\}.$$

Moreover, if $G^+(x, u) \subset A$ and $y \in A$, then also $G^+(y, u) \subset A$.
10. Each closed convex set $A \subset \mathbb{R}^n$ can be written in the form $A = \overline{A} \oplus V$, where V is a linear subspace and $\overline{A}$ is a line-free, closed convex set which is contained in a linear subspace complementary to V.
11. Let $A \subset \mathbb{R}^n$ be closed and convex. A ray $G^+(x, u) \subset A$ is called extreme if it is an extreme set in A. The union of all extreme rays of A is denoted by $\operatorname{extr}(A)$. By definition, it is clear that if $G^+(x, u)$ is an extreme ray of A, then $x \in \operatorname{ext}(A)$. Prove the following representation result.
    If $A \subset \mathbb{R}^n$ is line-free, closed and convex, then

$$A = \operatorname{conv}(\operatorname{ext}(A) \cup \operatorname{extr}(A)) = \operatorname{conv}(\operatorname{ext}(A)) + \operatorname{rec}(A).$$

12. A set of the form

$$\operatorname{pos}\{a_1, \dots, a_m\} = \left\{ \sum_{i=1}^m \lambda_i a_i : \lambda_i \geq 0 \text{ for } i = 1, \dots, m \right\}$$

with $a_1, \dots, a_m \in \mathbb{R}^n$ is called a *finitely generated convex cone*.
The convex hull of finitely many points is a polytope and hence a polyhedral set (as shown before). Show the following fact, which states that the positive hull of finitely many vectors is a polyhedral set: A finitely generated convex cone in $\mathbb{R}^n$ is a polyhedral set, in particular, it is a closed set.
13. Let $P \subset \mathbb{R}^n$ be a polyhedral set. Then there are points $a_1, \dots, a_m \in \mathbb{R}^n, m \geq 1$, and vectors $b_1, \dots, b_p \in \mathbb{R}^n$ with $P = \operatorname{conv}\{a_1, \dots, a_m\} + \operatorname{pos}\{b_1, \dots, b_p\}$. In particular, a bounded polyhedral set is a polytope.
14. Let points $a_1, \dots, a_m \in \mathbb{R}^n$ with $m \geq 1$ and vectors $b_1, \dots, b_p \in \mathbb{R}^n$ be given. Then $P := \operatorname{conv}\{a_1, \dots, a_m\} + \operatorname{pos}\{b_1, \dots, b_p\}$ is a polyhedral set.
15. Hints to the literature: Convex polytopes and polyhedral sets are treated in greater detail in [2, 23, 39, 40, 49, 64, 71, 93, 94]. Combinatorial aspects of convexity are in the focus of [5, 17, 29, 44, 60, 68, 71, 75, 90]. The connection between discrete and convex geometry is the subject of [17, 19, 21, 36, 38, 60, 71, 77, 95–97]. For algorithmic aspects and combinatorial geometry, see [16, 28, 29, 36, 49, 54, 74, 75].

Chapter 2
Convex Functions

In this chapter, we study convex functions, which are the analytic counterpart of convex sets. There are many relations between convex functions and sets. Our definition of a convex function via the convexity of its epigraph emphasizes this connection. For convex functions, the study of regularity properties such as continuity or differentiability is particularly natural, but in turn this suggests to consider smoothness of convex sets as well. A strong and very useful link between sets and functions is provided by the support function of a convex set. This tool will be crucial throughout the book.

2.1 Properties and Operations

In the following, we consider functions

$$f : \mathbb{R}^n \to [-\infty, \infty].$$

We assume the usual rules for addition and multiplication with ∞, namely:

$$
\begin{aligned}
\alpha + \infty &:= \infty, & &\text{for } \alpha \in (-\infty, \infty], \\
\alpha - \infty &:= -\infty, & &\text{for } \alpha \in [-\infty, \infty), \\
\alpha \infty &:= \infty, \ (-\alpha)\infty := -\infty, & &\text{for } \alpha \in (0, \infty].
\end{aligned}
$$

In the context of measure and integration theory, it is often convenient to use the convention $0\,\infty := 0$, where ∞ is the value of a function and 0 is the measure of the set of points where the function is infinite.

Definition 2.1 For a function $f : \mathbb{R}^n \to (-\infty, \infty]$, the set

$$\operatorname{epi} f := \{(x, \alpha) : x \in \mathbb{R}^n, \alpha \in \mathbb{R}, f(x) \le \alpha\} \subset \mathbb{R}^n \times \mathbb{R}$$

© Springer Nature Switzerland AG 2020
D. Hug, W. Weil, *Lectures on Convex Geometry*, Graduate Texts
in Mathematics 286, https://doi.org/10.1007/978-3-030-50180-8_2

is called the *epigraph* of f. A function f is *convex*, if epi f is a convex subset of $\mathbb{R}^n \times \mathbb{R} = \mathbb{R}^{n+1}$.

The following remarks contain further definitions and straightforward consequences.

Remark 2.1 A function $f : \mathbb{R}^n \to [-\infty, \infty)$ is said to be *concave* if $-f$ is convex. Thus, for a convex function f we exclude the value $-\infty$, whereas for a concave function we exclude ∞.

Remark 2.2 If $A \subset \mathbb{R}^n$ is a subset of $\mathbb{R}^n$, a function $f : A \to (-\infty, \infty)$ is called *convex* if the extended function $\tilde{f} : \mathbb{R}^n \to (-\infty, \infty]$, given by

$$\tilde{f}(x) := \begin{cases} f(x), & x \in A, \\ \infty, & x \in \mathbb{R}^n \setminus A, \end{cases}$$

is convex. This automatically requires that A is a convex set. In view of this construction, we usually need not consider convex functions defined on subsets of $\mathbb{R}^n$, but instead we can assume that convex functions are defined on all of $\mathbb{R}^n$.

Remark 2.3 On the other hand, we often are only interested in convex functions $f : \mathbb{R}^n \to (-\infty, \infty]$ at points where f is finite. We call

$$\mathrm{dom}\, f := \{x \in \mathbb{R}^n : f(x) < \infty\}$$

the *effective domain* of the function $f : \mathbb{R}^n \to (-\infty, \infty]$. For a convex function f, the effective domain dom f is a convex subset of $\mathbb{R}^n$.

Remark 2.4 The function $f \equiv \infty$ is convex, it is called the *improper* convex function; convex functions f with $f \not\equiv \infty$ are called *proper*. The improper convex function $f \equiv \infty$ has epi $f = \emptyset$ and dom $f = \emptyset$.

Theorem 2.1 *A function $f : \mathbb{R}^n \to (-\infty, \infty]$ is convex if and only if*

$$f(\alpha x + (1 - \alpha)y) \leq \alpha f(x) + (1 - \alpha)f(y)$$

for $x, y \in \mathbb{R}^n$ and $\alpha \in [0, 1]$. For a concave function, the inequality is reversed.

Proof By definition, f is convex if and only if epi $f = \{(x, \beta) \in \mathbb{R}^n \times \mathbb{R} : f(x) \leq \beta\}$ is convex. The latter condition means that

$$\alpha(x_1, \beta_1) + (1 - \alpha)(x_2, \beta_2) = (\alpha x_1 + (1 - \alpha)x_2, \alpha\beta_1 + (1 - \alpha)\beta_2) \in \text{epi } f$$

for $\alpha \in [0, 1]$ and whenever $(x_1, \beta_1), (x_2, \beta_2) \in \text{epi } f$, that is, whenever $f(x_1) \leq \beta_1$, $f(x_2) \leq \beta_2$.

Hence, f is convex if and only if

$$f(\alpha x_1 + (1 - \alpha)x_2) \leq \alpha\beta_1 + (1 - \alpha)\beta_2,$$

for $x_1, x_2 \in \mathbb{R}^n, \alpha \in [0, 1]$ and $\beta_1 \geq f(x_1), \beta_2 \geq f(x_2)$. Then, it is necessary and sufficient that this inequality is satisfied for $\beta_1 = f(x_1), \beta_2 = f(x_2)$, since the inequality is always satisfied if $f(x_1) = \infty$ or $f(x_2) = \infty$, and we obtain the assertion.

The assertion for a concave function f follows by applying to $-f$ the characterization just proved. $\qquad \square$

Remark 2.5 A function $f : \mathbb{R}^n \to \mathbb{R}$ is affine if and only if f is convex and concave. If f is affine, then epi f is a halfspace in $\mathbb{R}^{n+1}$ and dom $f = \mathbb{R}^n$.

Remark 2.6 For a convex function f, the sublevel sets $\{f < \alpha\}$ and $\{f \leq \alpha\}$ are convex subsets of $\mathbb{R}^n$. The converse is not true, that is, if all sublevel sets of a function are convex, the function need not be convex. A function $f : \mathbb{R}^n \to \mathbb{R}$ for which $\{f \leq \alpha\} \subset \mathbb{R}^n$ is convex for all $\alpha \in \mathbb{R}$ is called *quasi-convex*. Note that $f : \mathbb{R}^n \to \mathbb{R}$ is quasi-convex if and only if

$$f(\lambda x + (1 - \lambda)y) \leq \max\{f(x), f(y)\}$$

whenever $x, y \in \mathbb{R}^n$ and $\lambda \in [0, 1]$.

Remark 2.7 If f, g are convex and $\alpha, \beta \geq 0$, then $\alpha f + \beta g$ is convex.

Remark 2.8 If $(f_i)_{i \in I}$ is a family of convex functions, the (pointwise) supremum $\sup_{i \in I} f_i$ is convex. This follows since

$$\text{epi}\left(\sup_{i \in I} f_i\right) = \bigcap_{i \in I} \text{epi } f_i.$$

It can also be verified by using the analytic description given in Theorem 2.1.

Remark 2.9 As a generalization of Theorem 2.1, we obtain that f is convex if and only if

$$f(\alpha_1 x_1 + \cdots + \alpha_k x_k) \leq \alpha_1 f(x_1) + \cdots + \alpha_k f(x_k)$$

for $k \in \mathbb{N}, x_i \in \mathbb{R}^n$, and $\alpha_i \in [0, 1]$ with $\sum \alpha_i = 1$. This follows easily by induction over k or by arguing via the epigraph.

Remark 2.10 A function $f : \mathbb{R}^n \to (-\infty, \infty]$ is *positively homogeneous* (of degree 1) if

$$f(\alpha x) = \alpha f(x) \quad \text{for all } x \in \mathbb{R}^n, \alpha \geq 0.$$

Sometimes this is only required for $\alpha > 0$. If f is positively homogeneous, f is convex if and only if it is *subadditive*, that is, if

$$f(x + y) \leq f(x) + f(y)$$

for $x, y \in \mathbb{R}^n$.

The following simple result is useful for generating convex functions from convex sets in $\mathbb{R}^n \times \mathbb{R}$. Note that $\inf \emptyset = \infty$.

Theorem 2.2 *Let $A \subset \mathbb{R}^n \times \mathbb{R}$ be convex and suppose that*

$$f_A(x) := \inf\{\alpha \in \mathbb{R} : (x, \alpha) \in A\} > -\infty$$

for $x \in \mathbb{R}^n$. Then f_A is a convex function.

Proof The definition of $f_A(x)$ implies that

$$\text{epi } f_A = \{(x, \beta) : \exists \alpha \in \mathbb{R}, \alpha \le \beta, \text{ and a sequence } \gamma_i \searrow \alpha \text{ with } (x, \gamma_i) \in A\}.$$

It is easy to see that epi f_A is convex.

A variant of the proof is based on Theorem 2.1. For this, let $x, y \in \mathbb{R}^n$ and $\lambda \in [0, 1]$. We can assume that $f_A(x), f_A(y) < \infty$. Let $\varepsilon > 0$ be given. Then there are $\alpha_x, \alpha_y \in \mathbb{R}$ such that $f_A(x) + \varepsilon > \alpha_x$, $f_A(y) + \varepsilon > \alpha_y$ and $(x, \alpha_x), (y, \alpha_y) \in A$. Since

$$((1-\lambda)x + \lambda y, (1-\lambda)\alpha_x + \lambda\alpha_y) = (1-\lambda)(x, \alpha_x) + \lambda(y, \alpha_y) \in A,$$

we get

$$f_A((1-\lambda)x + \lambda y) \le (1-\lambda)\alpha_x + \lambda\alpha_y$$
$$\le (1-\lambda)(f_A(x) + \varepsilon) + \lambda(f_A(y) + \varepsilon)$$
$$= (1-\lambda)f_A(x) + \lambda f_A(y) + \varepsilon,$$

which yields the assertion, since $\varepsilon > 0$ was arbitrary. $\square$

Remark 2.11 The condition $f_A > -\infty$ is satisfied if and only if A does not contain a vertical half-line which is unbounded from below.

Remark 2.12 For $x \in \mathbb{R}^n$, let $\{x\} \times \mathbb{R} := \{(x, \alpha) : \alpha \in \mathbb{R}\}$ be the *vertical line* in $\mathbb{R}^n \times \mathbb{R}$ through x. Let $A \subset \mathbb{R}^n \times \mathbb{R}$ be closed and convex and assume that $f_A(x) > -\infty$ for $x \in \mathbb{R}^n$. Then $A = \text{epi } f_A$ if and only if

$$A \cap (\{x\} \times \mathbb{R}) = \{x\} \times [f_A(x), \infty) \quad \text{for } x \in \mathbb{R}^n. \tag{2.1}$$

Theorem 2.2 allows us to define operations on convex functions by applying corresponding operations on convex sets to the epigraphs of the functions. We give two examples of that kind.

Definition 2.2 A convex function $f : \mathbb{R}^n \to (-\infty, \infty]$ is *closed* if epi f is closed.

Example 2.1 Let $f : \mathbb{R}^n \to (-\infty, \infty]$ be convex. Then $A := \text{cl epi } f \subset \mathbb{R}^{n+1}$ is a closed convex set. We assert that $f_A > -\infty$. The case $f \equiv \infty$ is trivial, since then $f \equiv \infty$, epi $f = \emptyset$ and $f_A = f$. In particular, f is closed. If f

is proper, that is, epi $f \neq \emptyset$, then w.l.o.g. we may assume that $\dim \operatorname{dom} f = n$ (since it is sufficient to work in the affine hull of $\operatorname{dom} f$). We choose a point $x \in \operatorname{int} \operatorname{dom} f$. Then, $(x, f(x)) \in \operatorname{bd} \operatorname{epi} f$. Hence, there is a supporting hyperplane $E \subset \mathbb{R}^n \times \mathbb{R}$ of $\operatorname{cl} \operatorname{epi} f$ at $(x, f(x))$. Since $x \in \operatorname{int} \operatorname{dom} f$, E is not vertical. Hence, the corresponding supporting halfspace is the epigraph of an affine function $h \leq f$. Thus, we get $f_A \geq h > -\infty$.

In addition, it is easy to check that for $A = \operatorname{cl} \operatorname{epi} f$ condition (2.1) is satisfied. In fact, if $(x', b) \in A \cap (\{x\} \times \mathbb{R})$, then $x' = x$, $b \in \mathbb{R}$ and $(x, b) \in A$. But then $f_A(x) \leq b$ and therefore $(x', b) \in \{x\} \times [f_A(x), \infty)$. Conversely, suppose that $(x', b) \in \{x\} \times [f_A(x), \infty)$. Then $x' = x$ and $f_A(x) \leq b < \infty$. Then there exists a sequence $\alpha_i \downarrow f_A(x)$ with $(x, \alpha_i) \in A$. Since A is closed, we get $(x, f_A(x)) \in A = \operatorname{cl} \operatorname{epi} f$. If $b = f_A(x)$, nothing remains to be shown. If $b > f_A(x)$, then there are $(x_i, \gamma_i) \in \operatorname{epi} f$ for $i \in \mathbb{N}$ with $(x_i, \gamma_i) \to (x, f_A(x))$ as $i \to \infty$. If i is large enough, we have $\gamma_i < b$, hence $(x_i, b) \in \operatorname{epi} f$, and therefore $(x, b) \in \operatorname{cl} \operatorname{epi} f = A$.

Hence we can define $\operatorname{cl} f := f_A$ with $A = \operatorname{cl} \operatorname{epi} f$. Moreover, we have $A = \operatorname{epi} f_A$, which can be rewritten in the form $\operatorname{cl} \operatorname{epi} f = \operatorname{epi} \operatorname{cl} f$. In other words, $\operatorname{cl} \operatorname{epi} f$ is the epigraph of a closed convex function, which we denote by $\operatorname{cl} f$, and $\operatorname{cl} f$ is the largest closed convex function smaller than or equal to f. It is clear that $\operatorname{cl} f = f_{\operatorname{cl} \operatorname{epi} f}$ is convex and closed, since $\operatorname{epi}(\operatorname{cl} f) = \operatorname{cl}(\operatorname{epi} f)$ is closed. Let g be a closed, convex function with $g \leq f$. Then $\operatorname{epi} f \subset \operatorname{epi} g$, hence $\operatorname{epi} \operatorname{cl} f = \operatorname{cl} \operatorname{epi} f \subset \operatorname{cl} \operatorname{epi} g = \operatorname{epi} g$, since g is closed. But this implies that $g \leq \operatorname{cl} f$.

Example 2.2 Our second example is the convex hull operator. If $(f_i)_{i \in I}$ is a family of (arbitrary) functions $f_i : \mathbb{R}^n \to (-\infty, \infty]$, we consider $B := \bigcup_{i \in I} \operatorname{epi} f_i$. Suppose $A = \operatorname{conv} B$ does not contain any vertical half-line which is unbounded from below. Then, by Theorem 2.2, $\operatorname{conv}(f_i)_{i \in I} := f_A$ is a convex function, which we call the *convex hull* of the functions f_i, $i \in I$. Then $\operatorname{conv}(f_i)_{i \in I}$ is the largest convex function less than or equal to f_i for $i \in I$, that is,

$$\operatorname{conv}(f_i)_{i \in I} = \sup\{g : g \text{ convex}, \ g \leq f_i \text{ for } i \in I\} =: h.$$

To see this, we observe that $h \leq f_i$ for $i \in I$ implies that $\operatorname{epi}(f_i) \subset \operatorname{epi}(h)$ for $i \in I$, and hence $\operatorname{conv}\left(\bigcup_{i \in I} \operatorname{epi}(f_i)\right) \subset \operatorname{epi}(h)$, since h is a convex function (as a supremum of convex functions) and $\operatorname{epi}(h)$ is a convex set. This shows that $h \leq f_A = \operatorname{conv}(f_i)_{i \in I}$. For the reverse inequality, let $i \in I$ and suppose first that $f_i(x) < \infty$. Then $(x, f_i(x)) \in \operatorname{epi}(f_i) \subset \operatorname{conv}\left(\bigcup_{i \in I} \operatorname{epi}(f_i)\right)$, and therefore $f_A(x) \leq f_i(x)$, which remains true if $f_i(x) = \infty$. Since f_A is convex and $i \in I$ is arbitrary, we conclude that $f_A \leq h$.

Furthermore, $\operatorname{conv}(f_i)$ exists if and only if there is an affine function h with $h \leq f_i$ for $i \in I$. In fact, if h is affine and $h \leq f_i$ for $i \in I$, then $\operatorname{conv}\left(\bigcup_{i \in I} \operatorname{epi}(f_i)\right) \subset \operatorname{epi} h$, and hence $f_A(x) \geq h(x) > -\infty$ for $x \in \mathbb{R}^n$.

Now suppose that $f_A = \operatorname{conv}(f_i)$ exists. Let $(x, \beta) \in \operatorname{relint} A$ and $(x, \alpha) \in \operatorname{relbd} A = \operatorname{relbd} \operatorname{cl} A$. Then there is a supporting hyperplane H' of $\operatorname{cl} A$ through (x, α) in $\operatorname{aff} A$, which is not vertical, since $(x, \beta) \in \operatorname{relint} A = \operatorname{relint} \operatorname{cl} A$. If $L(A)$

denotes the linear subspace parallel to aff A, then $H := H' + L(A)^\perp$ is a supporting hyperplane of cl A in $\mathbb{R}^{n+1}$ which is not vertical. But then $H = \{(z, h(z)) : z \in \mathbb{R}^n\}$ with an affine function h and $f_i \geq h$ for $i \in I$.

Further applications of Theorem 2.2 are listed in the exercises.

The following representation of convex functions is a counterpart to the support theorem for convex sets. It provides a characterization of closed convex functions, since for any family $\mathcal{H}$ of affine functions the function $\sup\{h : h \in \mathcal{H}\}$ is closed and convex (recall that the epigraph of the sup is the intersection of the epigraphs).

Theorem 2.3 *Let* $f : \mathbb{R}^n \to (-\infty, \infty]$ *be closed and convex. Then*

$$f = \sup\{h : h \leq f, h \text{ affine}\}.$$

Proof By assumption, epi f is closed and convex. Moreover, we can assume that f is proper, i.e., epi $f \neq \emptyset$. By Corollary 1.4, epi f is the intersection of all closed halfspaces $H \subset \mathbb{R}^n \times \mathbb{R}$ which contain epi f.

There are three types of closed halfspaces in $\mathbb{R}^n \times \mathbb{R}$:

$$H_1 = \{(x, r) : r \geq l(x)\}, \quad l : \mathbb{R}^n \to \mathbb{R} \text{ affine,}$$

$$H_2 = \{(x, r) : r \leq l(x)\}, \quad l : \mathbb{R}^n \to \mathbb{R} \text{ affine,}$$

$$H_3 = \widetilde{H} \times \mathbb{R}, \quad\quad\quad \widetilde{H} \text{ halfspace in } \mathbb{R}^n.$$

Halfspaces of type H_2 cannot occur, due to the definition of epi f and since epi $f \neq \emptyset$. Halfspaces of type H_3 can occur, hence we have to show that these 'vertical' halfspaces can be avoided, i.e., epi f is the intersection of all halfspaces of type H_1 containing epi f. Then the proof will be finished since the intersection of halfspaces of type H_1 is the epigraph of the supremum of the corresponding affine functions l.

For the result just explained it is sufficient to show that any point $(x_0, r_0) \notin$ epi f can be separated by a non-vertical hyperplane E from epi f. Hence, let E_3 be a vertical hyperplane separating (x_0, r_0) and epi f, obtained from Theorem 1.14, and let H_3 be the corresponding vertical halfspace containing epi f. We may represent H_3 as

$$H_3 = \{(x, r) \in \mathbb{R}^n \times \mathbb{R} : l_0(x) \leq 0\}$$

for some affine function $l_0 : \mathbb{R}^n \to \mathbb{R}$, and we may assume that $l_0(x_0) > 0$ and $l_0(x) \leq 0$ for $x \in \text{dom } f$.

Next we show that there exists an affine function l_1 with $l_1 \leq f$. To verify this, observe that since $f > -\infty$ and $f \not\equiv \infty$, there is some $(x_1, f(x_1)) \in \text{bd epi } f$. Then $(x_1, f(x_1) - 1) \notin \text{epi } f$, and since epi f is closed and convex this point can be strongly separated from epi f by a hyperplane (see Exercise 1.4.4). Thus there are $(u, v) \in (\mathbb{R}^n \times \mathbb{R}) \setminus \{(0, 0)\}$ and $\alpha \in \mathbb{R}$ such that $\langle (x_1, f(x_1) - 1), (u, v) \rangle < \alpha$ and $\langle (x, f(x) + s), (u, v) \rangle \geq \alpha$ for $x \in \text{dom } f$ and $s \geq 0$. Choosing $x = x_1$ in the second condition and combining it with the first condition, we see that $v > 0$. But then the second condition implies that $f(x) \geq v^{-1}\alpha + \langle x, -v^{-1}u \rangle =: l_1(x)$ for $x \in \operatorname{dom} f$.

For $x \in \operatorname{dom} f$, we then have

$$l_0(x) \leq 0, \quad l_1(x) \leq f(x),$$

hence

$$al_0(x) + l_1(x) \leq f(x) \quad \text{for } a \geq 0.$$

For $x \notin \operatorname{dom} f$, this inequality holds trivially since then $f(x) = \infty$. Hence

$$m_a := al_0 + l_1$$

is an affine function for which $m_a \leq f$. Since $l_0(x_0) > 0$, we have $m_a(x_0) > r_0$ if a is chosen sufficiently large, for instance,

$$a := \max\{(r_0 - l_1(x_0))/l_0(x_0) + 1, 1\}$$

is a proper choice. $\qquad\square$

We now come to another important operation on convex functions, the construction of the conjugate function.

Definition 2.3 Let $f : \mathbb{R}^n \to (-\infty, \infty]$ be proper and convex. Then the function f^* defined by

$$f^*(y) := \sup\{\langle x, y \rangle - f(x) : x \in \mathbb{R}^n\}, \quad y \in \mathbb{R}^n,$$

is called the *conjugate function* of f.

Example 2.3 Let $f : \mathbb{R}^n \to (-\infty, \infty]$ be constant, $f \equiv -\alpha \in \mathbb{R}$. Then $f^*(y) = \infty$ if $y \neq 0$ and $f^*(0) = \alpha$.

Example 2.4 Consider $f : \mathbb{R}^n \to \mathbb{R}$ given by $f(x) := \frac{1}{2}\|x\|^2$. Then we obtain

$$f^*(y) = \frac{1}{2} \sup\left\{-\|x - y\|^2 + \|y\|^2 : x \in \mathbb{R}^n\right\} = \frac{1}{2}\|y\|^2 = f(y),$$

that is, $f^* = f$.

Let $f, g : \mathbb{R}^n \to (-\infty, \infty]$ be proper and convex. Then $f \leq g$ implies that $g^* \leq f^*$. This simple observation will be used repeatedly in the following.

Theorem 2.4 *Let* f^* *be the conjugate function of the proper convex function* $f : \mathbb{R}^n \to (-\infty, \infty]$. *Then*

(a) f^* *is proper, closed, and convex.*
(b) $f^{**} := (f^*)^* = \operatorname{cl} f$.

Proof (a) For $x \notin \operatorname{dom} f$, we have $\langle x, y \rangle - f(x) = -\infty$ for $y \in \mathbb{R}^n$, hence

$$f^* = \sup_{x \in \operatorname{dom} f} \left(\langle x, \cdot \rangle - f(x) \right).$$

For $x \in \operatorname{dom} f$, the function

$$g_x : y \mapsto \langle x, y \rangle - f(x)$$

is affine, therefore f^* is convex (as the supremum of affine functions).
Since

$$\operatorname{epi} f^* = \operatorname{epi} \left(\sup_{x \in \operatorname{dom} f} g_x \right) = \bigcap_{x \in \operatorname{dom} f} \operatorname{epi} g_x$$

and since $\operatorname{epi} g_x$ is a closed halfspace, $\operatorname{epi} f^*$ is closed, and hence f^* is closed.
In order to show that f^* is proper, we consider an affine function $h \leq f$. Such a function exists by Theorem 2.3 and it has a representation

$$h = \langle \cdot, y \rangle - \alpha \quad \text{with suitable } y \in \mathbb{R}^n, \alpha \in \mathbb{R}.$$

This implies $\langle \cdot, y \rangle - \alpha \leq f$, hence $\langle \cdot, y \rangle - f \leq \alpha$, and therefore $f^*(y) \leq \alpha$.
  (b) By Theorem 2.3,

$$\operatorname{cl} f = \sup\{h : h \leq \operatorname{cl} f, h \text{ affine}\}.$$

Writing h again as

$$h = \langle \cdot, y \rangle - \alpha, \quad y \in \mathbb{R}^n, \alpha \in \mathbb{R},$$

we obtain

$$\operatorname{cl} f = \sup_{(y,\alpha)} \left(\langle \cdot, y \rangle - \alpha \right),$$

where the supremum is taken over all (y, α) with

$$\langle \cdot, y \rangle - \alpha \leq \operatorname{cl} f.$$

The latter holds if and only if

$$\alpha \geq \sup\{\langle x, y \rangle - \mathrm{cl}\, f(x) : x \in \mathbb{R}^n\} = (\mathrm{cl}\, f)^*(y).$$

Consequently, we have

$$\mathrm{cl}\, f(x) \leq \sup\{\langle x, y \rangle - (\mathrm{cl}\, f)^*(y) : y \in \mathbb{R}^n\} = (\mathrm{cl}\, f)^{**}(x),$$

for $x \in \mathbb{R}^n$. Since $\mathrm{cl}\, f \leq f$, the definition of the conjugate function implies

$$(\mathrm{cl}\, f)^* \geq f^*,$$

and therefore

$$\mathrm{cl}\, f \leq (\mathrm{cl}\, f)^{**} \leq f^{**}.$$

On the other hand,

$$f^{**}(x) = (f^*)^*(x) = \sup\{\langle x, y \rangle - f^*(y) : y \in \mathbb{R}^n\},$$

where

$$f^*(y) = \sup\{\langle z, y \rangle - f(z) : z \in \mathbb{R}^n\} \geq \langle x, y \rangle - f(x).$$

Therefore,

$$f^{**}(x) \leq \sup\{\langle x, y \rangle - \langle x, y \rangle + f(x) : y \in \mathbb{R}^n\} = f(x),$$

which gives us $f^{**} \leq f$. By part (a), f^{**} is closed, hence $f^{**} \leq \mathrm{cl}\, f$. $\quad\square$

Finally, we mention a canonical description of convex sets $A \subset \mathbb{R}^n$ by convex functions. A common way to describe a set A is by the indicator function

$$\mathbf{1}_A(x) := \begin{cases} 1, & x \in A, \\ 0, & x \notin A. \end{cases}$$

However, $\mathbf{1}_A$ is neither convex nor concave. Therefore, we define the convex *indicator function* δ_A of an arbitrary set $A \subset \mathbb{R}^n$ by

$$\delta_A(x) := \begin{cases} 0, & x \in A, \\ \infty, & x \notin A. \end{cases}$$

Note that A is convex if and only if δ_A is convex. If A is convex, then δ_A is also called the convex characteristic function of A. Moreover, for $A \neq \emptyset$ we have

$$\delta_A^*(x) = \sup\{\langle y, x \rangle - \delta_A(y) : y \in \mathbb{R}^n\} = \sup\{\langle y, x \rangle : y \in A\}.$$

The expression on the right will be called the support function of A and studied in more detail in Sect. 2.3.

Exercises and Supplements for Sect. 2.1

1. Let $A \subset \mathbb{R}^n$ be nonempty, closed, convex, and line-free. Let further $f : \mathbb{R}^n \to \mathbb{R}$ be convex and assume there is a point $y \in A$ with

$$f(y) = \max_{x \in A} f(x).$$

Show that there is also a point $z \in \mathrm{ext}\, A$ with

$$f(z) = \max_{x \in A} f(x).$$

2.* Let $f : \mathbb{R}^n \to (-\infty, \infty]$ be convex. Show that the following assertions are equivalent:

(a) f is closed.
(b) f is lower semi-continuous, that is, for $x \in \mathbb{R}^n$ we have

$$f(x) \leq \liminf_{y \to x} f(y).$$

(c) All the sublevel sets $\{f \leq \alpha\}$, $\alpha \in \mathbb{R}$, are closed.

3. Let $f, f_1, \ldots, f_m : \mathbb{R}^n \to (-\infty, \infty]$ be convex functions and $\alpha \geq 0$. Suppose that epi $f_1 + \cdots +$ epi f_m does not contain a vertical line (that is, a vertical half-ray which is unbounded from below). Prove the following assertions.

(a) The function $\alpha \circ f : x \mapsto \inf\{\beta \in \mathbb{R} : (x, \beta) \in \alpha \cdot$ epi $f\}$ is convex.
(b) The function $f_1 \square \cdots \square f_m : x \mapsto \inf\{\beta \in \mathbb{R} : (x, \beta) \in$ epi $f_1 + \cdots +$ epi $f_m\}$ is convex, and we have

$$(f_1 \square \cdots \square f_m)(x)$$
$$= \inf\{f_1(x_1) + \cdots + f_m(x_m) : x_1, \ldots, x_m \in \mathbb{R}^n, x_1 + \cdots + x_m = x\}.$$

The function $f_1 \square \cdots \square f_m$ is called the *infimal convolution* of $f_1, \ldots, f_m$.

(c) Let $\{f_i : i \in I\}$ $(I \neq \emptyset)$ be a family of convex functions on $\mathbb{R}^n$ such that $\mathrm{conv}(f_i)_{i \in I}$ exists. Show that

$$\mathrm{conv}(f_i)_{i \in I}$$

$$= \inf \left\{ \alpha_1 \circ f_{i_1} \square \cdots \square \alpha_m \circ f_{i_m} : \alpha_j \geq 0, \sum_{j=1}^m \alpha_j = 1, i_j \in I, m \in \mathbb{N} \right\}.$$

4. Let $A \subset \mathbb{R}^n$ be convex and $0 \in A$. The *inner distance function* or *gauge function* $d_A : \mathbb{R}^n \to (-\infty, \infty]$ of A is defined by

$$d_A(x) = \inf\{\alpha \geq 0 : x \in \alpha A\}, \quad x \in \mathbb{R}^n.$$

Show that d_A has the following properties:

(a) d_A is positively homogeneous, nonnegative and convex.
(b) d_A is finite if and only if $0 \in \mathrm{int}\, A$.
(c) $\{d_A < 1\} \subset A \subset \{d_A \leq 1\} \subset \mathrm{cl}\, A$.
(d) If $0 \in \mathrm{int}\, A$, then $\mathrm{int}\, A = \{d_A < 1\}$ and $\mathrm{cl}\, A = \{d_A \leq 1\}$.
(e) $d_A(x) > 0$ if and only if $x \neq 0$ and $\beta x \notin A$ for some $\beta > 0$.
(f) Let A be closed. Then d_A is even (i.e., $d_A(x) = d_A(-x)$ for $x \in \mathbb{R}^n$) if and only if A is symmetric with respect to 0 (i.e., $A = -A$).
(g) Let A be closed. Then d_A is a norm on $\mathbb{R}^n$ if and only if A is symmetric, compact and contains 0 in its interior.
(h) If A is closed, then d_A is closed.

Note: For some parts of this exercise it is used that a real-valued convex function is continuous. This fact will be proved in Sect. 2.2.

5. Let $f : \mathbb{R}^n \to [0, \infty]$ be a proper, positively homogeneous, convex function.

(a) Show that there is a convex set $A \subset \mathbb{R}^n$ with $0 \in A$ such that $f = d_A$.
(b) Show that if f is closed, then A can be chosen as a closed set (and then A is uniquely determined).

6. Let $f : \mathbb{R}^n \to \mathbb{R}$ be a continuous function satisfying

$$f\left(\frac{x_1 + x_2}{2}\right) \leq \frac{1}{2}(f(x_1) + f(x_2)) \quad \text{for } x_1, x_2 \in \mathbb{R}^n.$$

Show that f is convex.

7. Let $f : \mathbb{R}^2 \to \mathbb{R}$ be defined by

$$f(x, y) := \begin{cases} y^2/x, & x > 0, y \geq 0, \\ \infty, & \text{otherwise,} \end{cases}$$

and consider the set

$$C := \{(x, y) : x, y \in [0, 1), y \leq \sqrt{x} - x^2\}.$$

(a) Show that f is a convex function and C is a convex set.
(b) Is f an upper semi-continuous function on C?
(c) Does f attain its maximum on $C \setminus \{(0, 0)^\top\}$?
(d) Determine cl f.

2.2 Regularity

The regularity of a function usually refers to its smoothness properties. We start with the (local Lipschitz) continuity of convex functions.

Theorem 2.5 *A convex function* $f : \mathbb{R}^n \to (-\infty, \infty]$ *is continuous in* int dom f *and Lipschitz continuous on compact subsets of* int dom f.

Proof Let $x \in$ int dom f. There exists an n-simplex P with $P \subset$ int dom f and $x \in$ int P. If $x_0, \ldots, x_n$ are the vertices of P and $y \in P$, we have

$$y = \alpha_0 x_0 + \cdots + \alpha_n x_n$$

with $\alpha_i \in [0, 1]$, $\sum \alpha_i = 1$, and hence

$$f(y) \leq \alpha_0 f(x_0) + \cdots + \alpha_n f(x_n) \leq \max_{i=0,\ldots,n} f(x_i) =: c < \infty.$$

Therefore, $f \leq c$ on P.

Let now $\alpha \in (0, 1)$ and choose a closed ball U centered at 0 such that $x + U \subset P$. Let $z = x + \alpha u$ with $u \in$ bd U. Then, $z = (1 - \alpha)x + \alpha(x + u)$ and

$$f(z) \leq (1 - \alpha)f(x) + \alpha f(x + u) \leq (1 - \alpha)f(x) + \alpha c.$$

This implies that

$$f(z) - f(x) \leq \alpha(c - f(x)).$$

On the other hand,

$$x = \frac{1}{1 + \alpha} z + \frac{\alpha}{1 + \alpha}(x - u),$$

and hence

$$f(x) \leq \frac{1}{1+\alpha} f(z) + \frac{\alpha}{1+\alpha} f(x-u),$$

which yields

$$f(x) \leq \frac{1}{1+\alpha} f(z) + \frac{\alpha}{1+\alpha} c.$$

We obtain

$$\alpha(f(x) - c) \leq f(z) - f(x).$$

Together, the two inequalities give

$$|f(z) - f(x)| \leq \alpha(c - f(x)),$$

for $z \in x + \alpha$ bd U. Let ϱ be the radius of U. Thus we have shown that

$$|f(z) - f(x)| \leq \frac{c - f(x)}{\varrho} \|z - x\|$$

for $z \in s + U$. In particular, f is continuous on int dom f.

Now let $A \subset$ int dom f be compact. Hence there is some $\varrho > 0$ such that also $A + \varrho B^n \subset$ int dom f. Let $x, z \in A$. Since f is continuous on $A + \varrho B^n$,

$$C := \max\{|f(y)| : y \in A + \varrho B^n\} < \infty.$$

By the preceding argument,

$$|f(z) - f(x)| \leq \frac{2C}{\varrho} \|z - x\|,$$

if $\|z - x\| \leq \varrho$. For $\|z - x\| \geq \varrho$, this is true as well. $\qquad \square$

Convex functions enjoy basic differentiability properties, which we discuss next. We first consider the case of a function on the real line with values in the extended real line $(-\infty, \infty]$.

Theorem 2.6 *Let $f : \mathbb{R}^1 \to (-\infty, \infty]$ be convex.*

(a) *In each point $x \in$ int dom f, the right derivative $f^+(x)$ and the left derivative $f^-(x)$ exist and satisfy $f^-(x) \leq f^+(x)$.*

(b) *On int dom f, the functions f^+ and f^- are increasing and, for almost all $x \in$ int dom f (with respect to the Lebesgue measure λ_1 on $\mathbb{R}^1$), we have $f^-(x) = f^+(x)$, hence f is almost everywhere differentiable on cl dom f.*

(c) *Moreover, f^+ is continuous from the right, f^- is continuous from the left, and f is the indefinite integral of f^+ (of f^- and of f') on int dom f.*

Proof W.l.o.g. we concentrate on the case dom $f = \mathbb{R}^1$.

(a) If $0 < m \leq l$ and $0 < h \leq k$, the convexity of f implies that

$$f(x - m) = f\left(\left(1 - \frac{m}{l}\right)x + \frac{m}{l}(x - l)\right) \leq \left(1 - \frac{m}{l}\right)f(x) + \frac{m}{l}f(x - l),$$

hence

$$\frac{f(x) - f(x - l)}{l} \leq \frac{f(x) - f(x - m)}{m}.$$

Similarly, we have

$$f(x) = f\left(\frac{h}{h + m}(x - m) + \frac{m}{h + m}(x + h)\right)$$

$$\leq \frac{h}{h + m}f(x - m) + \frac{m}{h + m}f(x + h),$$

which gives us

$$\frac{f(x) - f(x - m)}{m} \leq \frac{f(x + h) - f(x)}{h}.$$

Finally,

$$f(x + h) = f\left(\left(1 - \frac{h}{k}\right)x + \frac{h}{k}(x + k)\right) \leq \left(1 - \frac{h}{k}\right)f(x) + \frac{h}{k}f(x + k),$$

and therefore

$$\frac{f(x + h) - f(x)}{h} \leq \frac{f(x + k) - f(x)}{k}.$$

We obtain that the left difference quotients in x are increasing and bounded from above by the right difference quotients, which are decreasing and bounded from below by the left difference quotients. Therefore, the limits

$$f^+(x) = \lim_{h \searrow 0} \frac{f(x + h) - f(x)}{h}$$

and

$$f^-(x) = \lim_{m \searrow 0} \frac{f(x) - f(x - m)}{m} \quad \left(= \lim_{t \nearrow 0} \frac{f(x + t) - f(x)}{t}\right)$$

exist and satisfy $f^-(x) \leq f^+(x)$.

(b) For $x' > x$, we have just seen that

$$f^-(x) \le f^+(x) \le \frac{f(x') - f(x)}{x' - x} \le f^-(x') \le f^+(x'). \qquad (2.2)$$

Therefore, the functions f^- and f^+ are increasing. As is well known, an increasing function has at most countably many points of discontinuity (namely jumps), and therefore it is continuous almost everywhere. At the points x of continuity of f^-, (2.2) implies $f^-(x) = f^+(x)$.

(c) Suppose that $x < y$. From

$$\frac{f(y) - f(x)}{y - x} = \lim_{z \searrow x} \frac{f(y) - f(z)}{y - z} \ge \lim_{z \searrow x} f^+(z)$$

we obtain $f^+(x) \le \lim_{z \searrow x} f^+(z) \le f^+(x)$, hence $\lim_{z \searrow x} f^+(z) = f^+(x)$, since f^+ is increasing. For $y < x$, we get by a similar argument

$$\lim_{z \nearrow x} f^-(z) \ge \lim_{z \nearrow x} \frac{f(z) - f(y)}{z - y} = \frac{f(x) - f(y)}{x - y},$$

and hence $f^-(x) \le \lim_{z \nearrow x} f^-(z) \le f^-(x)$. Thus we also have the relation $\lim_{z \nearrow x} f^-(z) = f^-(x)$.

Finally, for arbitrary $a \in \mathbb{R}$, we define a function g by

$$g(x) := f(a) + \int_a^x f^-(s)\, ds.$$

We first show that g is convex, and then $g = f$.

For $z := \alpha x + (1 - \alpha)y, \alpha \in [0, 1], x < y$, we have

$$g(z) - g(x) = \int_x^z f^-(s)\, ds \le (z - x) f^-(z),$$

$$g(y) - g(z) = \int_z^y f^-(s)\, ds \ge (y - z) f^-(z).$$

It follows that

$$\alpha(g(z) - g(x)) + (1 - \alpha)(g(z) - g(y))$$
$$\le \alpha(z - x) f^-(z) + (1 - \alpha)(z - y) f^-(z)$$
$$= f^-(z)(z - [\alpha x + (1 - \alpha)y]) = 0,$$

and therefore

$$g(z) \leq \alpha g(x) + (1 - \alpha)g(y),$$

which shows that g is convex.

As a consequence, g^+ and g^- exist. For $y > x$,

$$\frac{g(y) - g(x)}{y - x} = \frac{1}{y - x} \int_x^y f^-(s) \, ds = \frac{1}{y - x} \int_x^y f^+(s) \, ds \geq f^+(x),$$

hence we obtain $g^+(x) \geq f^+(x)$. Analogously, for $y < x$ we obtain

$$\frac{g(x) - g(y)}{x - y} = \frac{1}{x - y} \int_y^x f^-(s) \, ds \leq f^-(x),$$

and thus we get $g^-(x) \leq f^-(x)$. Since $g^+ \geq f^+ \geq f^- \geq g^-$ and $g^+ = g^-$, except for at most countably many points, we have $g^+ = f^+$ and $g^- = f^-$, except for at most countably many points. By the continuity from the left of g^- and f^-, and the continuity from the right of g^+ and f^+, it follows that $g^+ = f^+$ and $g^- = f^-$ on $\mathbb{R}$. Hence, $h := g - f$ is differentiable everywhere and $h' \equiv 0$. Therefore, $h \equiv c = 0$ because we have $g(a) = f(a)$. $\qquad\square$

Now we consider the n-dimensional case. Basically, this is reduced to the one-dimensional case by restricting a convex function to lines. If $f : \mathbb{R}^n \to (-\infty, \infty]$ is convex and $x \in \operatorname{int} \operatorname{dom} f$, then for each $u \in \mathbb{R}^n$, $u \neq 0$, the equation

$$g_{(u)}(t) := f(x + tu), \quad t \in \mathbb{R},$$

defines a convex function $g_{(u)} : \mathbb{R}^1 \to (-\infty, \infty]$ and we have $0 \in \operatorname{int} \operatorname{dom} g_{(u)}$. By Theorem 2.6, the right derivative $g_{(u)}^+(0)$ exists. This is precisely the *directional derivative*

$$f'(x; u) := \lim_{t \searrow 0} \frac{f(x + tu) - f(x)}{t} \tag{2.3}$$

of f at $x \in \operatorname{int} \operatorname{dom} f$ in direction u. In fact, there is no reason for excluding the case $u = 0$. The right-hand side of (2.3) also makes sense for $u = 0$ and yields the value 0. We therefore define $f'(x; 0) := 0$ and obtain the following corollary to Theorem 2.6.

Corollary 2.1 *Let* $f : \mathbb{R}^n \to (-\infty, \infty]$ *be convex and* $x \in \operatorname{int} \operatorname{dom} f$. *Then, for each* $u \in \mathbb{R}^n$ *the directional derivative* $f'(x; u)$ *of* f *exists.*

The corollary does not imply that $f'(x; -u) = -f'(x; u)$ holds. In fact, since $g_{(u)}^-(0) = -f'(x; -u)$, the latter equation is true if and only if $g_{(u)}^-(0) = g_{(u)}^+(0)$.

Let $e_1, \ldots, e_n$ be an orthonormal basis of $\mathbb{R}^n$. Then the partial derivative of f at x with respect to e_i exists if and only if $f'(x; -e_i) = -f'(x; e_i)$, and we denote this by $f_i(x) = f'(x; e_i)$, for a fixed orthonormal basis. For a convex function f, the partial derivatives $f_1(x), \ldots, f_n(x)$ of f at x need not exist in each point $x \in \operatorname{int dom} f$. However, in analogy to Theorem 2.6 it can be can shown that $f_1, \ldots, f_n$ exist almost everywhere with respect to the Lebesgue measure λ_n in $\mathbb{R}^n$ (see Exercise 2.2.7) and that at points x where the partial derivatives $f_1(x), \ldots, f_n(x)$ exist, the function f is even differentiable (see Exercise 2.2.6). Even more, a convex function f on $\mathbb{R}^n$ is twice differentiable almost everywhere (in a suitable sense). We refer to the exercises, for these and a number of further results on derivatives of convex functions.

The map $u \mapsto f'(x; u)$ is positively homogeneous on $\mathbb{R}^n$. Moreover, if f is convex, then $f'(x; \cdot)$ is also convex. For support functions, we shall continue the discussion of directional derivatives in the next section.

For a function f which is differentiable or twice differentiable, the first or second derivatives can be used to characterize convexity of f.

Remark 2.13 (See Exercise 2.2.3) Let $A \subset \mathbb{R}$ be open and convex and let $f : A \to \mathbb{R}$ be a real function.

- If f is differentiable, then f is convex if and only if f' is monotone increasing (on A).
- If f is twice differentiable, then f is convex if and only if $f'' \geq 0$ (on A).

Remark 2.14 (See Exercise 2.2.4) Let $A \subset \mathbb{R}^n$ be open and convex and let $f : A \to \mathbb{R}$ be a real function.

- If f is differentiable, then f is convex if and only if

$$\langle \operatorname{grad} f(x) - \operatorname{grad} f(y), x - y \rangle \geq 0 \quad \text{for } x, y \in A.$$

(Here, $\operatorname{grad} f(x) := (f_1(x), \ldots, f_n(x))^\top$ is the *gradient* of f at x.)
- If f is twice differentiable, then f is convex if and only if the *Hessian matrix*

$$\partial^2 f(x) := (f_{ij}(x))_{i,j=1}^n \in \mathbb{R}^{n,n}$$

of f at x is positive semi-definite for $x \in A$. Here, $f_{ij}(x) = d^2 f(x)(e_i, e_j)$ are the second partial derivatives of f at x in the directions e_i and e_j and the second differential is considered to be a bilinear map on $\mathbb{R}^n$. Alternatively, we can view $d^2 f(x)$ as a linear map $d^2 f(x) : \mathbb{R}^n \to \mathbb{R}^n$ so that $f_{ij}(x) = \langle d^2 f(x)(e_i), e_j \rangle$.

Example 2.5 Remark 2.13 can be used to verify that the function $h : [0, 1] \to [0, 1]$ with $h(x) = \sqrt{x} - x^2$ is concave, and hence the set $C := \{(x, y) \in [0, 1)^2 : y \leq \sqrt{x} - x^2\}$ is convex. In fact, $h''(x) < 0$ for $x > 0$, and hence $-h$ is convex. (Or simply note that h is a sum of two concave functions.) Moreover, Remark 2.14

implies that the function $f : \mathbb{R}^2 \to \mathbb{R}$ defined by

$$f(x, y) := \begin{cases} \frac{y^2}{x}, & x > 0, y \geq 0, \\ \infty, & \text{otherwise,} \end{cases}$$

is convex. In fact, the Hessian matrix of f at $(x, y) \in \mathbb{R}^2$ with $x > 0$ and $y \geq 0$ is positive semi-definite, specifically

$$\partial^2 f(x, y) = 2 \begin{pmatrix} \frac{y^2}{x^3} & -\frac{y}{x} \\ -\frac{y}{x} & \frac{1}{x} \end{pmatrix}.$$

For a more direct argument, see the solution of Exercise 2.1.7.

Exercises and Supplements for Sect. 2.2

1. (a) Give an example of two convex functions $f, g : \mathbb{R}^n \to (-\infty, \infty]$ such that f and g both have minimal points (that is, points in $\mathbb{R}^n$ where the infimum of the function is attained), but $f + g$ does not have a minimal point.
    (b) Suppose $f, g : \mathbb{R}^n \to \mathbb{R}$ are convex functions which both have a unique minimal point in $\mathbb{R}^n$. Show that $f + g$ has a minimal point.
        Hint: Show first that the sets

$$\{x \in \mathbb{R}^n : f(x) \leq \alpha\} \quad \text{and} \quad \{x \in \mathbb{R}^n : g(x) \leq \alpha\}$$

are compact, for each $\alpha \in \mathbb{R}$.

2. Let $f : \mathbb{R} \to \mathbb{R}$ be a convex function. Show that

$$f(x) - f(0) = \int_0^x f^+(t)\, dt = \int_0^x f^-(t)\, dt$$

for $x \in \mathbb{R}$.

3. Let $A \subset \mathbb{R}$ be open and convex, and let $f : A \to \mathbb{R}$ be a real function.

    (a) Suppose that f is differentiable. Show that f is convex if and only if f' is increasing (on A).
    (b) Suppose that f is twice differentiable. Show that f is convex if and only if $f'' \geq 0$ (on A).

4.* Let $A \subset \mathbb{R}^n$ be open and convex, and let $f : A \to \mathbb{R}$ be a real function.

    (a) Suppose that f is differentiable. Show that f is convex if and only if

$$\langle \operatorname{grad} f(x) - \operatorname{grad} f(y), x - y \rangle \geq 0 \quad \text{for } x, y \in A.$$

(b) Suppose that f is twice differentiable. Show that f is convex if and only if

$$\partial^2 f(x) := (f_{ij}(x))_{i,j=1}^n \in \mathbb{R}^{n,n}$$

is positive semi-definite for $x \in A$.

5.* For a convex function $f : \mathbb{R}^n \to (-\infty, \infty]$ and $x \in \operatorname{dom} f$, the *subdifferential* of f at x is the set

$$\partial f(x) := \{v \in \mathbb{R}^n : f(y) \geq f(x) + \langle v, y - x \rangle \text{ for } y \in \mathbb{R}^n\}.$$

Each $v \in \partial f(x)$ is called a *subgradient* of f at x.
   Let $f \in \operatorname{int} \operatorname{dom} f$. Show that:

(a) $\partial f(x)$ is nonempty, compact and convex.
(b) We have

$$\partial f(x) = \{v \in \mathbb{R}^n : \langle v, u \rangle \leq f'(x; u) \text{ for } u \in \mathbb{R}^n\}.$$

(c) If f is differentiable in x, then

$$\partial f(x) = \{\operatorname{grad} f(x)\}.$$

6.* Let $f : \mathbb{R}^n \to (-\infty, \infty]$ be convex and $x \in \operatorname{int} \operatorname{dom} f$. Suppose that all partial derivatives $f_1(x), \ldots, f_n(x)$ at x exist. Show that f is differentiable at x.
7.* Let $f : \mathbb{R}^n \to \mathbb{R}$ be convex. Show that f is differentiable almost everywhere.
   Hint. Use Exercise 2.2.6.
8. Let $f : \mathbb{R}^n \to (-\infty, \infty]$ be convex, and let $x \in \operatorname{dom} f$. Show that

$$s \in \partial f(x) \iff f^*(s) + f(x) = \langle s, x \rangle.$$

Conclude that if f is closed, then

$$s \in \partial f(x) \iff x \in \partial f^*(s).$$

9. Let $b \in \mathbb{R}^n$ and let Q be a symmetric, positive definite $n \times n$ matrix. Define the function

$$f(x) := \frac{1}{2}\langle x, Qx \rangle + \langle b, x \rangle, \quad x \in \mathbb{R}^n.$$

Determine f^*.
10. Let $A \subset \mathbb{R}^n$ and let $U \subset \mathbb{R}^n$ be an open neighbourhood of A. Suppose that for each $x \in U$ there is a unique point $f(x) \in A$ such that

$$\|f(x) - x\| = \min\{\|y - x\| : y \in A\}.$$

Show that A is closed and f is continuous.

11. Let $f : \mathbb{R}^n \to \mathbb{R}$ be convex. A function $f : \mathbb{R}^n \to \mathbb{R}$ is said to be *strictly convex* if $f((1-t)x + ty) < (1-t)f(x) + tf(y)$ for $x, y \in \mathbb{R}^n$ with $x \neq y$ and $t \in (0, 1)$.

   (a) Show that f is strictly convex if and only if epi f does not contain a non-degenerate segment in its boundary.
   (b) Show that f is strictly convex if and only if $\partial f(a) \cap \partial f(b) = \emptyset$ for $a, b \in \mathbb{R}^n$ with $a \neq b$.

12. (a) Let $f : \mathbb{R}^n \to \mathbb{R}$ be a convex function which has a local minimum at $x_0 \in \mathbb{R}^n$. Show that $0 \in \partial f(x_0)$.
   (b) (Mean value theorem for convex functions) Let $f : \mathbb{R} \to \mathbb{R}$ be convex, and let $a, b \in \mathbb{R}$ with $a \neq b$. Show that there is some $c \in (a, b)$ such that

$$\frac{f(b) - f(a)}{b - a} \in \partial f(c).$$

13.* Let $f : \mathbb{R}^n \to (-\infty, \infty]$ be a convex function and $x \in \text{dom } f$. Then $v \in \partial f(x)$ if and only if $(v, -1) \in N(\text{epi } f, (x, f(x)))$.

14. Let $f_1, f_2 : \mathbb{R}^n \to \mathbb{R}$ be convex functions, and let $f := \max\{f_1, f_2\}$. Let $x \in \mathbb{R}^n$ be such that $f_1(x) = f_2(x)$. Show that

$$\partial f(x) = \text{conv}(\partial f_1(x) \cup \partial f_2(x)).$$

15. Let $\emptyset \neq A \subset \mathbb{R}^n$ be a closed convex set with (exterior) distance function $d_A := d(A, \cdot)$ in $\mathbb{R}^n$, where $d(A, x) = \|x - p(A, x)\|$ for $x \in \mathbb{R}^n$. Show that d_A is a convex function. For $x \in \text{bd } A$ we denote by $v(A, x)$ the set of exterior unit normal vectors of A at x, that is, $v(A, x) := \{u \in \mathbb{S}^{n-1} : \langle a - x, u \rangle \leq 0 \text{ for } a \in A\}$. Show that

$$\partial d_A(x) = \text{conv}(\{0\} \cup v(A, x)).$$

16. Let $g : [0, 1] \to \mathbb{R}$ be an increasing function. Show that the function $f : [0, 1] \to \mathbb{R}$ given by $f(t) := \int_0^t g(s)\, ds$ is convex.
   Let $(r_n)_{n \in \mathbb{N}}$ be an enumeration of $\mathbb{Q} \cap [0, 1]$. The function $g : [0, 1] \to [0, \infty)$ is defined by $g(x) := \sum_{r_n < x} 2^{-n}$.

   (a) Show that g is strictly increasing, continuous in $x \in [0, 1] \setminus \mathbb{Q}$, and continuous from the left, but not continuous from the right, in $x \in [0, 1] \cap \mathbb{Q}$.
   (b) Determine the right and left derivative of the function $f(t) := \int_0^t g(s)\, ds$ for $t \in [0, 1]$.
   Hint: Recall the proof of the fundamental theorem of calculus.

   The epigraph of f provides an example of a closed convex set having a countable dense set of singular boundary points (that is, points through which at least two different supporting hyperplanes pass).

17. (Jensen's inequality.) Let μ be a probability measure in a space X, let U be an open convex set in $\mathbb{R}^n$, and let φ be a convex real-valued function on U. Suppose that $g : X \to U$ is measurable and component-wise μ-integrable, and that $\varphi \circ g$ is μ-integrable. Let $z_0 = \int_X g(x)\, d\mu(x)$. Then $z_0 \in U$ and

$$\int_X \varphi(g(x))\, \mu(dx) \geq \varphi\left(\int_X g(x)\, \mu(dx) \right). \tag{2.4}$$

If φ is strictly convex, then equality holds if and only if $g(x) = z_0$ for μ-almost all $x \in X$.

If φ is concave, then the inequality in (2.4) is reversed, with the same equality condition if φ is strictly concave.

2.3 The Support Function

A particularly useful analytic description of (compact) convex sets can be given in terms of the support function, which we introduce and study in this section. It is also one of the basic tools in the following chapters. The support function of a set $A \subset \mathbb{R}^n$ with $0 \in A$ is in a certain sense dual to the (inner) distance function (or gauge function), which is considered in Exercises 2.1.4 and 2.1.5. The precise nature of this duality is shown in Exercise 2.3.2 (d). The support function turns out to be particularly useful in dealing with Minkowski combinations of convex sets.

Definition 2.4 Let $A \subset \mathbb{R}^n$ be nonempty and convex. The *support function* $h_A : \mathbb{R}^n \to (-\infty, \infty]$ of A is defined as

$$h_A(u) := \sup\{\langle x, u \rangle : x \in A\}, \quad u \in \mathbb{R}^n.$$

It is often convenient to use the notation $h(A, \cdot) := h_A$ for the support function of the set A. For instance, this is the case whenever we focus on the dependence of the support function on the underlying set or if the set A and the vector u in $h_A(u) = h(A, u)$ deserve equal attention.

Example 2.6 If $A = \{x_0\}$, then $h_A(u) = \langle x_0, u \rangle$ for $u \in \mathbb{R}^n$.

Example 2.7 The support function of the Euclidean ball of radius $r \geq 0$ centred at 0 is given by $h_{rB^n}(u) = r\|u\|$, $u \in \mathbb{R}^n$. Here we use that for $x \in rB^n$, we have $\langle x, u \rangle \leq \|x\|\|u\| \leq r\|u\|$ and $\langle ru/\|u\|, u \rangle = r\|u\|$ if $u \neq 0$ with $ru/\|u\| \in rB^n$.

Theorem 2.7 *Let $A, B \subset \mathbb{R}^n$ be nonempty convex sets. Then*

(a) h_A *is positively homogeneous, closed and convex (and hence subadditive).*
(b) $h_A = h_{\operatorname{cl} A}$ *and*

$$\operatorname{cl} A = \{x \in \mathbb{R}^n : \langle x, u \rangle \leq h_A(u) \text{ for } u \in \mathbb{R}^n\}.$$

(c) $A \subset B$ implies $h_A \leq h_B$; conversely, $h_A \leq h_B$ implies cl $A \subset$ cl B.
(d) h_A is finite (real-valued) if and only if A is bounded.
(e) $h_{\alpha A + \beta B} = \alpha h_A + \beta h_B$ for $\alpha, \beta \geq 0$.
(f) $h_{-A}(u) = h_A(-u)$ for $u \in \mathbb{R}^n$.
(g) If A_i, $i \in I$, are nonempty and convex and $A := \mathrm{conv}\left(\bigcup_{i \in I} A_i\right)$, then

$$h_A = \sup_{i \in I} h_{A_i}.$$

(h) If A_i, $i \in I$, are closed and convex, and if $A := \bigcap_{i \in I} A_i$ is nonempty, then

$$h_A = \mathrm{cl}\,\mathrm{conv}(h_{A_i})_{i \in I}.$$

(i) $\delta_A^* = h_A$.

Proof (a) For $\alpha \geq 0$ and $u, v \in \mathbb{R}^n$, we have

$$h_A(\alpha u) = \sup_{x \in A} \langle x, \alpha u \rangle = \alpha \sup_{x \in A} \langle x, u \rangle = \alpha h_A(u)$$

and

$$h_A(u + v) = \sup_{x \in A} \langle x, u + v \rangle \leq \sup_{x \in A} \langle x, u \rangle + \sup_{x \in A} \langle x, v \rangle = h_A(u) + h_A(v).$$

Furthermore, as a supremum of closed functions, h_A is closed.

(b) The first part follows from

$$\sup_{x \in A} \langle x, u \rangle = \sup_{x \in \mathrm{cl}\,A} \langle x, u \rangle, \quad u \in \mathbb{R}^n.$$

For $x \in \mathrm{cl}\,A$, we therefore have $\langle x, u \rangle \leq h_A(u)$ for $u \in \mathbb{R}^n$. Conversely, suppose $x \in \mathbb{R}^n$ is such that $\langle x, \cdot \rangle \leq h_A(\cdot)$, and assume $x \notin \mathrm{cl}\,A$. Then, by Theorem 1.14, there exists a (supporting) hyperplane separating x and cl A, that is, a direction $y \in \mathbb{S}^{n-1}$ and $\alpha \in \mathbb{R}$ such that

$$\langle x, y \rangle > \alpha \text{ and } \langle z, y \rangle \leq \alpha \text{ for } z \in \mathrm{cl}\,A.$$

This implies

$$h_{\mathrm{cl}\,A}(y) = h_A(y) \leq \alpha < \langle x, y \rangle,$$

a contradiction.

(c) The first part is obvious, the second follows from (b).

(d) If A is bounded, we have $A \subset B^n(r)$ for some $r > 0$. Then, (c) implies $h_A \leq h_{B^n(r)} = r\|\cdot\|$, hence $h_A < \infty$. Conversely, $h_A < \infty$ and Theorem 2.5 imply that h_A is continuous on $\mathbb{R}^n$. Therefore, h_A is bounded on $\mathbb{S}^{n-1}$, that is, $h_A \leq r = h_{B^n(r)}$ on $\mathbb{S}^{n-1}$, for some $r > 0$. The positive homogeneity, proved in (a), implies that $h_A \leq h_{B^n(r)}$ on all of $\mathbb{R}^n$, hence (c) shows that $\operatorname{cl} A \subset B^n(r)$, that is, A is bounded.

(e) For any $u \in \mathbb{R}^n$, we have

$$h_{\alpha A + \beta B}(u) = \sup_{x \in \alpha A + \beta B} \langle x, u \rangle = \sup_{y \in A, z \in B} \langle \alpha y + \beta z, u \rangle$$

$$= \sup_{y \in A} \langle \alpha y, u \rangle + \sup_{z \in B} \langle \beta z, u \rangle$$

$$= \alpha h_A(u) + \beta h_B(u).$$

(f) For any $u \in \mathbb{R}^n$, we have

$$h_{-A}(u) = \sup_{x \in -A} \langle x, u \rangle = \sup_{y \in A} \langle -y, u \rangle = \sup_{y \in A} \langle y, -u \rangle$$

$$= h_A(-u).$$

(g) Since $A_i \subset A$, we have $h_{A_i} \leq h_A$ (from (c)), hence

$$\sup_{i \in I} h_{A_i} \leq h_A.$$

Conversely, $y \in A$ has a representation

$$y = \alpha_1 y_{i_1} + \cdots + \alpha_k y_{i_k},$$

with $k \in \mathbb{N}$, $y_{i_j} \in A_{i_j}$, $\alpha_j \geq 0$, $\sum \alpha_j = 1$ and $i_j \in I$. Hence, for $u \in \mathbb{R}^n$ we get

$$\langle y, u \rangle = \langle \alpha_1 y_{i_1} + \cdots + \alpha_k y_{i_k}, u \rangle \leq \alpha_1 h_{A_{i_1}}(u) + \cdots + \alpha_k h_{A_{i_k}}(u) \leq \sup_{i \in I} h_{A_i}(u),$$

and therefore $h_A(u) = \sup_{y \in A} \langle y, u \rangle \leq \sup_{i \in I} h_{A_i}(u)$.

(h) Let $x_0 \in A$. Since $A \subset A_i$ for $i \in I$, we get from (c) that $\langle x_0, \cdot \rangle \leq h_A \leq h_{A_i}$ for $i \in I$. Using the inclusion of the epigraphs, the definition of cl and conv for functions and (a), we obtain

$$h_A \leq \operatorname{cl} \operatorname{conv}(h_{A_i})_{i \in I} =: g.$$

On the other hand, Theorem 2.3 shows that the closed convex function g is the supremum of all affine functions below g. Since

$$(a, b) \in \mathrm{cl\,conv}\left(\bigcup_{i \in I} \mathrm{epi}\, h_i\right) \iff (\lambda a, \lambda b) \in \mathrm{cl\,conv}\left(\bigcup_{i \in I} \mathrm{epi}\, h_i\right)$$

for $\lambda > 0$, g is positively homogeneous, and hence we can concentrate on linear functions. In fact, if $\langle \cdot, y \rangle + \alpha \leq g$, then $\alpha \leq 0$ since $0 + \alpha \leq g(0) = 0$. For $u \in \mathbb{R}^n$ and $\lambda > 0$, we have $\langle \lambda u, y \rangle + \alpha \leq g(\lambda u)$. Hence $\langle u, y \rangle + \alpha/\lambda \leq g(u)$, and therefore $\langle u, y \rangle \leq g(u)$. This shows that the given estimate can be replaced by the stronger estimate $\langle \cdot, y \rangle \leq g$.

Therefore, suppose that $\langle \cdot, y \rangle \leq g$, $y \in \mathbb{R}^n$, is such a function. Since $g \leq h_{A_i}$ for $i \in I$,

$$\langle \cdot, y \rangle \leq h_{A_i} \quad \text{for } i \in I.$$

Now (c) implies that $y \in A_i$, $i \in I$, hence $y \in \bigcap_{i \in I} A_i = A$. Therefore,

$$\langle \cdot, y \rangle \leq h_A,$$

from which we get

$$g = \mathrm{cl\,conv}(h_{A_i})_{i \in I} \leq h_A.$$

(i) For $x \in \mathbb{R}^n$, we have

$$\delta_A^*(x) = \sup_{y \in \mathbb{R}^n} \left(\langle x, y \rangle - \delta_A(y) \right) = \sup_{y \in A} \langle x, y \rangle = h_A(x),$$

hence $\delta_A^* = h_A$. $\qquad\square$

We mention without proof a couple of further properties of support functions, which are simple consequences of the definition or of Theorem 2.7. In the following remarks, A is always a nonempty closed convex subset of $\mathbb{R}^n$.

Remark 2.15 For $a \in \mathbb{R}^n$, we have $A = \{a\}$ if and only if $h_A = \langle a, \cdot \rangle$.

Remark 2.16 For a nonempty convex set $A \subset \mathbb{R}^n$ and $x \in \mathbb{R}^n$, we have $h_{A+x} = h_A + \langle x, \cdot \rangle$.

Remark 2.17 A is origin-symmetric (i.e., $A = -A$) if and only if h_A is *even*, i.e., $h_A(x) = h_A(-x)$ for $x \in \mathbb{R}^n$.

Remark 2.18 We have $0 \in A$ if and only if $h_A \geq 0$.

The following result is crucial for our subsequent considerations. It provides a characterization of support functions of closed convex sets.

Theorem 2.8 *Let* $h : \mathbb{R}^n \to (-\infty, \infty]$ *be positively homogeneous, closed and convex. Then there exists a unique nonempty, closed and convex set* $A \subset \mathbb{R}^n$ *such that*

$$h_A = h.$$

Proof If $h \equiv \infty$, we can choose $A = \mathbb{R}^n$. Hence we can assume that h is proper. Then there is some $x_0 \in \mathbb{R}^n$ with $h(x_0) < \infty$, and thus $h(0) = h(0 \cdot x_0) = 0h(x_0) = 0$, by the assumed positive homogeneity.

For $\alpha > 0$, we obtain from the positive homogeneity of h that

$$h^*(x) = \sup_{y \in \mathbb{R}^n} (\langle x, y \rangle - h(y)) = \sup_{y \in \mathbb{R}^n} (\langle x, \alpha y \rangle - h(\alpha y))$$

$$= \alpha \sup_{y \in \mathbb{R}^n} (\langle x, y \rangle - h(y)) = \alpha h^*(x).$$

Therefore, h^* can only attain the values 0 and ∞. Since h is convex and proper, so is h^*. We put $A := \operatorname{dom} h^*$. By Theorem 2.4 (a), A is nonempty, closed and convex, and

$$h^* = \delta_A.$$

Theorem 2.7 (i) implies that

$$h^{**} - \delta_A^* = h_A.$$

By Theorem 2.4 (b), we have $h^{**} = h$, hence $h_A = h$.

The uniqueness of A follows from Theorem 2.7 (b). □

Since the class of nonempty compact convex sets will play a major role, we provide the following definition.

Definition 2.5 A compact convex set $K \neq \emptyset$ is called a *convex body*. We denote by $\mathcal{K}^n$ the set of all convex bodies in $\mathbb{R}^n$.

In the literature, convex bodies are sometimes required to have nonempty interior, whereas here we only exclude empty (compact convex) sets.

Corollary 2.2 *Let* $h : \mathbb{R}^n \to \mathbb{R}$ *be positively homogeneous and convex. Then there exists a unique* $K \in \mathcal{K}^n$ *such that* $h_K = h$. *Conversely, for* $K \in \mathcal{K}^n$ *the support function* $h_K : \mathbb{R}^n \to \mathbb{R}$ *is real-valued, positively homogeneous, convex and continuous.*

Proof This follows as a consequence of Theorem 2.8 in connection with Theorem 2.7 (d). □

Let $A \subset \mathbb{R}^n$ be nonempty, closed and convex. For $u \in \mathbb{R}^n \setminus \{0\}$, we consider the sets

$$H(A, u) := \{x \in \mathbb{R}^n : \langle x, u \rangle = h_A(u)\}$$

and

$$A(u) := A \cap H(A, u) = \{x \in A : \langle x, u \rangle = h_A(u)\}.$$

If $h_A(u) = \infty$, both sets are empty. If $h_A(u) < \infty$, then $H(A, u)$ is a hyperplane which bounds A, but $H(A, u)$ need not be a supporting hyperplane (see the example in Sect. 1.4), namely if $A(u) = \emptyset$. If $A(u) \neq \emptyset$, then $H(A, u)$ is a supporting hyperplane of A (at each point $x \in A(u)$) and $A(u)$ is the corresponding support set. We discuss now the support function of $A(u)$. In order to simplify the considerations, we concentrate on the case where A is compact, and hence also $A(u)$ is nonempty and compact for $u \in \mathbb{S}^{n-1}$.

Next we provide a description of support functions of support sets of convex bodies.

Theorem 2.9 *Let $K \in \mathcal{K}^n$ and $u \in \mathbb{R}^n \setminus \{0\}$. Then*

$$h_{K(u)}(x) = h'_K(u; x), \quad x \in \mathbb{R}^n,$$

that is, the support function of $K(u)$ is given by the directional derivatives of h_K at the point u.

Proof First, we show that $h_{K(u)}(x) \leq h'_K(u; x)$ for $x \in \mathbb{R}^n$. For $y \in K(u)$ and $v \in \mathbb{R}^n$, we have

$$\langle y, v \rangle \leq h_K(v),$$

since $y \in K$, with equality for $v = u$. In particular, for $v := u + tx$, $x \in \mathbb{R}^n$ and $t > 0$, we thus get

$$\langle y, u \rangle + t \langle y, x \rangle \leq h_K(u + tx),$$

and hence

$$\langle y, x \rangle \leq \frac{h_K(u + tx) - h_K(u)}{t}.$$

For $t \searrow 0$, we obtain

$$\langle y, x \rangle \leq h'_K(u; x).$$

Since this holds for all $y \in K(u)$, we arrive at

$$h_{K(u)}(x) \leq h'_K(u; x). \tag{2.5}$$

Now we prove that $h_{K(u)}(x) \geq h'_K(u; x)$ for $x \in \mathbb{R}^n$. From the subadditivity of h_K we obtain, for $t > 0$,

$$\frac{h_K(u + tx) - h_K(u)}{t} \leq \frac{h_K(tx)}{t} = h_K(x),$$

and thus

$$h'_K(u; x) \leq h_K(x).$$

This shows that the function $x \mapsto h'_K(u; x)$ is finite. As we have mentioned in the last section, it is also convex and positively homogeneous. In fact,

$$h'_K(u; x + z) = \lim_{t \searrow 0} \frac{h_K(u + tx + tz) - h_K(u)}{t}$$

$$\leq \lim_{t \searrow 0} \frac{h_K(\frac{u}{2} + tx) - h_K(\frac{u}{2})}{t} + \lim_{t \searrow 0} \frac{h_K(\frac{u}{2} + tz) - h_K(\frac{u}{2})}{t}$$

$$\leq \lim_{t \searrow 0} \frac{h_K(u + 2tx) - h_K(u)}{2t} + \lim_{t \searrow 0} \frac{h_K(u + 2tz) - h_K(u)}{2t}$$

$$= h'_K(u; x) + h'_K(u, z)$$

and

$$h'_K(u; \alpha x) = \lim_{t \searrow 0} \frac{h_K(u + t\alpha x) - h_K(u)}{t} = \alpha h'_K(u; x),$$

for $x, z \in \mathbb{R}^n$ and $\alpha \geq 0$. By Corollary 2.2, there exists an $L \in \mathcal{K}^n$ with

$$h_L(x) = h'_K(u; x), \quad x \in \mathbb{R}^n.$$

For $y \in L$, we have

$$\langle y, x \rangle \leq h'_K(u; x) \leq h_K(x), \quad x \in \mathbb{R}^n,$$

hence $y \in K$. Furthermore, the definition of the directional derivative and the positive homogeneity of support functions imply that

$$\langle y, u \rangle \leq h'_K(u; u) = h_K(u)$$

and

$$\langle y, -u \rangle \le h'_K(u; -u) = -h_K(u),$$

from which we obtain

$$\langle y, u \rangle = h_K(u),$$

and thus $y \in K \cap H(K, u) = K(u)$. It follows that $L \subset K(u)$, and therefore (again by Theorem 2.7)

$$h'_K(u; x) = h_L(x) \le h_{K(u)}(x). \tag{2.6}$$

Combining the inequalities (2.5) and (2.6), we obtain the assertion.   □

Remark 2.19 As a consequence, we obtain that $K(u)$ consists of one point if and only if $h'_K(u; \cdot)$ is linear. In this case, the unique boundary point of K with exterior unit normal u is grad $h_K(u) =: \nabla h_K(u)$. In view of Exercise 2.2.6, the latter is equivalent to the differentiability of h_K at u. If all the support sets $K(u)$, $u \in \mathbb{S}^{n-1}$, of a nonempty, compact convex set K consist of points, the boundary bd K does not contain any segments. Such sets K are called *strictly convex*. Hence, K is strictly convex if and only if h_K is differentiable on $\mathbb{R}^n \setminus \{0\}$.

We finally consider the support functions of polytopes. A set $\emptyset \ne C \subset \mathbb{R}^n$ is called a *convex cone* if $\alpha C \subset C$ for $\alpha \ge 0$ and if C is convex. Hence, C is closed under nonnegative combinations of vectors in C.

We call a (positively homogeneous) function h on $\mathbb{R}^n$ *piecewise linear* if there are finitely many convex cones $A_1, \dots, A_m \subset \mathbb{R}^n$ such that $\mathbb{R}^n = \bigcup_{i=1}^m A_i$ and h is linear on A_i, $i = 1, \dots, m$.

Theorem 2.10 *Let $K \in \mathcal{K}^n$. Then K is a polytope if and only if h_K is piecewise linear.*

Proof The convex body K is a polytope if and only if

$$K = \mathrm{conv}\{x_1, \dots, x_k\},$$

for some $x_1, \dots, x_k \in \mathbb{R}^n$. In view of Theorem 2.7, the latter is equivalent to

$$h_K = \max_{i=1,\dots,k} \langle x_i, \cdot \rangle,$$

which holds if and only if h_K is piecewise linear.

In fact, if h_K has the above form, we define

$$A_i := \{x \in \mathbb{R}^n : \max_{j=1,\dots,k} \langle x_j, x \rangle = \langle x_i, x \rangle\}, \quad i = 1, \dots, k.$$

It is easy to check that A_i is a closed convex cone, and clearly $h_K = \langle x_i, \cdot \rangle$ is linear on A_i. Thus h_K is piecewise linear.

Conversely, suppose that h_K is linear on the convex cones $A_1, \ldots, A_k$ which cover $\mathbb{R}^n$. Then we may assume that A_i is closed, since we can replace A_i by cl A_i, which is still a convex cone, and use that h_K remains linear on cl A_i. Moreover, we can assume that all closed convex cones $A_1, \ldots, A_k$ have interior points, since lower-dimensional closed sets A_i can be omitted (if a point $x_0 \in \mathbb{R}^n$ lies only in lower-dimensional closed convex cones, then there is a point in a neighbourhood of x_0 which is not covered by the union of all cones, a contradiction). Then, for $i = 1, \ldots, k$ there is some (uniquely determined) $x_i \in \mathbb{R}^n$ such that $\langle x_i, \cdot \rangle = h_K$ on A_i. This already implies that

$$\max\{\langle x_i, \cdot \rangle : i = 1, \ldots, k\} \geq h_K.$$

For the reverse inequality, let $x \in \mathbb{R}^n$. For $i \in \{1, \ldots, k\}$, we can choose $z \in \text{int}(A_i) \setminus \{x\}$. Then there are $y \in A_i$ and $\lambda \in (0, 1)$ such that $z = \lambda x + (1 - \lambda)y$. Hence,

$$\langle x_i, z \rangle = h_K(z) = h_K(\lambda x + (1 - \lambda)y)$$
$$\leq \lambda h_K(x) + (1 - \lambda)h_K(y) = \lambda h_K(x) + (1 - \lambda)\langle x_i, y \rangle,$$

and thus $\langle x_i, x \rangle \leq h_K(x)$. Since this holds for $i = 1, \ldots, k$, we get

$$\max\{\langle x_i, x \rangle : i = 1, \ldots, k\} \leq h_K(x)$$

for $x \in \mathbb{R}^n$. This finally yields that $h_K = \max\{\langle x_i, \cdot \rangle : i = 1, \ldots, k\}$. □

Exercises and Supplements for Sect. 2.3

1.* Let $f : \mathbb{R}^n \to \mathbb{R}$ be positively homogeneous and twice continuously partially differentiable on $\mathbb{R}^n \setminus \{0\}$. Show that there are $K, L \in \mathcal{K}^n$ such that

$$f = h_K - h_L.$$

Hint: Use Exercise 2.2.4 (b).

2. Let $K \in \mathcal{K}^n$ with $0 \in \text{int } K$, and let K° be the polar of K (see Exercise 1.1.14). Show that

(a) K° is compact and convex with $0 \in \text{int } K^\circ$,
(b) $K^{\circ\circ} := (K^\circ)^\circ = K$,
(c) K is a polytope if and only if K° is a polytope,
(d) $h_K = d_{K^\circ}$.

3. Let $K \in \mathcal{K}^n$ with $0 \in \operatorname{int} K$. Then the *radial function* of K is defined by

$$\rho(K, x) := \rho_K(x) := \max\{\lambda \geq 0 : \lambda x \in K\}, \quad x \in \mathbb{R}^n \setminus \{0\}.$$

Show that $\rho(K, x) = h(K^\circ, x)^{-1}$ for $x \in \mathbb{R}^n \setminus \{0\}$. See Exercise 1.1.14 for the definition of the polar body K°.

4. Let $f : \mathbb{R}^n \to \mathbb{R}$ be sublinear. Then there is a uniquely determined $K \in \mathcal{K}^n$ satisfying $f = h(K, \cdot)$. Prove the existence of K by using Helly's theorem.

5.* Let $K \subset \mathbb{R}^n$ be a compact set, and for $k \in \mathbb{N}$ let $f_k : \mathbb{R}^n \to \mathbb{R}$ be a convex function. Assume that for $x \in \mathbb{R}^n$ the limit $f(x) := \lim_{k \to \infty} f_k(x)$ exists. Prove the following assertions.

   (a) $f : \mathbb{R}^n \to \mathbb{R}$ is a convex function.
   (b) There is a constant $M \in \mathbb{R}$ such that $|f_k(x)| \leq M$ for $x \in K$ and $k \in \mathbb{N}$.
   (c) The sequence $(f_k)_{k \in \mathbb{N}}$ converges uniformly on K to f.

6. Let $K, L \in \mathcal{K}^n$ and let $C > 0$ be a constant such that $K, L \subset B^n(0, C)$. Show that

   (a) h_K is Lipschitz continuous with

   $$|h_K(x) - h_K(y)| \leq C \|x - y\|, \quad x, y \in \mathbb{R}^n.$$

   (b)

   $$|h_K(u) - h_L(v)| \leq C \|u - v\| + \|h_K - h_L\|, \quad u, v \in \mathbb{S}^{n-1},$$

   where $\|f\| := \sup\{|f(u)| : u \in \mathbb{S}^{n-1}\}$ for a function $f : \mathbb{S}^{n-1} \to \mathbb{R}$.

7. Let $0 < r \leq R < \infty$, $K, L \in \mathcal{K}^n$ with $B^n(0, r) \subset K, L \subset B^n(0, R)$ and $u, v \in \mathbb{S}^{n-1}$. Prove that

$$|\rho(K, u) - \rho(L, v)| \leq \frac{R}{r} \|h_K - h_L\| + \frac{R^2}{r} \|u - v\|.$$

8. Let $K \in \mathcal{K}^n$ with $\operatorname{int} K \neq \emptyset$ and

$$l(K, u) := \sup\{\|x - y\| : x, y \in K, x - y \in \mathbb{R}u\}, \quad u \in \mathbb{S}^{n-1}.$$

Prove the following assertions.

   (a) The supremum in the definition of $l(K, u)$ is a maximum. (This is the maximal length of a segment in K having direction u.)
   (b) $l(K, u) = \rho(K - K, u)$ for $u \in \mathbb{S}^{n-1}$.
   (c) $\min\{l(K, u) : u \in \mathbb{S}^{n-1}\} = \min\{h(K, u) + h(K, -u) : u \in \mathbb{S}^{n-1}\}$.

9. Determine the support functions of the following convex bodies (cube and cross-polytope).

   (a) $K_1 := \{(x_1, \ldots, x_n)^\top \in \mathbb{R}^n : |x_i| \leq r \text{ for } i = 1, \ldots, n\}, r \geq 0$.
   (b) $K_2 := \{(x_1, \ldots, x_n)^\top \in \mathbb{R}^n : |x_1| + \cdots + |x_n| \leq r\}, r > 0$.

10. Let $K_1, K_2 \subset \mathbb{R}^n$ be convex bodies. Determine the support function of the convex body

$$\operatorname{conv}(K_1 \cup K_2).$$

Suppose in addition that $K_1 \cap K_2 \neq \emptyset$. Prove that the function

$$g(u) := \inf\{h(K_1, u_1) + h(K_2, u_2) : u_1 + u_2 = u\}, \quad u \in \mathbb{R}^n,$$

is sublinear. In fact, g is the support function of $K_1 \cap K_2$.

11. Let $f, g : \mathbb{R}^n \to \mathbb{R}$ be convex functions, and let $x \in \mathbb{R}^n$. Show that

$$\partial(f + g)(x) = \partial f(x) + \partial g(x).$$

12. Hints to the literature: Analytic aspects of convexity and convex functions are the subject of [37, 45, 55, 56, 65, 76, 87]. There exists a vast literature on optimization related to convexity. In this respect, we mention only the classical text [89].

Chapter 3
Brunn–Minkowski Theory

The classical Steiner formula states that for a convex body $K \in \mathcal{K}^n$ the volume $V_n(K + \varrho B^n)$ of the parallel set $K + \varrho B^n$ (see Fig. 3.1 for an illustration), for $\varrho \geq 0$, is a polynomial in $\varrho \geq 0$ whose coefficients (if suitably normalized) are the intrinsic volumes (or Minkowski functionals) $V_0(K), \ldots, V_n(K)$ evaluated at the given convex body K. Among these functionals are the volume V_n and the surface area V_{n-1} (up to a constant factor), but also less known functionals such as the mean width (which is proportional to V_1) and the Euler characteristic V_0. The intrinsic volumes are continuous, additive and motion invariant functionals on convex bodies. In Chap. 5 we shall see that these properties characterize the intrinsic volumes. In addition, the intrinsic volumes are subject to various inequalities, of which the celebrated isoperimetric inequality is a particular example.

The Steiner formula is a special case of a much more general polynomial expansion available for the volume of Minkowski combinations of convex bodies. The coefficient functionals arising from such an expansion are the mixed volumes of convex bodies, which provide a far-reaching generalization of the intrinsic volumes. It is a main goal of this chapter to introduce the mixed volume, which is a functional on n-tuples of convex bodies in $\mathbb{R}^n$. Mixed volumes of convex bodies satisfy deep geometric inequalities and can be localized by the introduction of mixed area measures. The latter will be studied in detail in Chap. 4.

3.1 The Space of Convex Bodies

In the following, we mostly concentrate on *convex bodies* (nonempty compact convex sets) K in $\mathbb{R}^n$ and first discuss the space $\mathcal{K}^n$ of convex bodies, which we endow with a metric and hence a topology. We emphasize that we do not require that a convex body has interior points; hence lower-dimensional bodies are included

© Springer Nature Switzerland AG 2020
D. Hug, W. Weil, *Lectures on Convex Geometry*, Graduate Texts
in Mathematics 286, https://doi.org/10.1007/978-3-030-50180-8_3

Fig. 3.1 Illustration of the
parallel set of a polygon and
its polynomial area growth
(Steiner's formula in the
plane)

in $\mathcal{K}^n$. The set $\mathcal{K}^n$ is closed under Minkowski addition, that is, we have

$$K, L \in \mathcal{K}^n \implies K + L \in \mathcal{K}^n,$$

and multiplication by nonnegative scalars,

$$K \in \mathcal{K}^n, \alpha \geq 0 \implies \alpha K \in \mathcal{K}^n.$$

In fact, we even have $\alpha K \in \mathcal{K}^n$ for all $\alpha \in \mathbb{R}$, since the reflection $-K$ of a convex
body K is again a convex body. Thus, $\mathcal{K}^n$ is a convex cone and the question arises
whether we can embed this cone into a suitable vector space. Since $(\mathcal{K}^n, +)$ is a
(commutative) semi-group, the problem reduces to the question of whether this
semi-group can be embedded into a group. A simple algebraic criterion (which
is necessary and sufficient) is that the cancellation rule must be valid. Although
this can be checked directly for convex bodies (see Exercise 3.1.3), we use now
the support function for a direct embedding, which has a number of additional
advantages.

For this purpose, we consider the support function h_K of a convex body as a
function on the unit sphere $\mathbb{S}^{n-1}$ (because of the positive homogeneity of h_K, the
values on $\mathbb{S}^{n-1}$ determine h_K completely). Let $\mathbf{C}(\mathbb{S}^{n-1})$ be the vector space of
continuous functions on $\mathbb{S}^{n-1}$. This is a Banach space with respect to the maximum
norm

$$\|f\| := \max_{u \in \mathbb{S}^{n-1}} |f(u)|, \quad f \in \mathbf{C}(\mathbb{S}^{n-1}).$$

We call a function $f : \mathbb{S}^{n-1} \to \mathbb{R}$ *convex* if the homogeneous extension

$$\tilde{f}(x) := \begin{cases} \|x\| f\left(\frac{x}{\|x\|}\right), & x \neq 0, \\ 0, & x = 0, \end{cases}$$

is convex on $\mathbb{R}^n$. Let $\mathcal{H}^n$ be the set of all convex functions on $\mathbb{S}^{n-1}$. By Theorem 2.5
and Remark 2.7, $\mathcal{H}^n$ is a convex cone in $\mathbf{C}(\mathbb{S}^{n-1})$.

Theorem 3.1 *The mapping*

$$T : \mathcal{K}^n \to \mathcal{H}^n, \quad K \mapsto h_K,$$

is (positively) linear on $\mathcal{K}^n$ and maps the convex cone $\mathcal{K}^n$ one-to-one onto the convex cone $\mathcal{H}^n$. Moreover, T is compatible with the inclusion order on $\mathcal{K}^n$ and the pointwise order $\leq$ on $\mathcal{H}^n$.

In particular, T embeds the (ordered) convex cone $\mathcal{K}^n$ into the (ordered) vector space $\mathbf{C}(\mathbb{S}^{n-1})$.

Proof The positive linearity of T follows from Theorem 2.7 (e) and the injectivity from Theorem 2.7 (b). The fact that $T(\mathcal{K}^n) = \mathcal{H}^n$ is a consequence of Theorem 2.8. The compatibility with respect to the orders on $\mathcal{K}^n$ and $\mathcal{H}^n$, respectively, follows from Theorem 2.7 (c). □

Remark 3.1 Positive linearity of T on the convex cone $\mathcal{K}^n$ means that

$$T(\alpha K + \beta L) = \alpha T(K) + \beta T(L)$$

for $K, L \in \mathcal{K}^n$ and $\alpha, \beta \geq 0$. In this case, positive linearity does not extend to negative α, β, in particular not to difference bodies $K - L = K + (-L)$. One reason is that the function $h_K - h_L$ is in general not convex, but even if it is, which means that

$$h_K - h_L = h_M,$$

for some $M \in \mathcal{K}^n$, then the body M is in general different from the difference body $K - L$. We write $K \ominus L := M$ and call this body the *Minkowski difference* of K and L. While the difference body $K - L$ exists for all $K, L \in \mathcal{K}^n$, the Minkowski difference $K \ominus L$ exists only in special cases, namely if K can be decomposed as $K = M + L$ (then $M = K \ominus L$).

With respect to the norm topology provided by the maximum norm in $\mathbf{C}(\mathbb{S}^{n-1})$, the cone $\mathcal{H}^n$ is closed (see Exercise 3.1.11 for a more general statement). Our next goal is to define a natural metric on $\mathcal{K}^n$ such that T becomes even an isometry. Thus we obtain an isometric embedding of $\mathcal{K}^n$ into the Banach space $\mathbf{C}(\mathbb{S}^{n-1})$.

Definition 3.1 For $K, L \in \mathcal{K}^n$, let

$$d(K, L) := \inf\{\varepsilon \geq 0 : K \subset L + B^n(\varepsilon), L \subset K + B^n(\varepsilon)\}.$$

It is easy to see that the infimum is attained, hence it is in fact a minimum (see Exercise 3.1.1). The Euclidean metric on $\mathbb{R}^n$ is also denoted by d. Since the arguments $K, L \in \mathcal{K}^n$ of $d(K, L)$ in the preceding definition are convex bodies, hence subsets of $\mathbb{R}^n$, there is no danger of confusing this quantity with the metric of the underlying space.

Theorem 3.2 *For $K, L \in \mathcal{K}^n$, we have*

$$d(K, L) = \|h_K - h_L\|.$$

Therefore, d is a metric on $\mathcal{K}^n$ and

$$d(K + M, L + M) = d(K, L),$$

for $K, L, M \in \mathcal{K}^n$.

Proof From Theorem 2.7 we obtain

$$K \subset L + B^n(\varepsilon) \iff h_K \leq h_L + \varepsilon h_{B^n}$$

and

$$L \subset K + B^n(\varepsilon) \iff h_L \leq h_K + \varepsilon h_{B^n}.$$

Since $h_{B^n} \equiv 1$ on $\mathbb{S}^{n-1}$, this implies that

$$K \subset L + B^n(\varepsilon), L \subset K + B^n(\varepsilon) \iff \|h_K - h_L\| \leq \varepsilon,$$

and the assertions follow.   □

In an arbitrary metric space (X, d_0), the class $\mathcal{C}(X)$ of nonempty compact subsets of X can be endowed with the *Hausdorff metric* $\widetilde{d}$ which is defined by

$$\widetilde{d}(A, B) := \max\{\max_{x \in A} d_0(x, B), \max_{y \in B} d_0(y, A)\}$$

for $A, B \in \mathcal{C}(X)$, where we have used the abbreviation

$$d_0(u, C) := \min_{v \in C} d_0(u, v), \quad u \in X, C \in \mathcal{C}(X).$$

The minimal and maximal values exist due to the compactness of the sets and the continuity of the metric.

We now show that on $\mathcal{K}^n \subset \mathcal{C}(\mathbb{R}^n)$, the Hausdorff metric $\widetilde{d}$ coincides with the metric d introduced in Definition 3.1.

Theorem 3.3 *If $K, L \in \mathcal{K}^n$, then*

$$d(K, L) = \widetilde{d}(K, L).$$

Proof We have

$$d(K, L) = \max\{\inf\{\varepsilon \geq 0 : K \subset L + B^n(\varepsilon)\}, \inf\{\varepsilon \geq 0 : L \subset K + B^n(\varepsilon)\}\}.$$

Recall that $d : \mathbb{R}^n \times \mathbb{R}^n \to [0, \infty)$ denotes the Euclidean metric. Then we have

$$K \subset L + B^n(\varepsilon) \iff d(x, L) \leq \varepsilon \quad \text{for } x \in K$$

$$\iff \max_{x \in K} d(x, L) \leq \varepsilon,$$

and hence

$$\inf\{\varepsilon \geq 0 : K \subset L + B^n(\varepsilon)\} = \max_{x \in K} d(x, L),$$

which yields the assertion.   □

The preceding results can be summarized by saying that a sequence of convex bodies $(K_j)_{j \in \mathbb{N}}$ converges to $K_0 \in \mathcal{K}^n$ in the metric space $(\mathcal{K}^n, d)$, that is, with respect to the Hausdorff metric, if and only if the sequence of support functions $(h_{K_j})_{j \in \mathbb{N}}$ converges uniformly to the support function of K_0 on $\mathbb{S}^{n-1}$. The latter in turn is equivalent to the pointwise convergence of the support functions, as shown in Exercise 2.3.6.

We now come to an important topological property of the metric space $(\mathcal{K}^n, d)$ which states that every bounded subset $\mathcal{M} \subset \mathcal{K}^n$ is relatively compact. Clearly, also in the metric space $(\mathbb{R}^n, d)$ every bounded subset is relatively compact, but this property is not available in general metric spaces.

In $(\mathcal{K}^n, d)$, a subset $\mathcal{M} \subset \mathcal{K}^n$ is bounded, if there exists a $c > 0$ such that

$$d(K, L) \leq c \quad \text{for } K, L \in \mathcal{M}.$$

This is equivalent to

$$K \subset B^n(c') \quad \text{for } K \in \mathcal{M},$$

for some constant $c' > 0$. Here we can replace the ball $B^n(c')$ by any compact set, in particular by a cube $W \subset \mathbb{R}^n$. The subset $\mathcal{M}$ is relatively compact if every sequence $K_1, K_2, \ldots$ with $K_k \in \mathcal{M}$ has a convergent subsequence (note that in metric spaces, sequential compactness and compactness are equivalent). Therefore, the mentioned topological property is a consequence of Theorem 3.4.

It is a crucial step in the proof of Theorem 3.4 to show that the metric space $(\mathcal{K}^n, d)$ is complete, that is, Cauchy sequences are convergent. We extract this part of the argument as a lemma, since it is of independent interest.

Lemma 3.1 *Cauchy sequences in $(\mathcal{K}^n, d)$ are convergent.*

Proof Suppose that $(K_k)_{k \in \mathbb{N}}$ is a Cauchy sequence, that is, for each $\varepsilon > 0$ there is some $m = m(\varepsilon) \in \mathbb{N}$ such that

$$d(K_k, K_l) \leq \varepsilon \quad \text{for } k, l \geq m. \tag{3.1}$$

Consider

$$\widetilde{K}_k := \text{cl conv}(K_k \cup K_{k+1} \cup \cdots) \in \mathcal{K}^n$$

and

$$K_0 := \bigcap_{k=1}^{\infty} \widetilde{K}_k \in \mathcal{K}^n.$$

We claim that $K_k \to K_0$ as $k \to \infty$.

First, by construction we have $\widetilde{K}_k \in \mathcal{K}^n$ and $\widetilde{K}_{k+1} \subset \widetilde{K}_k$ for $k \in \mathbb{N}$. Therefore, $K_0 \neq \emptyset$, and hence indeed $K_0 \in \mathcal{K}^n$.

For $\varepsilon > 0$, the Cauchy property (3.1) implies that there is some $m = m(\varepsilon) \in \mathbb{N}$ such that

$$K_l \subset K_k + B^n(\varepsilon) \quad \text{for } k, l \geq m,$$

therefore

$$\widetilde{K}_r \subset K_k + B^n(\varepsilon) \quad \text{for } k, r \geq m,$$

and thus in particular

$$K_0 \subset K_k + B^n(\varepsilon) \quad \text{for } k \geq m.$$

Next we show that for each $\varepsilon > 0$, there exists an $m' = m'(\varepsilon) \in \mathbb{N}$ such that

$$\widetilde{K}_k \subset K_0 + B^n(\varepsilon) \quad \text{for } k \geq m'.$$

To verify this, assume on the contrary that (for some $\varepsilon > 0$)

$$\widetilde{K}_k \not\subset K_0 + B^n(\varepsilon) \quad \text{for infinitely many } k \in \mathbb{N}.$$

Then

$$\widetilde{K}_{k_i} \cap [W \setminus \text{int}(K_0 + B^n(\varepsilon))] \neq \emptyset$$

for a subsequence $(k_i)_{i \in \mathbb{N}}$ in $\mathbb{N}$. Since $\widetilde{K}_{k_i}$ and $W \setminus \mathrm{int}(K_0 + B^n(\varepsilon))$ are compact, this implies that

$$\emptyset \neq \bigcap_{i=1}^{\infty} (\widetilde{K}_{k_i} \cap [W \setminus \mathrm{int}(K_0 + B^n(\varepsilon))]) = K_0 \cap [W \setminus \mathrm{int}(K_0 + B^n(\varepsilon))] = \emptyset,$$

a contradiction.

Since $\widetilde{K}_k \subset K_0 + B^n(\varepsilon)$ implies that $K_k \subset K_0 + B^n(\varepsilon)$, we finally obtain that $d(K_0, K_k) \leq \varepsilon$ for $k \geq \max\{m, m'\}$. This proves the lemma. $\square$

Now we show that the metric space $(\mathcal{K}^n, d)$ is sequentially compact.

Theorem 3.4 (Blaschke's Selection Theorem) *Every bounded sequence in the metric space $(\mathcal{K}^n, d)$ has a convergent subsequence with limit in $\mathcal{K}^n$.*

Proof Let $(K_j)_{j \in \mathbb{N}}$ be a given bounded sequence in $\mathcal{K}^n$. By repeated selection of subsequences and a subsequent choice of a diagonal sequence, we construct a subsequence of $(K_j)_{j \in \mathbb{N}}$ which is a Cauchy sequence. Then the proof is completed by an application of Lemma 3.1.

W.l.o.g. we may assume that $K_j \subset W$, $j \in \mathbb{N}$, and W is the unit cube. For each $i \in \mathbb{N}$, we divide W into 2^{in} closed subcubes of edge length 2^{-i}. For $K \in \mathcal{K}^n$, let $W_i(K)$ be the union of all subcubes of the ith dissection of W which intersect K.

Since for fixed $i \in \mathbb{N}$ there are only finitely many different sets $W_i(K_j)$, $j \in \mathbb{N}$, there is a subsequence $(K_j^{(1)})_{j \in \mathbb{N}}$ of $(K_j)_{j \in \mathbb{N}}$ such that

$$W_1(K_1^{(1)}) = W_1(K_2^{(1)}) = \cdots .$$

Then there is a subsequence $(K_j^{(2)})_{j \in \mathbb{N}}$ of $(K_j^{(1)})_{j \in \mathbb{N}}$ such that

$$W_2(K_1^{(2)}) = W_2(K_2^{(2)}) = \cdots .$$

This can be iterated (for $i \geq 2$) and leads to a subsequence $(K_j^{(i)})_{j \in \mathbb{N}}$ of $(K_j^{(i-1)})_{j \in \mathbb{N}}$ such that

$$W_i(K_1^{(i)}) = W_i(K_2^{(i)}) = \cdots .$$

Let $k, l \in \mathbb{N}$ and $i \in \mathbb{N}$. Since

$$d(x, K_l^{(i)}) = \min_{y \in K_l^{(i)}} \|x - y\| \leq \sqrt{n}\, 2^{-i} \quad \text{for } x \in K_k^{(i)},$$

we have

$$d(K_k^{(i)}, K_l^{(i)}) \leq \sqrt{n}\, 2^{-i} \quad \text{for } k, l \in \mathbb{N},\ i \in \mathbb{N}.$$

By the subsequence property, we deduce

$$d(K_k^{(j)}, K_l^{(i)}) \leq \sqrt{n}\, 2^{-i} \quad \text{for } k, l \in \mathbb{N}, \ j, i \in \mathbb{N}, \ j \geq i.$$

In particular, the 'diagonal sequence' $K_k^{(k)}$, $k \in \mathbb{N}$, satisfies

$$d(K_k^{(k)}, K_l^{(l)}) \leq \sqrt{n}\, 2^{-l} \quad \text{for } k, l \in \mathbb{N}, \ k \geq l.$$

Hence $(K_k^{(k)})_{k \in \mathbb{N}}$ is a Cauchy sequence in $(\mathcal{K}^n, d)$.

An application of Lemma 3.1 completes the proof. $\qquad\qquad\qquad\qquad\square$

Remark 3.2 It is clear that Theorem 3.4 implies that $(\mathcal{K}^n, d)$ is a complete metric space, which is just the assertion of Lemma 3.1.

Remark 3.3 Blaschke's selection theorem can also be proved by using support functions and the Arzelá–Ascoli theorem for continuous functions on the unit sphere. In fact, if $(K_k)_{k \in \mathbb{N}}$ is a bounded sequence of convex bodies in $\mathbb{R}^n$, then the sequence of support functions $h_i = h_{K_i}$, $i \in \mathbb{N}$, is pointwise bounded and the sequence is equicontinuous on $\mathbb{S}^{n-1}$ (see Exercise 2.3.6). Hence the sequence has a convergent subsequence (with respect to the sup norm). The limit function is again positively homogenous and convex, hence the support function of a convex body K_0. It follows that the convex bodies corresponding to the subsequence converge to K_0.

The topology on $\mathcal{K}^n$ given by the Hausdorff metric allows us to introduce and study geometric functionals on convex bodies by first defining them for a special subclass, for example the class $\mathcal{P}^n$ of polytopes. Such a program requires that the geometric functionals under consideration have a continuity or monotonicity property and also that the class $\mathcal{P}^n$ of polytopes is dense in $\mathcal{K}^n$. We now discuss the latter aspect. Geometric functionals will then be investigated in the next section.

Theorem 3.5 *Let $K \in \mathcal{K}^n$ and $\varepsilon > 0$.*

(a) *There exists a polytope $P \in \mathcal{P}^n$ with $P \subset K$ and $d(K, P) \leq \varepsilon$.*
(b) *There exists a polytope $P \in \mathcal{P}^n$ with $K \subset P$ and $d(K, P) \leq \varepsilon$.*
(c) *If $0 \in \operatorname{relint} K$, then there exists a convex polytope $P \in \mathcal{P}^n$ which satisfies*
$$P \subset K \subset (1 + \varepsilon) P.$$
   There is even a convex polytope $\widetilde{P} \in \mathcal{P}^n$ with $\widetilde{P} \subset \operatorname{relint} K$ and which satisfies $K \subset \operatorname{relint}((1 + \varepsilon)\widetilde{P})$.

Proof Let $K \in \mathcal{K}^n$ and $\varepsilon > 0$ be fixed.

(a) The family $\{\operatorname{int} B^n(x, \varepsilon) : x \in \operatorname{bd} K\}$ is an open covering of the compact set $\operatorname{bd} K$. Therefore there exist $x_1, \ldots, x_m \in \operatorname{bd} K$ with

$$\operatorname{bd} K \subset \bigcup_{i=1}^{m} \operatorname{int} B^n(x_i, \varepsilon).$$

Defining

$$P := \operatorname{conv}\{x_1, \ldots, x_m\},$$

we obtain

$$P \subset K \quad \text{and} \quad \operatorname{bd} K \subset P + B^n(\varepsilon).$$

The latter implies that $K \subset P + B^n(\varepsilon)$ and therefore $d(K, P) \leq \varepsilon$.

(b) For each $u \in \mathbb{S}^{n-1}$, $H(K, u) = \{x \in \mathbb{R}^n : \langle x, u \rangle = h_K(u)\}$ is the supporting hyperplane of K (in direction u) and

$$H^-(K, u) := \{x \in \mathbb{R}^n : \langle x, u \rangle \leq h_K(u)\}$$

is a halfspace bounded by $H(K, u)$ which contains K. Then

$$A(K, u) := \{x \in \mathbb{R}^n : \langle x, u \rangle > h_K(u)\}$$

is the complement of $H^-(K, u)$. The family $\{A(K, u) : u \in \mathbb{S}^{n-1}\}$ is an open covering of the compact set $\operatorname{bd}(K + B^n(\varepsilon))$. In fact, for $y \in \operatorname{bd}(K + B^n(\varepsilon))$ we have $y \notin K$, and hence y can be separated from K by a supporting hyperplane $H(K, u)$ of K such that $y \in A(K, u)$. Therefore there exist $u_1, \ldots, u_m \in \mathbb{S}^{n-1}$ with

$$\operatorname{bd}(K + B^n(\varepsilon)) \subset \bigcup_{i=1}^m A(K, u_i).$$

We define a polyhedral set by

$$P := \bigcap_{i=1}^m (\mathbb{R}^n \setminus A(K, u_i)) = \bigcap_{i=1}^m H^-(K, u_i),$$

and hence

$$K \subset P \quad \text{and} \quad \mathbb{R}^n \setminus P = \bigcup_{i=1}^m A(K, u_i).$$

Finally, we claim that

$$P \subset K + B^n(\varepsilon).$$

In fact, if there is some $x \in P \setminus (K + B^n(\varepsilon))$, we choose any $p \in K \subset P$ and then find some

$$q \in [x, p] \cap \mathrm{bd}(K + B^n(\varepsilon)) \subset P \cap \bigcup_{i=1}^{m} A(K, u_i) = P \cap (\mathbb{R}^n \setminus P) = \emptyset,$$

a contradiction. In particular, the polyhedral set P is bounded and hence a polytope. Thus we conclude that $d(K, P) \leq \varepsilon$.

(c) W.l.o.g. we may assume that $\dim K = n$, hence $0 \in \mathrm{int}\, K$. If we copy the proof of (b) with $B^n(\varepsilon) = \varepsilon B^n$ replaced by εK, we obtain a polytope P' with

$$K \subset P' \subset (1 + \varepsilon)K.$$

The polytope $P := \frac{1}{1+\varepsilon} P'$ then satisfies $0 \in \mathrm{int}\, P$ and

$$P \subset K \subset (1 + \varepsilon)P.$$

In particular, we get a polytope $\overline{P}$ with $0 \in \mathrm{int}\, \overline{P}$ and

$$\overline{P} \subset K \subset \left(1 + \frac{\varepsilon}{2}\right) \overline{P}.$$

We choose $\widetilde{P} := \delta \overline{P}$ with $0 < \delta < 1$. Then

$$\widetilde{P} \subset \mathrm{int}\, \overline{P} \subset \mathrm{int}\, K.$$

If δ is close to 1, such that $(1 + \frac{\varepsilon}{2})\frac{1}{\delta} < 1 + \varepsilon$, then

$$K \subset \left(1 + \frac{\varepsilon}{2}\right) \frac{1}{\delta} \widetilde{P} \subset \mathrm{int}((1 + \varepsilon)\widetilde{P}),$$

which completes the proof.

$\square$

Remark 3.4 Theorem 3.5 shows that $\mathrm{cl}\, \mathcal{P}^n = \mathcal{K}^n$. The metric space $\mathcal{K}^n$ is also separable, since there is a countable dense set $\widetilde{\mathcal{P}}^n$ of polytopes. To see this, observe that a given polytope can be approximated by polytopes having vertices with rational coordinates.

Remark 3.5 The polytope P which was constructed in the course of the proof of Theorem 3.5 (a) has its vertices on bd K. If we use the open covering $\{\mathrm{int}\, B^n(x, \varepsilon) : x \in \mathrm{relint}\, K\}$ of K instead, we obtain a polytope P with $d(K, P) \leq \varepsilon$ and $P \subset \mathrm{relint}\, K$.

Remark 3.6 There is also a simultaneous proof of Theorem 3.5 (b) and the first part of Theorem 3.5 (c), which uses Theorem 3.5 (a). Namely, assuming $\dim K = n$ and

$0 \in \operatorname{int} K$, the body K contains a ball $B^n(\alpha)$, $\alpha > 0$. For given $\varepsilon \in (0, 1)$, by (a) there is some $P \in \mathcal{P}^n$, $P \subset K$, such that $d(K, P) < \alpha\varepsilon/2$. Note that P depends on ε and α. Hence

$$h_P(u) \geq h_K(u) - \frac{\alpha\varepsilon}{2} \geq \alpha\left(1 - \frac{\varepsilon}{2}\right) > 0, \quad u \in \mathbb{S}^{n-1},$$

and therefore $\alpha(1 - \varepsilon/2)B^n \subset P$. This shows that

$$P \subset K \subset P + \frac{\alpha\varepsilon}{2}B^n \subset P + \frac{\alpha\varepsilon}{2}\frac{1}{\alpha(1 - \varepsilon/2)}P = \left(1 + \frac{\varepsilon/2}{1 - \varepsilon/2}\right)P \subset (1 + \varepsilon)P.$$

Hence we obtain (c) and also get

$$\|h_{(1+\varepsilon)P} - h_K\| \leq \varepsilon\|h_P\| \leq \varepsilon\|h_K\|.$$

To deduce (b), we may assume that $0 \in \operatorname{int} K$. Let $\varepsilon \in (0, 1)$ be given. Define $\varepsilon' := \varepsilon/(1 + \|h_K\|) \in (0, 1)$. Then by what we have already shown there exists a polytope P' such that $0 \in \operatorname{int} P'$, $P' \subset K \subset (1 + \varepsilon')P'$ and

$$\|h_{(1+\varepsilon')P'} - h_K\| \leq \varepsilon'\|h_K\| \leq \varepsilon.$$

Thus the polytope $(1 + \varepsilon')P'$ satisfies all requirements.

Exercises and Supplements for Sect. 3.1

1. For $K, L \in \mathcal{K}^n$ show that

    $$d(K, L) = \min\{\varepsilon \geq 0 : K \subset L + B^n(\varepsilon), L \subset K + B^n(\varepsilon)\},$$

    that is, the infimum in the definition of $d(K, L)$ is attained.
2. Let $\mathcal{K}_c^n$ denote the set of all $K \in \mathcal{K}^n$ for which there is some point $c \in \mathbb{R}^n$ such that $K - c = -(K - c)$. Show that for each $K \in \mathcal{K}_c^n$ such a point c is uniquely determined. It is denoted by $c(K)$ and called the *centre of symmetry* of K. Show that the map $c : \mathcal{K}_c^n \to \mathbb{R}^n$ is continuous.
3. Let $K, L, M \in \mathcal{K}^n$. Show that if $K + L \subset M + L$, then $K \subset M$ (generalized cancellation rule). Can you avoid the use of support functions?
    More generally, let A, B, C be sets in $\mathbb{R}^n$. Suppose that A is nonempty and bounded, that C is closed and convex, and that $A + B \subset A + C$. Show that then $B \subset C$.
4. Let $K, L, M \in \mathcal{K}^n$. Let $u \in \mathbb{S}^{n-1}$. Show that $K(u) + M(u) = (K + M)(u)$. Can you avoid the use of support functions?

5. Let $(K_i)_{i \in \mathbb{N}}$ be a sequence in $\mathcal{K}^n$ and $K \in \mathcal{K}^n$. Show that $K_i \to K$ (in the Hausdorff metric) if and only if the following two conditions are satisfied:

   (a) Every $x \in K$ is a limit point of a suitable sequence $(x_i)_{i \in \mathbb{N}}$ with $x_i \in K_i$ for $i \in \mathbb{N}$.
   (b) For each sequence $(x_i)_{i \in \mathbb{N}}$ with $x_i \in K_i$, for $i \in \mathbb{N}$, every accumulation point lies in K.

6. Let $K_i, K \in \mathcal{K}^n, i \in \mathbb{N}$. Then the following statements are equivalent.

   (i) If $U \subset \mathbb{R}^n$ is open and $K \cap U \neq \emptyset$, then $K_i \cap U \neq \emptyset$ for almost all $i \in \mathbb{N}$.
   (ii) For $x \in K$, there are $x_i \in K_i, i \in \mathbb{N}$, such that $x_i \to x$ as $i \to \infty$.

7. Let $K_i, K \in \mathcal{K}^n, i \in \mathbb{N}$. Suppose that $\operatorname{int} K \neq \emptyset$. Then the following statements (a), (b), and (c) are equivalent.

   (a) $K_i \to K$ as $i \to \infty$.
   (b) (i) If $x \in \operatorname{int} K$, then $x \in \operatorname{int} K_i$ for almost all $i \in \mathbb{N}$,
       (ii) If $I \subset \mathbb{N}$ with $|I| = \infty$, $x_i \in K_i$ for $i \in I$, and $x_i \to x$, as $i \to \infty$, then $x \in K$.
   (c) (i') If $x \in \operatorname{int} K$, then $x \in K_i$ for almost all $i \in \mathbb{N}$,
       (ii) If $I \subset \mathbb{N}$ with $|I| = \infty$, $x_i \in K_i$ for $i \in I$, and $x_i \to x$, as $i \to \infty$, then $x \in K$.

8.* (a) Let $K, M \in \mathcal{K}^n$ be convex bodies, which cannot be separated by a hyperplane (i.e., there is no hyperplane $\{f = \alpha\}$ with $K \subset \{f \leq \alpha\}$ and $M \subset \{f \geq \alpha\}$). Further, let $(K_i)_{i \in \mathbb{N}}$ and $(M_i)_{i \in \mathbb{N}}$ be sequences in $\mathcal{K}^n$. Show that

$$K_i \to K, \ M_i \to M \Longrightarrow K_i \cap M_i \to K \cap M.$$

   (b) Let $K \in \mathcal{K}^n$ be a convex body, and let $E \subset \mathbb{R}^n$ be an affine subspace with $E \cap \operatorname{int} K \neq \emptyset$. Further, let $(K_i)_{i \in \mathbb{N}}$ be a sequence in $\mathcal{K}^n$. Show that

$$K_i \to K \Longrightarrow E \cap K_i \to E \cap K.$$

   Hint: Use Exercise 3.1.5.

9. Let $K \subset \mathbb{R}^n$ be compact. Show that:

   (a) There is a unique Euclidean ball $B_c(K)$ of smallest diameter with $K \subset B_c(K)$ (which is called the *circumball* of K). The map $K \mapsto B_c(K)$ is continuous.
   (b) If $\operatorname{int} K \neq \emptyset$, then there exists a Euclidean ball $B_i(K)$ of maximal diameter with $B_i(K) \subset K$ (which is called an *inball* of K). In general, inballs of convex bodies are not uniquely determined.

10. A body $K \in \mathcal{K}^n$ is *strictly convex*, if it does not contain any segments in the boundary.

(a) Show that the set of all strictly convex bodies in $\mathbb{R}^n$ is a G_δ-set in $\mathcal{K}^n$, that is, an intersection of countably many open sets in $\mathcal{K}^n$.

(b) Show that the set of all strictly convex bodies in $\mathbb{R}^n$ is dense in $\mathcal{K}^n$.

11. Let $(K_i)_{i \in \mathbb{N}}$ be a sequence in $\mathcal{K}^n$ for which the support functions $h_{K_i}(u)$ converge to the values $h(u)$ of a function $h : \mathbb{S}^{n-1} \to \mathbb{R}$, for each $u \in \mathbb{S}^{n-1}$. Show that h is the support function of a convex body and that $h_{K_i} \to h$ uniformly on $\mathbb{S}^{n-1}$.

12. Let P be a convex polygon in $\mathbb{R}^2$ with int $P \neq \emptyset$. Show that:

   (a) There is a polygon P_1 and a triangle (or a segment) Δ with $P = P_1 + \Delta$.
   (b) P has a representation $P = \Delta_1 + \cdots + \Delta_m$, with triangles (segments) Δ_j which are pairwise not homothetic.
   (c) P is a triangle if and only if $m = 1$.

13. A body $K \in \mathcal{K}^n, n \geq 2$, is *indecomposable* if $K = M + L$ implies that $M = \alpha K + x$ and $L = \beta K + y$, for some $\alpha, \beta \geq 0$ and $x, y \in \mathbb{R}^n$. Show that:

   (a) If $P \in \mathcal{K}^n$ is a polytope and all 2-faces of P are triangles, then P is indecomposable.
   (b) For $n \geq 3$, the set of indecomposable convex bodies is a dense G_δ-set in $\mathcal{K}^n$.

14. Let $\mathcal{I}^n$ be the set of convex bodies $K \in \mathcal{K}^n$ which are strictly convex **and** indecomposable. Let $n \geq 3$. Show that $\mathcal{I}^n$ is dense in $\mathcal{K}^n$
   It seems to be an open problem to explicitly construct an element of $\mathcal{I}^n$

15.* (Simultaneous approximation of convex sets, their unions and nonempty intersections) Let $K_1, \ldots, K_m \in \mathcal{K}^n$ be convex bodies such that $K := K_1 \cup \cdots \cup K_m$ is convex. Let $\varepsilon > 0$. Then there are polytopes $P_1, \ldots, P_m \in \mathcal{P}^n$ with $K_i \subset P_i \subset K_i + \varepsilon B^n$ for $i = 1, \ldots, m$ and such that $P := P_1 \cup \cdots \cup P_m$ is convex.

16. Let $K \in \mathcal{K}^n$ with dim $K = n$. Suppose that $H(u, t) = \{x \in \mathbb{R}^n : \langle x, u \rangle = t\}$, $u \in \mathbb{R}^n \setminus \{0\}$ and $t \in \mathbb{R}$, is a supporting hyperplane of K with $K \subset H^-(u, t)$. Show that

$$\lim_{s \nearrow t}(H(u, s) \cap K) = K \cap H(u, t),$$

where the convergence is with respect to the Hausdorff metric.

17. Let $K \in \mathcal{K}^2$. A convex body M is called a rotation average of K if there is some $m \in \mathbb{N}$ and there are proper rotations $\rho_1, \ldots, \rho_m \in SO(2)$ such that $M = \frac{1}{m}(\rho_1 K + \cdots + \rho_m K)$. Show that there is a sequence of rotation averages of K which converges to a two-dimensional ball (a disc) (which might be a point).

3.2   Volume and Surface Area

The volume of a convex body $K \in \mathcal{K}^n$ can be defined as the Lebesgue measure $\lambda_n(K)$ of K. However, the convexity of K implies that the volume also exists in an elementary sense and, moreover, that also the surface area of K exists. Therefore, we now introduce both notions in an elementary way, first for polytopes and then for arbitrary convex bodies by approximation.

Since we shall use a recursive definition with respect to the dimension n, we first remark that the support set $K(u)$, $u \in \mathbb{S}^{n-1}$, of a convex body K lies in a hyperplane parallel to $u^\perp$ (the orthogonal complement of u). Therefore, the orthogonal projection $K(u)|u^\perp$ of K to $u^\perp$ is a translate of $K(u)$, and we can consider $K(u)|u^\perp$ as a convex body in $\mathbb{R}^{n-1}$ if we identify $u^\perp$ with $\mathbb{R}^{n-1}$. Assuming that the volume is already defined in $(n-1)$-dimensional linear subspaces, we then denote by $V^{(n-1)}(K(u)|u^\perp)$ the $(n-1)$-dimensional volume of this projection. In principle, the identification of $u^\perp$ with $\mathbb{R}^{n-1}$ requires that an orthonormal basis in $u^\perp$ is given. However, it will be apparent that the quantities we define depend only on the Euclidean metric in $u^\perp$, hence they are independent of the choice of a basis.

Definition 3.2 Let $P \in \mathcal{P}^n$ be a polytope.
   If $n = 1$, then $P = [a, b]$ with $a \le b$ and we define

$$V^{(1)}(P) := b - a \quad \text{and} \quad F^{(1)}(P) := 2.$$

For $n \ge 2$, let

$$V^{(n)}(P) := \begin{cases} \dfrac{1}{n} \displaystyle\sum_{(*)} h_P(u) V^{(n-1)}(P(u)|u^\perp), & \text{if } \dim P \ge n - 1, \\[2mm] 0, & \text{if } \dim P \le n - 2, \end{cases}$$

and

$$F^{(n)}(P) := \begin{cases} \displaystyle\sum_{(*)} V^{(n-1)}(P(u)|u^\perp), & \text{if } \dim P \ge n - 1, \\[2mm] 0, & \text{if } \dim P \le n - 2, \end{cases}$$

where the summation $(*)$ is over all $u \in \mathbb{S}^{n-1}$ for which $P(u)$ is a facet of P; here, in $\mathbb{R}^n$, by a facet we mean a face (= support set) of dimension $n - 1$. In $\mathbb{R}^n$, we shortly write $V(P)$ for $V^{(n)}(P)$ and call this the *volume* of P. Similarly, we write $F(P)$ instead of $F^{(n)}(P)$ and call this the *surface area* of P.

   For $\dim P = n - 1$, there are two support sets of P which are facets, namely $P = P(u_0)$ and $P = P(-u_0)$, where u_0 is a normal vector to P. Since then $V^{(n-1)}(P(u_0)|u_0^\perp) = V^{(n-1)}(P(-u_0)|u_0^\perp)$ and $h_P(u_0) = -h_P(-u_0)$, we obtain

$V(P) = 0$, which coincides with the Lebesgue measure of P. In this case, we also have $F(P) = 2V^{(n-1)}(P(u_0)|u_0^\perp)$. For dim $P \leq n-2$, the polytope P does not have any facets, hence $V(P) = 0$ and $F(P) = 0$ (but we already defined this explicitly).

Proposition 3.1 *The volume V and surface area F of polytopes $P, Q \in \mathcal{P}^n$ have the following properties:*

(a) $V(P) = \lambda_n(P)$,
(b) V and F are invariant with respect to rigid motions,
(c) $V(\alpha P) = \alpha^n V(P)$ and $F(\alpha P) = \alpha^{n-1} F(P)$ for $\alpha \geq 0$,
(d) $V(P) = 0$ if and only if dim $P \leq n - 1$,
(e) if $P \subset Q$, then $V(P) \leq V(Q)$ and $F(P) \leq F(Q)$.

Proof (a) We proceed by induction on n. The result is clear for $n = 1$. Let $n \geq 2$. We have already mentioned that $V^{(n)}(P) = 0 = \lambda_n(P)$, if dim $P \leq n - 1$. Hence, suppose that dim $P = n$ and $P(u_1), \ldots, P(u_k)$ are the facets of P. If $0 \in \operatorname{int}(P)$, then first using Definition 3.2, then the induction hypothesis and finally basic geometry and calculus (for determining the volume of the pyramid $\operatorname{conv}(\{0\} \cup P(u_i)))$, we get

$$V^{(n)}(P) = \frac{1}{n} \sum_{i=1}^{k} h_P(u_i) V^{(n-1)}(P(u_i)|u_i^\perp)$$

$$= \frac{1}{n} \sum_{i=1}^{k} h_P(u_i) \lambda_{(n-1)}(P(u_i)|u_i^\perp)$$

$$= \sum_{i=1}^{k} \lambda_n(\operatorname{conv}(\{0\} \cup P(u_i)))$$

$$= \lambda_n(P), \tag{3.2}$$

where we also used that any two of the convex hulls in the final sum intersect in a set of λ_n-measure zero. We still assume that $0 \in \operatorname{int}(P)$. If $t \in \mathbb{R}^n$ and $\|t\|$ is small enough, then $-t \in \operatorname{int} P$, hence $0 \in \operatorname{int}(P + t)$. Using (3.2), the translation invariance of λ_n and again (3.2), we obtain

$$V^{(n)}(P + t) = \lambda_n(P + t) = \lambda_n(P) = V^{(n)}(P). \tag{3.3}$$

For any $t \in \mathbb{R}^n$, we have $h_{P+t}(u_i) = h_P(u_i) + \langle t, u_i \rangle$. Using that $(P + t)(u_i) = P(u_i) + t$ and again the induction hypothesis, we get

$$V^{(n-1)}((P + t)(u_i)|u_i^\perp) = V^{(n-1)}(P(u_i)|u_i^\perp).$$

Hence,

$$V^{(n)}(P+t) = \frac{1}{n}\sum_{i=1}^{k}(h_P(u_i) + \langle t, u_i\rangle)V^{(n-1)}(P(u_i)|u_i^{\perp})$$

$$= V^{(n)}(P) + \frac{1}{n}\left\langle t, \sum_{i=1}^{k}u_i V^{(n-1)}(P(u_i)|u_i)\right\rangle. \qquad (3.4)$$

Since $V^{(n)}(P+t) = V^{(n)}(P)$ if $\|t\|$ is small enough, it follows that

$$\sum_{i=1}^{k}u_i V^{(n-1)}(P(u_i)|u_i) = 0. \qquad (3.5)$$

But now (3.4) shows that $V^{(n)}(P) = V^{(n)}(P+t)$ for all $t \in \mathbb{R}^n$, and first for a polytope P with $0 \in \mathrm{int}(P)$, but then clearly for any full-dimensional polytope.

Now let $P \in \mathcal{P}^n$ be n-dimensional. Then there is some $t_0 \in \mathbb{R}^n$ such that $0 \in \mathrm{int}(P + t_0)$, and hence

$$V^{(n)}(P) = V^{(n)}(P+t_0) = \lambda_n(P+t_0) = \lambda_n(P),$$

which proves (a), and also the translation invariance of $V^{(n)}$.

The assertions (b), (c), (d) and the first part of (e) now follow directly from (a) (and the corresponding properties of the Lebesgue measure).

It remains to show $F(P) \le F(Q)$ for polytopes $P \subset Q$. We may assume that $\dim Q = n$. Again we denote the facets of P by $P(u_1), \ldots, P(u_k)$. We make use of the inequality (a generalization of the triangle inequality),

$$V^{(n-1)}(P(u_i)|u_i^{\perp}) \le \sum_{j\neq i}V^{(n-1)}(P(u_j)|u_j^{\perp}), \quad i = 1, \ldots, k. \qquad (3.6)$$

To see this, we rewrite (3.5) in the form

$$u_i V^{(n-1)}(P(u_i)|u_i^{\perp}) = -\sum_{j:j\neq i}u_j V^{(n-1)}(P(u_j)|u_j^{\perp}).$$

Taking the norm of both sides and using the triangle inequality on the right-hand side, we obtain the requested estimate, since $\|u_j\| = 1$.

The estimate (3.6) implies that

$$F(Q \cap H) \le F(Q),$$

for any closed halfspace $H \subset \mathbb{R}^n$. Since $P \subset Q$ is a finite intersection of halfspaces, we obtain $F(P) \le F(Q)$ by successive truncation. $\qquad \square$

Remark 3.7 In the proof of Proposition 3.1 (a), we first considered P such that $0 \in \operatorname{int} P$. Alternatively, we can take into account signed volumes and proceed as follows. By Definition 3.2, we have

$$V(P) = \frac{1}{n} \sum_{i=1}^{k} h_P(u_i) V^{(n-1)}(P(u_i)|u_i^{\perp}).$$

By the induction hypothesis, $V^{(n-1)}(P(u_i)|u_i^{\perp})$ equals the $(n-1)$-dimensional Lebesgue measure (in $u_i^{\perp}$) of $P(u_i)|u_i^{\perp}$. We can assume without loss of generality that $h_P(u_1), \ldots, h_P(u_m) \geq 0$ and $h_P(u_{m+1}), \ldots, h_P(u_k) < 0$, and consider the pyramid-shaped polytopes $P_i := \operatorname{conv}(P(u_i) \cup \{0\})$ for $i = 1, \ldots k$. Then

$$V(P_i) = \frac{1}{n} h_P(u_i) V^{(n-1)}(P(u_i)|u_i^{\perp}), \qquad i = 1, \ldots, m,$$

and

$$V(P_i) = -\frac{1}{n} h_P(u_i) V^{(n-1)}(P(u_i)|u_i^{\perp}), \qquad i = m+1, \ldots, k.$$

Hence,

$$V(P) = \sum_{i=1}^{m} V(P_i) - \sum_{i=m+1}^{k} V(P_i) = \sum_{i=1}^{m} \lambda_n(P_i) \sum_{i=m+1}^{k} \lambda_n(P_i') - \lambda_n(P).$$

Here, we have used that the Lebesgue measure of the pyramid parts outside P cancel out, and the pyramid parts inside P yield a dissection of P (into sets with disjoint interiors).

Remark 3.8 For another proof of (3.6), we can project $P(u_j)$, $j \neq i$, orthogonally onto the hyperplane $u_i^{\perp}$. The projections then cover $P(u_i)|u_i^{\perp}$. Since the projection does not increase the $(n-1)$-dimensional Lebesgue measure, (3.6) follows.

Remark 3.9 We can now simplify our formulas for the volume $V(P)$ and the surface area $F(P)$ of a polytope P. First, since we have shown that our elementary definition of volume equals the Lebesgue measure and is thus translation invariant, we do not need the orthogonal projection of the facets anymore. Second, since $V^{(n-1)}(P(u)) = 0$, for $\dim P(u) \leq n-2$, we can sum over all $u \in \mathbb{S}^{n-1}$. If we write, in addition, v instead of $V^{(n-1)}$, we obtain

$$V(P) = \frac{1}{n} \sum_{u \in \mathbb{S}^{n-1}} h_P(u) v(P(u))$$

and

$$F(P) = \sum_{u \in \mathbb{S}^{n-1}} v(P(u)).$$

Definition 3.3 For a convex body $K \in \mathcal{K}^n$, we define

$$V_+(K) := \inf_{P \supset K} V(P), \quad V_-(K) := \sup_{P \subset K} V(P),$$

and

$$F_+(K) := \inf_{P \supset K} F(P), \quad F_-(K) := \sup_{P \subset K} F(P),$$

where $P \in \mathcal{P}^n$. If $V_+(K) = V_-(K) =: V(K)$, we call this value the *volume* of K. Moreover, if $F_+(K) = F_-(K) =: F(K)$, we call this value the *surface area* of K.

The following result shows that volume and surface area of general convex bodies exist and summarizes important properties of these functionals.

Theorem 3.6 *Let $K, L \in \mathcal{K}^n$.*

(a) *Then*

$$V_+(K) = V_-(K) = V(K)$$

and

$$F_+(K) = F_-(K) = F(K).$$

(b) *Volume and surface area have the following properties:*

  (b1) $V(K) = \lambda_n(K)$,

  (b2) *V and F are invariant with respect to rigid motions,*

  (b3) $V(\alpha K) = \alpha^n V(K)$ *and* $F(\alpha K) = \alpha^{n-1} F(K)$ *for $\alpha \geq 0$,*

  (b4) $V(K) = 0$ *if and only if* $\dim K \leq n - 1$,

  (b5) *if $K \subset L$, then $V(K) \leq V(L)$ and $F(K) \leq F(L)$,*

  (b6) $K \mapsto V(K)$ *is continuous.*

Proof (a) We first remark that for a polytope P the monotonicity of V and F (which was proved in Proposition 3.1 (e)) shows that $V^+(P) = V(P) = V^-(P)$ and $F^+(P) = F(P) = F^-(P)$. In particular, the new definition of $V(P)$ and $F(P)$ is consistent with the preliminary definition.

For an arbitrary convex body $K \in \mathcal{K}^n$, we get from Proposition 3.1 (e)

$$V_-(K) \le V_+(K) \quad \text{and} \quad F_-(K) \le F_+(K),$$

and by Proposition 3.1 (b), $V_-(K), V_+(K), F_-(K)$ and $F_+(K)$ are motion invariant. After a suitable translation, we may therefore assume that $0 \in \operatorname{relint} K$. For $\varepsilon \in (0, 1)$, we then use Theorem 3.5 (c) and find a polytope P (depending on ε) with $P \subset K \subset (1 + \varepsilon)P$. From Proposition 3.1 (c), we get

$$V(P) \le V_-(K) \le V_+(K) \le V((1 + \varepsilon)P) = (1 + \varepsilon)^n V(P)$$

and

$$F(P) \le F_-(K) \le F_+(K) \le F((1 + \varepsilon)P) = (1 + \varepsilon)^{n-1} F(P).$$

Let Q be an arbitrary polytope with $K \subset Q$, hence $V(P) \le V(Q)$ and $F(P) \le F(Q)$. Since

$$(1 + \varepsilon)^n V(P) - V(P) = \left[(1 + \varepsilon)^n - 1\right] V(P) \le \varepsilon 2^n V(Q) \to 0 \quad \text{as } \varepsilon \searrow 0,$$

the assertion in (a) follows for the volume functional, and in a similar way we conclude the argument for the surface area.

(b) The assertions (b1)–(b5) now follow directly for convex bodies $K \in \mathcal{K}^n$ ((b1) by approximation with polytopes; (b2)–(b5) by approximation with polytopes or from the corresponding properties of Lebesgue measure).

It remains to prove (b6). Consider a convergent sequence $K_i \to K$, $K_i, K \in \mathcal{K}^n$. In view of (b2), we can assume $0 \in \operatorname{relint} K$. Using again Theorem 3.5 (c), we find a polytope P with $P \subset \operatorname{relint} K$ and $K \subset \operatorname{relint}((1 + \varepsilon)P)$. If $\dim K = n$, then $r := \min\{h_{(1+\varepsilon)P}(u) - h_K(u) : u \in \mathbb{S}^{n-1}\} > 0$ so that $K + B^n(r) \subset (1 + \varepsilon)P$. Then $K_i \subset K + B^n(r) \subset (1 + \varepsilon)P$ for $i \ge i_0$. Analogously, we can choose some $r' > 0$ such that $P + B^n(r') \subset K \subset K_i + B^n(r')$ for $i \ge i_1$. This implies that $P \subset K_i$, by the cancellation property of Minkowski addition. For $i \ge \max\{i_0, i_1\}$, we therefore obtain

$$V(P) - V((1 + \varepsilon)P) \le V(K_i) - V(K) \le V((1 + \varepsilon)P) - V(P),$$

and hence

$$|V(K_i) - V(K)| \le (1 + \varepsilon)^n V(P) - V(P) \le [(1 + \varepsilon)^n - 1]V(K) \to 0$$

as $\varepsilon \searrow 0$. If $\dim K = j \le n - 1$, hence $V(K) = 0$, we have

$$K \subset \operatorname{int}((1 + \varepsilon)P + \varepsilon W),$$

where W is a cube, centred at 0, with edge length 1 and dimension $n - j$, lying in the orthogonal space $(\text{aff } K)^\perp$. As above, we obtain $K_i \subset (1+\varepsilon)P + \varepsilon W$ for $i \geq i_0$. Since

$$V((1 + \varepsilon)P + \varepsilon W) \leq \varepsilon^{n-j}(1 + \varepsilon)^j C$$

(where we can choose the constant C to be the j-dimensional Lebesgue measure of K), this gives us $V(K_i) \to 0 = V(K)$, as $\varepsilon \searrow 0$. $\qquad \square$

Remark 3.10 We shall see in the next section that the surface area F is also continuous.

Exercises and Supplements for Sect. 3.2

1. Show that among all convex bodies in $\mathbb{R}^n$ having a fixed diameter, there is one having maximal volume.
2. A convex body $K \in \mathcal{K}^2$ is called a *universal cover* if for each $L \in \mathcal{K}^2$ with diameter ≤ 1 there is a rigid motion g_L of $\mathbb{R}^2$ with $L \subset g_L K$.

   (a) Show that there is a universal cover K_0 with minimal area.
   (b) Find the shape and the area of K_0.

3. Let $K \in \mathcal{K}^n$ with $\dim K = n$. Show:

   (a) Among all simplices containing K there is one having minimal volume.
   (b) Among all simplices contained in K there is one having maximal volume.

4.* Let $K \in \mathcal{K}^n$ with $\dim K = n$. Show that there is a simplex S with centre c such that $S \subset K \subset c - n(S - c)$.
5. If f is an affine invariant, continuous functional on convex bodies with nonempty interiors, then f attains its extremal values.
6. Let $\mathcal{K}_0^n$ denote the set of convex bodies having nonempty interiors. Suppose that $f : \mathcal{K}_0^n \to [0, \infty)$ is increasing (with respect to inclusion), translation invariant and positively homogeneous of degree $r \geq 0$. Show that f is continuous.
7. Let $\mathbb{R}^2$ be an infinitely large baking tray and F the shape of a symmetric cutout cookie having centre $m \in \mathbb{R}^2$, that is, $F - m = -F + m$, which has area strictly larger than 4. At each integer point of the tray there is a raisin (which can be assumed to be a point). Show that if we use a translate of F to cut out a cookie such that the centre of the cutout is 0, then the cookie contains at least three raisins.

3.3 Mixed Volumes

There is another common definition of the surface area of a set $K \subset \mathbb{R}^n$, which describes the surface area as the derivative of the volume functional of the outer parallel sets of K, in the sense that

$$F(K) = \lim_{\varepsilon \searrow 0} \frac{1}{\varepsilon}(V(K + B^n(\varepsilon)) - V(K)).$$

In this section, we shall see that our notion of surface area of a convex body K satisfies this limit relation. More generally, we shall show that $V(K + B^n(\varepsilon))$ is a polynomial in $\varepsilon \geq 0$, which is the famous *Steiner formula*. In this way, we get a whole family of important geometric functionals of which volume and surface area are just two prominent examples. We start with an even more general problem and investigate how the volume

$$V(\alpha_1 K_1 + \cdots + \alpha_m K_m)$$

for $K_i \in \mathcal{K}^n$ and $\alpha_i \geq 0$, $i = 1, \ldots, n$, depends on the variables $\alpha_1, \ldots, \alpha_m$. This will lead us to a family of mixed functionals of convex bodies, which are called *mixed volumes*.

As in Sect. 3.2, we first consider the case of polytopes. Since the recursive representation of the volume of a polytope P was based on the support sets (facets) of P, we now discuss how support sets behave under linear combinations and intersections. The following lemma holds for arbitrary convex bodies with the same proof.

Proposition 3.2 *Let* $m \in \mathbb{N}$, *let* $\alpha_1, \ldots, \alpha_m > 0$, *let* $P_1, \ldots, P_m \in \mathcal{P}^n$ *be polytopes, and let* $u, v \in \mathbb{S}^{n-1}$. *Then,*

(a) $(\alpha_1 P_1 + \cdots + \alpha_m P_m)(u) = \alpha_1 P_1(u) + \cdots + \alpha_m P_m(u)$,
(b) $\dim(\alpha_1 P_1 + \cdots + \alpha_m P_m)(u) = \dim(P_1 + \cdots + P_m)(u)$,
(c) *if* $(P_1 + \cdots + P_m)(u) \cap (P_1 + \cdots + P_m)(v) \neq \emptyset$, *then*

$$(P_1 + \cdots + P_m)(u) \cap (P_1 + \cdots + P_m)(v)$$

$$= (P_1(u) \cap P_1(v)) + \cdots + (P_m(u) \cap P_m(v)).$$

Proof (a) By Theorems 2.7 and 2.9, for $x \in \mathbb{R}^n$ we have

$$h_{(\alpha_1 P_1 + \cdots + \alpha_m P_m)(u)}(x) = h'_{\alpha_1 P_1 + \cdots + \alpha_m P_m}(u; x)$$

$$= \alpha_1 h'_{P_1}(u; x) + \cdots + \alpha_m h'_{P_m}(u; x)$$

$$= \alpha_1 h_{P_1(u)}(x) + \cdots + \alpha_m h_{P_m(u)}(x)$$

$$= h_{\alpha_1 P_1(u) + \cdots + \alpha_m P_m(u)}(x).$$

Theorem 2.7 now yields the assertion.

(b) Let $P := P_1 + \cdots + P_m$ and $\widetilde{P} := \alpha_1 P_1 + \cdots + \alpha_m P_m$. W.l.o.g. we may assume $0 \in P_i(u)$ for $i = 1, \ldots, m$, and hence $0 \in P(u)$. (We may even assume $0 \in \operatorname{relint} P_i(u)$ for $i = 1, \ldots, m$. By Exercise 1.3.3 (a) it then follows that $0 \in \operatorname{relint} P(u)$. But this is not needed.) We put

$$\alpha := \min_{i=1,\ldots,m} \alpha_i, \quad \beta := \max_{i=1,\ldots,m} \alpha_i.$$

Then, $0 < \alpha \le \beta$ and (in view of (a))

$$\alpha P(u) \subset \widetilde{P}(u) \subset \beta P(u),$$

that is, $\dim P(u) = \dim \widetilde{P}(u)$.

(c) Using the notation introduced above, we assume $P(u) \cap P(v) \ne \emptyset$. Consider $x \in P(u) \cap P(v)$. Since $x \in P$, we have $x = x_1 + \cdots + x_m$ with $x_i \in P_i$. Because of

$$h_P(u) = \langle x, u \rangle = \sum_{i=1}^{m} \langle x_i, u \rangle \le \sum_{i=1}^{m} h_{P_i}(u) = h_P(u),$$

it follows that $\langle x_i, u \rangle = h_{P_i}(u)$ and thus $x_i \in P_i(u)$ for $i = 1, \ldots, m$. In the same way, we obtain $x_i \in P_i(v)$ for $i = 1, \ldots, m$.

Conversely, it is clear that any $x \in (P_1(u) \cap P_1(v)) + \cdots + (P_m(u) \cap P_m(v))$ satisfies $x \in P_1(u) + \cdots + P_m(u) = P(u)$ and $x \in P_1(v) + \cdots + P_m(v) = P(v)$, by (a). $\qquad\square$

In the proof of an important symmetry property of mixed volumes, we shall use the following lemma.

Lemma 3.2 *Let $K \in \mathcal{K}^n$, let $u, v \in \mathbb{S}^{n-1}$ be linearly independent unit vectors, and let $w = \lambda u + \mu v$ with some $\lambda \in \mathbb{R}$ and $\mu > 0$. Then $K(u) \cap K(v) \ne \emptyset$ implies that $K(u) \cap K(v) = K(u)(w)$.*

Proof Let $z \in K(u) \cap K(v)$ and $w = \lambda u + \mu v$ with some $\lambda \in \mathbb{R}$ and $\mu > 0$. Then $z \in K(u)$, hence $\langle z, u \rangle = h_K(u) = h_{K(u)}(u)$ and

$$\begin{aligned}
h_{K(u)}(-u) &= \max\{\langle x, -u \rangle : x \in K(u)\} = \max\{-\langle x, u \rangle : x \in K(u)\} \\
&= \max\{-h_{K(u)}(u) : x \in K(u)\} = -h_{K(u)}(u) = -\langle z, u \rangle \\
&= \langle z, -u \rangle.
\end{aligned}$$

Therefore we have $\langle z, \lambda u \rangle = h_{K(u)}(\lambda u)$ for $\lambda \in \mathbb{R}$. We deduce

$$\begin{aligned}
\langle z, w \rangle &= \langle z, \lambda u \rangle + \langle z, \mu v \rangle = h_{K(u)}(\lambda u) + h_K(\mu v) \ge h_{K(u)}(\lambda u) + h_{K(u)}(\mu v) \\
&\ge h_{K(u)}(\lambda u + \mu v) = h_{K(u)}(w) \ge \langle z, w \rangle,
\end{aligned}$$

which yields $z \in K(u)(w)$.

Now let $z \in K(u)(w)$. There is some $x_0 \in K(u) \cap K(v) \neq \emptyset$. Then $\langle x_0, u \rangle = h_K(u) = \langle z, u \rangle$, since $z \in K(u)$, and $\langle x_0, v \rangle = h_K(v)$. By the preceding argument, $x_0 \in K(u)(w)$, and therefore

$$\lambda \langle z, u \rangle + \mu \langle z, v \rangle = \langle z, w \rangle = \langle x_0, w \rangle = \lambda \langle x_0, u \rangle + \mu \langle x_0, v \rangle,$$

hence $\langle z, v \rangle = \langle x_0, v \rangle = h_K(v)$, that is, $z \in K(v)$.

Thus it follows that $z \in K(u) \cap K(v)$.   □

In analogy to the recursive definition of the volume of a polytope, we now define the mixed volume of polytopes. Again, we use projections of the support sets (faces) in order to make the definition rigorous. After we have shown translation invariance of the functionals, the corresponding formulas simplify.

For polytopes $P_1, \ldots, P_k \in \mathcal{P}^n$, let $N(P_1, \ldots, P_k)$ denote the set of all facet unit normals of the convex polytope $P_1 + \cdots + P_k$ in $\mathbb{R}^n$. As in Sect. 3.2, a facet of a convex body in $\mathbb{R}^n$ is a face (support set) of dimension $n - 1$.

Definition 3.4 For polytopes $P_1, \ldots, P_n \in \mathcal{P}^n$, we define the *mixed volume* $V^{(n)}(P_1, \ldots, P_n)$ of $P_1, \ldots, P_n$ recursively.

For $n = 1$ and $P_1 = [a, b] \subset \mathbb{R}$ with $a \leq b$,

$$V^{(1)}(P_1) := V(P_1) = h_{P_1}(1) + h_{P_1}(-1) = b - a,$$

and, for $n \geq 2$,

$$V^{(n)}(P_1, \ldots, P_n) := \frac{1}{n} \sum_{(*)} h_{P_n}(u) V^{(n-1)}(P_1(u)|u^\perp, \ldots, P_{n-1}(u)|u^\perp),$$

where the summation $(*)$ extends over all $u \in N(P_1, \ldots, P_{n-1})$.

This recursive definition of the mixed volumes leads to a functional satisfying various properties and relationships. A first collection of basic properties is provided in the following theorem.

Theorem 3.7 *The mixed volume* $V^{(n)}(P_1, \ldots, P_n)$ *of* n *convex polytopes* $P_1, \ldots, P_n \in \mathcal{P}^n$ *has the following properties.*

(a) *It is symmetric in the arguments* $P_1, \ldots, P_n$.
(b) *It is independent of individual translations of the polytopes* $P_1, \ldots, P_n$.
(c) *If* $\dim(P_1 + \cdots + P_n) \leq n - 1$, *then* $V^{(n)}(P_1, \ldots, P_n) = 0$.
(d) *If* $m \in \mathbb{N}$, $P_1, \ldots, P_m \in \mathcal{P}^n$, *and* $\alpha_1, \ldots, \alpha_m \geq 0$, *then*

$$V(\alpha_1 P_1 + \cdots + \alpha_m P_m) = \sum_{i_1=1}^{m} \cdots \sum_{i_n=1}^{m} \alpha_{i_1} \cdots \alpha_{i_n} V^{(n)}(P_{i_1}, \ldots, P_{i_n}). \quad (3.7)$$

(e) *The mixed volume is linear with respect to positive Minkowski combinations in each argument.*

For the proof, it is convenient to extend the k-dimensional mixed volume $V^{(k)}(Q_1, \ldots, Q_k)$, which is defined for convex polytopes $Q_1, \ldots, Q_k$ in a k-dimensional linear subspace $E \subset \mathbb{R}^d$, to convex polytopes $Q_1, \ldots, Q_k \in \mathcal{P}^n$ for which $\dim(Q_1 + \cdots + Q_k) \leq k$, by

$$V^{(k)}(Q_1, \ldots, Q_k) := V^{(k)}(Q_1|E, \ldots, Q_k|E),$$

where E is a k-dimensional subspace parallel to $Q_1 + \cdots + Q_k$, $1 \leq k \leq n-1$. The translation invariance and the dimensional condition, which we shall prove, show that this extension is consistent and independent of E in case $\dim(Q_1 + \cdots + Q_k) < k$. In the following proof, which proceeds by induction, we already make use of this extension in order to simplify the presentation. In particular, in the induction step, we use the mixed volume $V^{(n-1)}(P_1(u), \ldots, P_{n-1}(u))$.

In addition, we extend the definition of the mixed volume of convex polytopes to the empty set by setting $V^{(n)}(P_1, \ldots, P_n) := 0$, if one of the sets P_i is empty.

Proof of Theorem 3.7 We use induction on the dimension n.

For $n = 1$, the polytopes P_i are intervals and the mixed volume equals the (one-dimensional) volume $V^{(1)}$ (the length functional), which is linear

$$V^{(1)}(\alpha_1 P_1 + \cdots + \alpha_m P_m) = \sum_{i=1}^{m} \alpha_i V^{(1)}(P_i).$$

Here we use that if $P_i = [a_i, b_i]$ with $a_1 \leq b_i$, then

$$\sum_{i=1}^{m} \alpha_i [a_i, b_i] = [\alpha_1 a_1 + \cdots + \alpha_m a_m, \alpha_1 b_1 + \cdots + \alpha_m b_m].$$

Hence we get (3.7), and the remaining assertions hold as well.

Now we assume that the assertions of the theorem are true for all dimensions $\leq n - 1$, and we consider dimension $n \geq 2$. We first discuss the dimensional statement. If $\dim(P_1 + \cdots + P_n) \leq n - 1$, then either $N(P_1, \ldots, P_{n-1}) = \emptyset$ or $N(P_1, \ldots, P_{n-1}) = \{-u, u\}$, where u is a unit normal on $\mathrm{aff}(P_1 + \cdots + P_n)$. In the first case, we have $V^{(n)}(P_1, \ldots, P_n) = 0$ by definition; in the second case, we have

$$V^{(n)}(P_1, \ldots, P_n) = \frac{1}{n} h_{P_n}(u) V^{(n-1)}(P_1(u), \ldots, P_{n-1}(u))$$

$$+ \frac{1}{n} h_{P_n}(-u) V^{(n-1)}(P_1(-u), \ldots, P_{n-1}(-u))$$

$$= \frac{1}{n} h_{P_n}(u) V^{(n-1)}(P_1(u), \ldots, P_{n-1}(u))$$

$$- \frac{1}{n} h_{P_n}(u) V^{(n-1)}(P_1(u), \ldots, P_{n-1}(u))$$

$$= 0.$$

Here we used that $P_i(-u)$ is a translate of $P_i(u)$ for $i = 1, \ldots, n-1$.

Next, we prove (3.7). If $\alpha_i = 0$, for a certain index i, the corresponding summand $\alpha_i P_i$ on the left-hand side can be deleted, as well as all summands on the right-hand side which contain this particular index i. Therefore it is sufficient to consider the case where $\alpha_1 > 0, \ldots, \alpha_m > 0$. By the definition of volume and Proposition 3.2,

$$V(\alpha_1 P_1 + \cdots + \alpha_m P_m)$$

$$= \frac{1}{n} \sum_{u \in N(P_1, \ldots, P_m)} h_{\sum_{i=1}^m \alpha_i P_i}(u) \, v\left(\left(\sum_{i=1}^m \alpha_i P_i\right)(u)\right)$$

$$= \sum_{i_n=1}^m \alpha_{i_n} \frac{1}{n} \sum_{u \in N(P_1, \ldots, P_m)} h_{P_{i_n}}(u) \, v\left(\sum_{i=1}^m \alpha_i (P_i(u)|u^\perp)\right).$$

The induction hypothesis implies that

$$v\left(\sum_{i=1}^m \alpha_i (P_i(u)|u^\perp)\right)$$

$$= \sum_{i_1=1}^m \cdots \sum_{i_{n-1}=1}^m \alpha_{i_1} \cdots \alpha_{i_{n-1}} V^{(n-1)}(P_{i_1}(u), \ldots, P_{i_{n-1}}(u)).$$

Hence we obtain

$$V(\alpha_1 P_1 + \cdots + \alpha_m P_m)$$

$$= \sum_{i_1=1}^m \cdots \sum_{i_n=1}^m \alpha_{i_1} \cdots \alpha_{i_{n-1}} \alpha_{i_n}$$

$$\times \frac{1}{n} \sum_{u \in N(P_1, \ldots, P_m)} h_{P_{i_n}}(u) \, V^{(n-1)}(P_{i_1}(u), \ldots, P_{i_{n-1}}(u))$$

$$= \sum_{i_1=1}^m \cdots \sum_{i_n=1}^m \alpha_{i_1} \cdots \alpha_{i_n} V^{(n)}(P_{i_1}, \ldots, P_{i_n}).$$

Here we have used that for a given set of indices $\{i_1, \ldots, i_n\}$, the summation over the set $N(P_1, \ldots, P_m)$ can be replaced by the summation over the subset $N(P_{i_1}, \ldots, P_{i_{n-1}})$. Namely, for a unit vector $u \notin N(P_{i_1}, \ldots, P_{i_{n-1}})$ the support set $P_{i_1}(u) + \cdots + P_{i_{n-1}}(u) = (P_{i_1} + \cdots + P_{i_{n-1}})(u)$ has dimension $\leq n - 2$ and hence $V^{(n-1)}(P_{i_1}(u), \ldots, P_{i_{n-1}}(u)) = 0$. We will also use this fact in the following parts of the proof.

We now prove the symmetry property. Since $V^{(n-1)}(P_1(u), \ldots, P_{n-1}(u))$ is symmetric (in the indices), by the induction hypothesis, it suffices to show that

$$V^{(n)}(P_1, \ldots, P_{n-2}, P_{n-1}, P_n) = V^{(n)}(P_1, \ldots, P_{n-2}, P_n, P_{n-1}).$$

Moreover, we may assume that $P := P_1 + \cdots + P_n$ is n-dimensional. By definition,

$$V^{(n-1)}(P_1(u), \ldots, P_{n-1}(u))$$
$$= \frac{1}{n-1} \sum_{\tilde{v} \in \widetilde{N}} h_{P_{n-1}(u)}(\tilde{v}) V^{(n-2)}(P_1(u)(\tilde{v}), \ldots, P_{n-2}(u)(\tilde{v})),$$

where the sum extends over the set $\widetilde{N}$ of facet normals of $P(u)$ (in $u^\perp$). Formally, we would have to work with the projections (that is, the shifted support sets) $P_1(u)|u^\perp, \ldots, P_{n-1}(u)|u^\perp$, but here we make use of our extended definition of the $(n - 2)$-dimensional mixed volume and of the fact that

$$h_{P_{n-1}(u)|u^\perp}(\tilde{v}) = h_{P_{n-1}(u)}(\tilde{v}),$$

for all $\tilde{v} \perp u$. The facets of $P(u)$ are $(n - 2)$-dimensional faces of P, thus they arise (because of $\dim P = n$) as intersections $P(u) \cap P(v)$ of the facet $P(u)$ with another facet $P(v)$ of P. Since $\dim P = n$, the case $v = -u$ cannot occur. If $P(u) \cap P(v)$ is a $(n - 2)$-face of P, hence a facet of $P(u)$, the corresponding normal (in $u^\perp$) is given by $\tilde{v} := \|v|u^\perp\|^{-1}(v|u^\perp)$, hence it is of the form $\tilde{v} = \lambda u + \mu v$ with some $\lambda \in \mathbb{R}$ and $\mu > 0$.

By Proposition 3.2(c), we have

$$P(u) \cap P(v) = (P_1(u) \cap P_1(v)) + \cdots + (P_n(u) \cap P_n(v));$$

in particular, $P_i(u) \cap P_i(v) \neq \emptyset$ for $i = 1, \ldots, n$. For an $(n - 2)$-face $P(u) \cap P(v)$ of P, we therefore obtain by Lemma 3.2

$$(P_i(u))(\tilde{v}) = P_i(u) \cap P_i(v), \quad i = 1, \ldots, n - 2,$$

which implies that

$$V^{(n-1)}(P_1(u), \ldots, P_{n-1}(u)) \qquad (3.8)$$

$$= \frac{1}{n-1} \sum_{(*)} h_{P_{n-1}(u)}(\tilde{v}) \, V^{(n-2)}(P_1(u) \cap P_1(v), \ldots, P_{n-2}(u) \cap P_{n-2}(v)),$$

where the sum $(*)$ extends over all $v \in N(P_1, \ldots, P_n)$ with $P(u) \cap P(v) \neq \emptyset$, since if v is a unit vector for which $P(u) \cap P(v) \neq \emptyset$ is not an $(n-2)$-face of P, then the mixed volume $V^{(n-2)}(P_1(u) \cap P_1(v), \ldots, P_{n-2}(u) \cap P_{n-2}(v))$ vanishes by the induction hypothesis. Also, for $n = 2$, the mixed volume $V^{(n-2)}(P_1(u) \cap P_1(v), \ldots, P_{n-2}(u) \cap P_{n-2}(v))$ is defined to be 1.

Let $\gamma(u, v) \in (0, \pi)$ denote the (outer) angle between u and v. Then

$$\| v | u^{\perp} \| = \sin \gamma(u, v), \quad \langle u, v \rangle = \cos \gamma(u, v),$$

and hence

$$\frac{v | u^{\perp}}{\| v | u^{\perp} \|} = \frac{1}{\sin \gamma(u, v)} v - \frac{1}{\tan \gamma(u, v)} u.$$

For $x \in P_{n-1}(u) \cap P_{n-1}(v)$, we have

$$h_{P_{n-1}(u)}(\tilde{v}) = \langle x, \tilde{v} \rangle = \frac{1}{\sin \gamma(u, v)} \langle x, v \rangle - \frac{1}{\tan \gamma(u, v)} \langle x, u \rangle$$

$$= \frac{1}{\sin \gamma(u, v)} h_{P_{n-1}}(v) - \frac{1}{\tan \gamma(u, v)} h_{P_{n-1}}(u). \qquad (3.9)$$

Hence, altogether we obtain

$$V^{(n)}(P_1, \ldots, P_{n-2}, P_{n-1}, P_n)$$

$$= \frac{1}{n} \sum_{u \in N(P_1, \ldots, P_n)} h_{P_n}(u) \, V^{(n-1)}(P_1(u), \ldots, P_{n-1}(u))$$

$$= \frac{1}{n(n-1)} \sum_{u, v \in N(P_1, \ldots, P_n), v \neq \pm u} \left[\frac{1}{\sin \gamma(u, v)} h_{P_n}(u) h_{P_{n-1}}(v) \right.$$

$$\left. - \frac{1}{\tan \gamma(u, v)} h_{P_n}(u) h_{P_{n-1}}(u) \right]$$

$$\times V^{(n-2)}(P_1(u) \cap P_1(v), \ldots, P_{n-2}(u) \cap P_{n-2}(v))$$

$$= V^{(n)}(P_1, \ldots, P_{n-2}, P_n, P_{n-1}),$$

which proves the symmetry property.

For the remaining assertion, observe that by the recursive definition and the translation invariance of the mixed volume in dimension $\leq n-1$, we obtain the translation invariance with respect to P_1 from the induction hypothesis. Here we also use that $(P_1 + x)(u) = P_1(u) + x$. The translation invariance with respect to the other arguments then follows from the symmetry.

Finally, property (e) follows for $n = 1$ from the polynomial expansion (which is linear for $n = 1$) which has already been shown. For $n \geq 2$ it follows by an induction argument and the symmetry. Alternatively, one can use the linearity properties of the support function and the symmetry.   □

Remark 3.11 In the following, we use similar abbreviations as in the case of volume,

$$V(P_1, \ldots, P_n) := V^{(n)}(P_1, \ldots, P_n)$$

and

$$v(P_1(u), \ldots, P_{n-1}(u)) := V^{(n-1)}(P_1(u), \ldots, P_{n-1}(u)).$$

As a special case of the polynomial expansion of volumes, we obtain

$$V(P_1 + \cdots + P_m) = \sum_{i_1=1}^{m} \cdots \sum_{i_n=1}^{m} V(P_{i_1}, \ldots, P_{i_n}).$$

The question arises whether this expansion can be inverted.

Example 3.1 To get an idea of what an inversion could look like, we first consider the case $n = 2$. In this case, we get

$$\frac{1}{2}\left(-V(K_1) - V(K_2) + V(K_1 + K_2)\right)$$

$$= \frac{1}{2}\left(-V(K_1) - V(K_2) + V(K_1, K_1) + 2V(K_1, K_2) + V(K_2, K_2)\right)$$

$$= V(K_1, K_2).$$

The case $n = 3$ is already more involved. Here we have

$$\frac{1}{6}\left(V(K_1) + V(K_2) + V(K_3) - [V(K_1 + K_2) + V(K_1 + K_3) + V(K_2 + K_3)]\right.$$

$$\left. + V(K_1 + K_2 + K_3)\right)$$

$$= \frac{1}{6}\left(V(K_1) + V(K_2) + V(K_3)\right.$$

$$\left. - [V(K_1) + 3V(K_1, K_1, K_2) + 3V(K_1, K_2, K_2) + V(K_2)]\right.$$

$$-[V(K_1) + 3V(K_1, K_1, K_3) + 3V(K_1, K_3, K_3) + V(K_3)]$$
$$-[V(K_2) + 3V(K_2, K_2, K_3) + 3V(K_2, K_3, K_3) + V(K_3)]$$
$$+[V(K_1) + V(K_2) + V(K_3) + 3(V(K_1, K_1, K_2) + V(K_1, K_2, K_2))$$
$$+3(V(K_1, K_1, K_3) + V(K_1, K_3, K_3))$$
$$+3(V(K_2, K_2, K_3) + V(K_2, K_3, K_3)) + 6V(K_1, K_2, K_3)])$$
$$= V(K_1, K_2, K_3).$$

The proof of the following theorem contains a general argument which explains why the cancellation always works in the right way.

Theorem 3.8 (Inversion Formula) *If* $P_1, \ldots, P_n \in \mathcal{P}^n$, *then*

$$V(P_1, \ldots, P_n) = \frac{1}{n!} \sum_{k=1}^{n} (-1)^{n+k} \sum_{1 \leq r_1 < \cdots < r_k \leq n} V(P_{r_1} + \cdots + P_{r_k}). \tag{3.10}$$

Proof We denote the right-hand side by $f(P_1, \ldots, P_n)$. Then formula (3.7) in Theorem 3.7 implies that $f(\alpha_1 P_1, \ldots, \alpha_n P_n)$ is a homogeneous polynomial of degree n in the variables $\alpha_1 \geq 0, \ldots, \alpha_n \geq 0$, or it is the zero polynomial. Replacing P_1 by $\{0\}$ and writing $\check{P}_j$ to indicate that P_j is omitted, we have

$$(-1)^{n+1} n! f(\{0\}, P_2, \ldots, P_n)$$

$$= \sum_{2 \leq r \leq n} V(P_r) - \left[\sum_{2 \leq r \leq n} V(\{0\} + P_r) + \sum_{2 \leq r < s \leq n} V(P_r + P_s) \right]$$

$$+ \left[\sum_{2 \leq r < s \leq n} V(\{0\} + P_r + P_s) + \sum_{2 \leq r < s < t \leq n} V(P_r + P_s + P_t) \right]$$

$$- \cdots$$

$$+ (-1)^{n-2} \left[\sum_{j=2}^{n} V(\{0\} + P_2 + \cdots + \check{P}_j + \cdots + P_n) + V(P_2 + \cdots + P_n) \right]$$

$$+ (-1)^{n-1} V(\{0\} + P_2 + \cdots + P_n)$$

$$= 0,$$

which means that $f(\{0\}, \alpha_2 P_2, \ldots, \alpha_n P_n) = f(0 \cdot P_1, \alpha_2 P_2, \ldots, \alpha_n P_n)$ is the zero polynomial. Consequently, in the polynomial $f(\alpha_1 P_1, \ldots, \alpha_n P_n)$ all monomials $\alpha_{i_1} \cdots \alpha_{i_n}$ with $1 \notin \{i_1, \ldots, i_n\}$ have zero coefficients. Replacing 1 by $2, \ldots, n$, we obtain finally that only the coefficient of $\alpha_1 \cdots \alpha_n$ can be non-zero. This coefficient occurs only once in the representation of f, namely for $k = n$ with $(r_1, \ldots, r_n) = (1, \ldots, n)$. Therefore, by Theorem 3.7, this coefficient must coincide with $V(P_1, \ldots, P_n)$. ☐

Remark 3.12 Relation (3.10) can be briefly written in the form

$$V(P_1, \ldots, P_n) = \frac{1}{n!} \sum_{\emptyset \neq I \subset [n]} (-1)^{n+|I|} V\left(\sum_{i \in I} P_i\right),$$

where the summation extends over all nonempty subsets I of $[n] = \{1, \ldots, n\}$ and $|I|$ denotes the cardinality of I.

Remark 3.13 We specialize (3.10) by choosing $P = P_1 = \cdots = P_n$. Then the result takes the form

$$V(P, \ldots, P) = \frac{1}{n!} \sum_{k=1}^{n} (-1)^{n+k} \binom{n}{k} k^n V(P) =: c_n V(P),$$

which yields $V(P, \ldots, P) = V(P)$, since $c_n = 1$. This can be seen by using the inclusion-exclusion formula for determining the number of surjective maps from an n-element set onto itself.

On the other hand, the equality $V(P, \ldots, P) = V(P)$ can be seen directly from the recursive definition of mixed volumes of polytopes, using an induction argument and the expression we had obtained for $V(P)$ previously.

Remark 3.14 Another way to see that $c_n = 1$ is as follows. Let $P_i = s_i$, $i = 1, \ldots, n$, be n non-degenerate segments with linearly independent directions. Put $C_n := s_1 + \cdots + s_n$. Then (3.10) yields

$$V(s_1, \ldots, s_n) = \frac{1}{n!} V(s_1 + \cdots + s_n) = \frac{1}{n!} V(C_n),$$

since $V(s_{r_1} + \cdots + s_{r_k}) = 0$ if $k < n$. On the other hand, using that $V(C_n, \ldots, C_n)$ can be linearly expanded in each of the n arguments (for the first equality) and the fact that the mixed volume is zero if the sum of the polytopes in the n arguments is lower-dimensional and that the mixed volume is symmetric (for the second equality), we obtain

$$V(C_n, \ldots, C_n) = \sum_{i_1=1}^{n} \cdots \sum_{i_n=1}^{n} V(s_{i_1}, \ldots, s_{i_n}) = n! V(s_1, \ldots, s_n),$$

where we used the dimensional and the symmetry property of mixed volumes. This shows that

$$c_n V(C_n) = V(C_n, \ldots, C_n) = n! V(s_1, \ldots, s_n) = V(C_n).$$

Since $V(C_n) > 0$, we get $c_n = 1$.

Theorem 3.9 *Let* $K_1, \ldots, K_n \in \mathcal{K}^n$ *be convex bodies. Let* $(P_i^{(j)})_{j\in\mathbb{N}}$ *for* $i \in \{1, \ldots, n\}$ *be arbitrary approximating sequences of polytopes such that* $P_i^{(j)} \to K_i$ *as* $j \to \infty$, *for* $i = 1, \ldots, n$. *Then the limit*

$$V(K_1, \ldots, K_n) = \lim_{j\to\infty} V(P_1^{(j)}, \ldots, P_n^{(j)})$$

exists and is independent of the choice of the approximating sequences $(P_i^{(j)})_{j\in\mathbb{N}}$. *The number* $V(K_1, \ldots, K_n)$ *is called the* mixed volume *of the convex bodies* $K_1, \ldots, K_n$. *The mapping* $V : (\mathcal{K}^n)^n \to \mathbb{R}$, *which is defined by* $(K_1, \ldots, K_n) \mapsto V(K_1, \ldots, K_n)$, *is called* mixed volume.
In particular,

$$V(K_1, \ldots, K_n) = \frac{1}{n!} \sum_{k=1}^{n} (-1)^{n+k} \sum_{1 \le r_1 < \cdots < r_k \le n} V(K_{r_1} + \cdots + K_{r_k}), \qquad (3.11)$$

and, for $m \in \mathbb{N}$, $K_1, \ldots, K_m \in \mathcal{K}^n$, *and* $\alpha_1, \ldots, \alpha_m \ge 0$,

$$V(\alpha_1 K_1 + \cdots + \alpha_m K_m) = \sum_{i_1=1}^{m} \cdots \sum_{i_n=1}^{m} \alpha_{i_1} \cdots \alpha_{i_n} V(K_{i_1}, \ldots, K_{i_n}). \qquad (3.12)$$

Furthermore, for $K, L, K_1, \ldots, K_n \in \mathcal{K}^n$,

(a) $V(K, \ldots, K) = V(K)$ *and* $nV(K, \ldots, K, B^n) = F(K)$.
(b) V *is symmetric.*
(c) V *is multilinear, that is, if* $\alpha, \beta \ge 0$, *then*

$$V(\alpha K + \beta L, K_2, \ldots, K_n) = \alpha V(K, K_2, \ldots, K_n) + \beta V(L, K_2, \ldots, K_n).$$

(d) $V(K_1 + x_1, \ldots, K_n + x_n) = V(K_1, \ldots, K_n)$ *for* $x_1, \ldots, x_n \in \mathbb{R}^n$.
(e) $V(gK_1, \ldots, gK_n) = V(K_1, \ldots, K_n)$ *for rigid motions* g *of* $\mathbb{R}^n$.
(f) V *is continuous, that is,*

$$V(K_1^{(j)}, \ldots, K_n^{(j)}) \to V(K_1, \ldots, K_n),$$

whenever $K_i^{(j)} \to K_i$ *as* $j \to \infty$, *for* $i = 1, \ldots, n$.
(g) $V \ge 0$ *and* V *is increasing in each argument.*

Proof The existence of the limit

$$V(K_1, \ldots, K_n) = \lim_{j \to \infty} V(P_1^{(j)}, \ldots, P_n^{(j)}),$$

the independence of the approximating sequences and formula (3.11) follow from
Theorem 3.8 and the continuity of the addition of convex bodies and of the volume
functional. Equation (3.12) is a consequence of (3.7).

(d), (e) and (f) now follow directly from (3.11).

(a) For polytopes the relation $V(K, \ldots, K) = V(K)$ has already been shown and
for general convex bodies it follows by approximation with polytopes; alternatively,
the relation follows from the inversion formula (see Remark 3.13).

Concerning the relation $nV(K, \ldots, K, B^n) = F(K)$, again we first discuss the
case $K \in \mathcal{P}^n$. Let $(Q_j)_{j \in \mathbb{N}}$ be a sequence of polytopes with $Q_j \to B^n$. Then,

$$nV(K, \ldots, K, Q_j) \to nV(K, \ldots, K, B^n)$$

and also

$$nV(K, \ldots, K, Q_j) = \sum_{u \in N(K)} h_{Q_j}(u) v(K(u))$$

$$\to \sum_{u \in N(K)} h_{B^n}(u) v(K(u)) = \sum_{u \in N(K)} v(K(u)) = F(K).$$

For the generalization to arbitrary bodies $K \in \mathcal{K}^n$, we approximate K from inside
and outside by polytopes $P_i' \subset K \subset P_i'', i \in \mathbb{N}$, and use (f) and the monotonicity of
F. Thus we get

$$nV(P_i', \ldots, P_i', B^n) = F(P_i') \le F(K) \le F(P_i'') = nV(P_i'', \ldots, P_i'', B^n).$$

If the polytopes $P_i', P_i'' \to K$ as $i \to \infty$, then the assertion follows, since

$$nV(P_i', \ldots, P_i', B^n) \to nV(K, \ldots, K, B^n)$$

and

$$nV(P_i'', \ldots, P_i'', B^n) \to nV(K, \ldots, K, B^n)$$

as $i \to \infty$. Note that only the monotonicity of the surface area is needed, *not* the
continuity (which has not been established up to this point, but now follows as a
consequence).

(b) follows from the corresponding property for polytopes or from (3.11).

(c) is a consequence of (3.12), if we apply it to the linear combination

$$\alpha_1(\alpha K + \beta L) + \alpha_2 K_2 + \cdots + \alpha_m K_m$$
$$= \alpha_1 \alpha K + \alpha_1 \beta L + \alpha_2 K_2 + \cdots + \alpha_m K_m$$

twice (once as a combination of m bodies and once as a combination of $m + 1$ bodies), and then compare the coefficients. Alternately, the result has already been shown for polytopes, the general case follows by approximation.

(g) Again it is sufficient to prove this for polytopes. Then $V \geq 0$ follows by induction and the formula

$$V(P_1, \ldots, P_n) = \frac{1}{n} \sum_{u \in N(P_1, \ldots, P_{n-1})} h_{P_n}(u) v(P_1(u), \ldots, P_{n-1}(u)),$$

where in view of (d) we may assume that $0 \in \text{relint } P_n$, hence $h_{P_n} \geq 0$. If $P_n \subset Q_n$, then $h_{P_n} \leq h_{Q_n}$, hence

$$V(P_1, \ldots, P_n) \leq V(P_1, \ldots, P_{n-1}, Q_n),$$

by the same formula and since the mixed volume is nonnegative.   □

Remark 3.15 In addition to $V \geq 0$, one can show that $V(K_1, \ldots, K_n) > 0$ if and only if there exist segments $s_1 \subset K_1, \ldots, s_n \subset K_n$ with linearly independent directions (see Exercise 3.3.1).

Remark 3.16 Theorem 3.9 (a) and (f) now imply the continuity of the surface area F.

The polynomial expansion in Theorem 3.9 can be written in a more economic way. Using the symmetry of the mixed volumes, we obtain for $K_1, \ldots, K_m \in \mathcal{K}^n$ and $\alpha_1, \ldots, \alpha_m \geq 0$ that

$$V_n \left(\sum_{i=1}^{m} \alpha_i K_i \right)$$

$$= \sum_{r_1, \ldots, r_m = 0}^{n} \binom{n}{r_1, \ldots, r_m} \alpha_1^{r_1} \cdots \alpha_m^{r_m} V(K_1[r_1], \ldots, K_m[r_m]), \quad (3.13)$$

where the number in brackets is the multiplicity r_i of the convex body K_i for $i = 1, \ldots, m$.

Now we consider the special case of the parallel body $K + B^n(\alpha)$, $\alpha \geq 0$, of a body $K \in \mathcal{K}^n$. With the choices $m = 2$, $\alpha_1 = 1$, $\alpha_2 = \alpha$ and $K_1 = K$, $K_2 = B^n$, relation (3.13) (or again Theorem 3.9) yields

$$V(K + B^n(\alpha)) = V(K + \alpha B^n) = V(\alpha_1 K_1 + \alpha_2 K_2)$$

$$= \sum_{i_1=1}^{2} \cdots \sum_{i_n=1}^{2} \alpha_{i_1} \cdots \alpha_{i_n} V(K_{i_1}, \ldots, K_{i_n}) \qquad (3.14)$$

$$= \sum_{i=0}^{n} \alpha^i \binom{n}{i} V(\underbrace{K, \ldots, K}_{n-i}, \underbrace{B^n, \ldots, B^n}_{i}).$$

The coefficients in this particular polynomial expansion deserve special attention. Recall that κ_k denotes the volume of the k-dimensional unit ball.

Definition 3.5 For $K \in \mathcal{K}^n$,

$$W_i(K) := V(\underbrace{K, \ldots, K}_{n-i}, \underbrace{B^n, \ldots, B^n}_{i})$$

is called the ith *quermassintegral* of K, $i \in \{0, \ldots, n\}$, and

$$V_j(K) = V_j^{(n)}(K) := \frac{\binom{n}{j}}{\kappa_{n-j}} W_{n-j}(K) = \frac{\binom{n}{j}}{\kappa_{n-j}} V(\underbrace{K, \ldots, K}_{j}, \underbrace{B^n, \ldots, B^n}_{n-j})$$

is called the jth *intrinsic volume* of K for $j \in \{0, \ldots, n\}$. In addition, we define

$$W_i(\emptyset) := V_j(\emptyset) := 0, \quad i, j = 0, \ldots, n.$$

The functional $W_i : \mathcal{K}^n \to \mathbb{R}$, for $i \in \{0, \ldots, n\}$, is called the ith quermassintegral, the functional $V_j : \mathcal{K}^n \to \mathbb{R}$, for $j \in \{0, \ldots, n\}$, is the jth intrinsic volume.

Formula (3.14) directly yields the following result.

Theorem 3.10 (Steiner Formula) *For* $K \in \mathcal{K}^n$ *and* $\alpha \geq 0$,

$$V(K + B^n(\alpha)) = \sum_{i=0}^{n} \alpha^i \binom{n}{i} W_i(K),$$

respectively

$$V(K + B^n(\alpha)) = \sum_{j=0}^{n} \alpha^{n-j} \kappa_{n-j} V_j(K).$$

Remark 3.17 In particular, we get

$$F(K) = nW_1(K) = 2V_{n-1}(K) = \lim_{\alpha \searrow 0} \frac{1}{\alpha}(V(K + B^n(\alpha)) - V(K)),$$

hence the surface area is the "derivative" of the volume functional.

Remark 3.18 As a generalization of the Steiner formula (3.14), one can show that

$$V_k(K + B^n(\alpha)) = \sum_{j=0}^{k} \alpha^{k-j} \binom{n-j}{n-k} \frac{\kappa_{n-j}}{\kappa_{n-k}} V_j(K),$$

for $k = 0, \ldots, n-1$ (see Exercise 3.3.7).

Remark 3.19 Here we deduced the Steiner formula as a special case of the polynomial expansion of the volume of a general Minkowski combination of convex bodies, that is via the introduction of mixed volumes. It is possible to follow a more direct approach by decomposing the outer parallel set of a convex polytope P by the inverse images under the projection map of the relative interiors of the faces of P (see Fig. 3.2). The result for a general convex body then follows again by approximation with polytopes. The details are the subject of Exercise 3.3.8.

Remark 3.20 The name "quermassintegral" is of German origin. Usually, a "Quermaß" of an object refers to a measure (= Maß) across (=quer) the body. This is deliberately vague and could mean for instance that one considers a section of the body and determines some kind of measure of this section or one might consider a projection of the body and a measure of this projection. The second part of the term "quermass**integral**" refers to striking integral-geometric properties of the quermassintegrals which for instance make it possible to introduce the

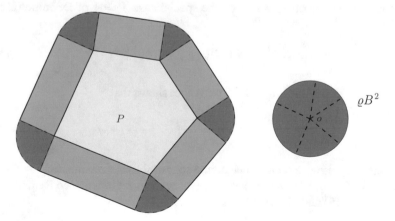

Fig. 3.2 Illustration of the parallel set of a polygon, which indicates the polynomial growth of the area $V_2(P + \varrho B^2) = V_2(P) + S_1(P) \cdot \varrho + V_2(B^2) \cdot \varrho^2$ as a function of $\varrho \geq 0$, where $S_1(P)$ is the boundary length of P

quermassintegrals of a body as integral averages of volumes of projections of the given body. We will discuss integral-geometric properties of quermassintegrals in Chap. 5, where we derive some integral-geometric projection and intersection formulas for these functionals. The name "quermassintegral" already appears in the classical book by Bonnesen and Fenchel [18] and has been in use since then.

Remark 3.21 The intrinsic volumes were introduced by Peter McMullen [69]. Their obvious advantages are that the index j of V_j equals the degree of homogeneity,

$$V_j(\alpha K) = \alpha^j V_j(K), \quad K \in \mathcal{K}^n, \alpha \geq 0,$$

and that they are independent of the dimension of the ambient Euclidean space, that is, for a body $K \in \mathcal{K}^n$ with $\dim K = k < n$, we have

$$V_j^{(n)}(K) = V_j^{(k)}(K), \quad j = 0, \ldots, k,$$

(see Exercise 3.3.5). Of course, this latter property justifies their name.

The intrinsic volumes $V_j : \mathcal{K}^n \to [0, \infty)$, $j \in \{0, \ldots, n\}$, are important geometric functionals of a convex body. First,

$$V_n(K) = V(K, \ldots, K) = V(K)$$

equals the volume of K. Second,

$$2V_{n-1}(K) = nV(K, \ldots, K, B^n) = F(K)$$

is the surface area of K. For a body K of dimension $n - 1$, $V_{n-1}(K)$ is the $(n - 1)$-dimensional content of K. Furthermore, $V_1(K)$ is proportional to the *mean width* of K. In order to introduce the mean width and to see how it is related to the first intrinsic volume, we observe that by the special case $j = 1$ of Definition 3.5 we have

$$\frac{\kappa_{n-1}}{n} V_1(K) = V(K, B^n, \ldots, B^n). \tag{3.15}$$

An approximation of the unit ball by polytopes shows that

$$V(K, B^n, \ldots, B^n) = \frac{1}{n} \int_{\mathbb{S}^{n-1}} h_K(u) \, \sigma(du), \tag{3.16}$$

where the integration is with respect to the spherical Lebesgue measure. A rigorous proof of (3.16) will be given in Sect. 4.1 (for another argument, see Remark 3.23). Combining (3.15) and (3.16), we get

$$\kappa_{n-1} V_1(K) = \int_{\mathbb{S}^{n-1}} h_K(u) \, \sigma(du). \tag{3.17}$$

Since $b_K(u) := h_K(u) + h_K(-u)$ gives the *width* of K in direction u (the distance between the two parallel supporting hyperplanes with common unit normal u), we obtain

$$\frac{1}{n} \int_{\mathbb{S}^{n-1}} h_K(u)\, \sigma(du) = \frac{1}{2n} \int_{\mathbb{S}^{n-1}} b_K(u)\, \sigma(du) = \frac{\kappa_n}{2}\overline{B}(K),$$

where

$$\overline{B}(K) := \frac{1}{n\kappa_n} \int_{\mathbb{S}^{n-1}} b_K(u)\, \sigma(du)$$

denotes the mean width of K (the letter B stands for the German word "Breite" (meaning "width"). Hence, we arrive at

$$V_1(K) = \frac{n\kappa_n}{2\kappa_{n-1}}\overline{B}(K). \tag{3.18}$$

Finally,

$$V_0(K) = \frac{1}{\kappa_n} W_n(K) = \begin{cases} 1, & K \neq \emptyset, \\ 0, & K = \emptyset, \end{cases}$$

is the *Euler–Poincaré characteristic* of K, which plays an important role in integral geometry (see Chap. 5). The other intrinsic volumes $V_j(K)$, $1 < j < n - 1$, have interpretations as integrals of curvature functions, if the boundary of K is smooth. For instance, $V_{n-2}(K)$ is proportional to the integral mean curvature of K. In general, V_j is obtained as an integral of the elementary symmetric function of order $n - 1 - j$ of the principal curvatures of K over the boundary of K. For a polytope P, an interpretation of $V_j(P)$ is provided in Exercise 3.3.9, as the sum of the weighted j-volumes of the j-faces of P where for each face the weight of the face is proportional to the volume of the spherical image of that face (the normal cone, taken at a point in the relative interior of the face, intersected with the unit sphere).

Remark 3.22 From Theorem 3.9 we obtain the following additional properties of the intrinsic volumes V_j:

- $K \mapsto V_j(K)$ is continuous,
- V_j is motion invariant,
- $V_j \geq 0$ and V_j is increasing.

In Sect. 4.5 we shall discuss an important additivity property of intrinsic volumes. The intrinsic volume V_j is *additive* in the sense that

$$V_j(K \cup M) + V_j(K \cap M) = V_j(K) + V_j(M)$$

for all $K, M \in \mathcal{K}^n$ such that $K \cup M \in \mathcal{K}^n$. Then we shall also discuss and prove Hadwiger's celebrated characterization theorem, which states that intrinsic volumes (or linear combinations thereof) can be characterized by some of their properties such as additivity, continuity (or monotonicity) and isometry invariance (see Exercise 3.3.8). As a consequence of this result, simple proofs of integral-geometric formulas can be given.

Finally, in this section we discuss a characterization (see Theorem 3.11) of the first intrinsic volume and of the mean width. From this result it follows that these functionals differ only by a constant factor. Theorem 3.11 is based on the notion of Minkowski additivity, which is more restrictive than additivity as defined above.

Definition 3.6 A functional $\varphi : \mathcal{K}^n \to \mathbb{R}$ is *Minkowski additive* if

$$\varphi(K + L) = \varphi(K) + \varphi(L) \qquad \text{for } K, L \in \mathcal{K}^n.$$

The map $K \mapsto h_K(u)$, for some fixed $u \in \mathbb{S}^{n-1}$, is Minkowski additive. Further examples are the first intrinsic volume V_1 and the mean width functional $\overline{B}$. A map which is Minkowski additive is also additive. This can be seen by observing that $(K \cup L) + (K \cap L) = K + L$ for $K, L \in \mathcal{K}^n$ such that also $K \cup L \in \mathcal{K}^n$.

Note that if φ is Minkowski additive, then $\varphi(\lambda K) = \lambda \varphi(K)$ for $K \in \mathcal{K}^n$ and rational $\lambda \geq 0$. In fact, by induction we get

$$\varphi(K_1 + \cdots + K_m) = \sum_{i=1}^{m} \varphi(K_i)$$

for $K_1, \ldots, K_m \in \mathcal{K}^n$. Hence $\varphi(mK) = m\varphi(K)$ for $K \in \mathcal{K}^n$ and $m \in \mathbb{N}$. But then, for $r, s \in \mathbb{N}$ we deduce that

$$\frac{r}{s}\varphi(K) = \frac{1}{s}\varphi(rK) = \frac{1}{s}\varphi\left(s\left(\frac{r}{s}K\right)\right) = \varphi\left(\frac{r}{s}K\right).$$

The functionals $V_1, \overline{B}$ further share the properties of being rigid motion invariant and continuous. In fact, fewer properties suffice for the following characterization theorem.

Theorem 3.11 *Let $n \geq 2$. If $\varphi : \mathcal{K}^n \to \mathbb{R}$ is Minkowski additive, invariant under proper rotations, and continuous at B^n, then $\varphi = c\overline{B}$ for some nonnegative constant $c \geq 0$.*

Remark 3.23 From the theorem it follows that $V_1 = c\overline{B}$ for some constant $c \geq 0$. Since $V_1(B^n) = (n\kappa_n)/\kappa_{n-1}$ and $\overline{B}(B^n) = 2$, it follows that $c = (n\kappa_n)/(2\kappa_{n-1})$. This proves (3.18) and thus also (3.17). Another proof will be provided in Sect. 4.1.

For the proof of Theorem 3.11, we use a method which can be described as rotation averaging. For $K \in \mathcal{K}^n$, $m \in \mathbb{N}$, and $\varrho_1, \ldots, \varrho_m \in SO(n)$, we define

$$K' := \frac{1}{m} (\varrho_1 K + \cdots + \varrho_m K) \in \mathcal{K}^n$$

and call this a *rotation average* (or rotation mean) of K. For this we have

$$\overline{B}(K') = \frac{1}{m} \sum_{i=1}^{m} \overline{B}(\varrho_i K) = \frac{1}{m} \sum_{i=1}^{m} \overline{B}(K) = \overline{B}(K),$$

where we used that $\overline{B}$ is Minkowski additive and invariant under proper rotations. This shows that the mean width is preserved under rotation averaging. Moreover, it is easy to check that a rotational average of K' is a rotation average of the original K. In the following, we use the fact that $\overline{B}(K) > 0$ if and only if $\mathrm{diam}(K) > 0$ for $K \in \mathcal{K}^n$.

Theorem 3.12 *Let $n \geq 2$. For each $K \in \mathcal{K}^n$ with $\mathrm{diam}(K) > 0$, there exists a sequence of rotation averages of K which converges to a nondegenerate Euclidean ball.*

Proof For $L \in \mathcal{K}^n$, let $R_0(L) := \min\{\lambda \geq 0 : L \subset \lambda B^n\}$ denote the centred circumradius of L. Clearly, R_0 is continuous on $\mathcal{K}^n$. Let $\mathcal{R}(K)$ denote the set of rotation averages of K. If $K \subset \lambda B^n$, then also $K' \subset \lambda B^n$ for $K' \in \mathcal{R}(K)$. Therefore, $\mathcal{R}(K)$ is bounded and $\mathrm{cl}\,\mathcal{R}(K)$ is compact (by Blaschke's selection theorem). Then R_0 attains its minimum r_0 in $L_0 \in \mathrm{cl}\,\mathcal{R}(K)$. Since $\mathrm{diam}(K) > 0$ and thence $\overline{B}(L_0) = \overline{B}(K) > 0$, it follows that $r_0 = R_0(L_0) > 0$. Clearly, we have $L_0 \subset r_0 B^n$. Suppose that $L_0 \neq r_0 B^n$. Then there is some $u_0 \in \mathbb{S}^{n-1}$ such that $h_{L_0}(u_0) < r_0$, and therefore there is an open set $U \subset \mathbb{S}^{n-1}$ such that $h_{L_0}(v) < r_0$ for $v \in U$. Hence there are $\varrho_1, \ldots, \varrho_m \in SO(n)$ such that $\bigcup_{i=1}^{m} \varrho_i U = \mathbb{S}^{n-1}$.

We define $L' := \frac{1}{m} \sum_{i=1}^{m} \varrho_i L_0$. Let $u \in \mathbb{S}^{n-1}$ be given. There is some $i \in \{1, \ldots, m\}$ such that $u \in \varrho_i U$, so that $\varrho_i^{-1} u \in U$ and $h_{L_0}(\varrho_i^{-1} u) < r_0$. But then we get

$$h_{L'}(u) = \frac{1}{m} \sum_{i=1}^{m} h_{\varrho_i L_0}(u) = \frac{1}{m} \sum_{i=1}^{m} h_{L_0}(\varrho_i^{-1} u) < r_0.$$

This shows that $R_0(L') < r_0$. Let $K_j \in \mathcal{R}(K)$, $j \in \mathbb{N}$, with $K_j \to L_0$ as $j \to \infty$. Then

$$K'_j := \frac{1}{m}(\varrho_1 K_j + \cdots + \varrho_m K_j) \to L',$$

which yields that $R_0(K'_j) < r_0$ if $j \in \mathbb{N}$ is large enough, since R_0 is continuous. This is a contradiction, since $K'_j \in \mathcal{R}(K)$. $\square$

Proof of Theorem 3.11 Suppose that $\varphi : \mathcal{K}^n \to \mathbb{R}$ is as in the statement of the theorem. For $x \in \mathbb{R}^n$, there is some $\varrho \in \mathrm{SO}(n)$ such that $\varrho x = -x$. Then we conclude that

$$\varphi(\{0\}) = \varphi(\{x\} + \varrho(\{x\})) = 2\varphi(\{x\}).$$

Taking first $x = 0$, we get $\varphi(\{0\}) = 0$, and then $\varphi(\{x\}) = 0$ for all $x \in \mathbb{R}^n$. Now suppose that $K \in \mathcal{K}^n$ with $\mathrm{diam}(K) > 0$. Let $m \in \mathbb{N}$, let λ_i be a positive rational number, let $\varrho_i \in \mathrm{SO}(n)$, for $i = 1, \ldots, m$, and define $K' := \lambda_1 \varrho_1 K + \cdots + \lambda_m \varrho_m K$. Then

$$\varphi(K') = \sum_{i=1}^m \varphi(\lambda_i \varrho_i K) = \sum_{i=1}^m \lambda_i \varphi(\varrho_i K) = \left(\sum_{i=1}^m \lambda_i \right) \varphi(K).$$

The same is true for $\overline{B}$. Hence, we deduce that

$$\frac{\varphi(K)}{\overline{B}(K)} = \frac{\varphi(K')}{\overline{B}(K')} \longrightarrow \frac{\varphi(B^n)}{\overline{B}(B^n)},$$

for a suitably chosen sequence of scaled rotational averages. The existence of such a sequence follows from Theorem 3.12. Since the left-hand side is independent of this sequence, the assertion follows.   □

Exercises and Supplements for Sect. 3.3

1.* (a) Let $s_1, \ldots, s_n \in \mathcal{K}^n$ be segments of the form $s_i = [0, x_i]$ with $x_i \in \mathbb{R}^n$. Show that

$$n!\, V(s_1, \ldots, s_n) = |\det(x_1, \ldots, x_n)|.$$

   (b) Let $K_1, \ldots, K_n \in \mathcal{K}^n$. Show that $V(K_1, \ldots, K_n) > 0$ if and only if there exist segments $s_i \subset K_i, i = 1, \ldots, n$, with linearly independent directions.

2. The following result has been shown in [13, Theorem 1] by U. Betke and W. Weil. There it has been proved by using Choquet's theorem (an integral version of the Krein–Milman theorem) and the result of Exercise 3.1.12. In Chap. 4 we shall provide an alternative and more elementary argument.

   (a) Let $K, M \in \mathcal{K}^2$. Prove the inequality

$$V(K, M) \le \frac{1}{8} F(K)\, F(M).$$

(b) Show that equality holds in the above inequality if and only if K, M are orthogonal segments (or if one of the bodies is a point).

3. The following result has been shown by U. Betke and W. Weil in [13, Theorem 2].

(a) Let $K \in \mathcal{K}^2$. Prove the inequality

$$V(K, -K) \leq \frac{\sqrt{3}}{18} F^2(K).$$

(b) Show that equality holds in the above inequality for a polygon K if and only if K is an equilateral triangle (or a point). It seems to be unknown whether this is the only equality case for general convex sets.

4. Let $K, L \in \mathcal{K}^n$. Show that

$$\lambda_n(D(K, L)) = \int_{\mathbb{R}^n} V_0(K \cap (L + x)) \lambda_n(dx)$$

$$= \sum_{j=0}^{n} \binom{n}{j} V(\underbrace{K, \ldots, K}_{n-j}, \underbrace{-L, \ldots, -L}_{j}),$$

where $D(K, L) := \{z \in \mathbb{R}^n : K \cap (L + z) \neq \emptyset\}$.

5. Let $K \in \mathcal{K}^n$. Show that the intrinsic volume $V_j(K) = V_j^{(n)}(K)$ is independent of the dimension n, that is, if $\dim K = k < n$, then

$$V_j^{(k)}(K) = V_j^{(n)}(K) \quad \text{for } 0 \leq j \leq k.$$

6. Suppose that $K \in \mathcal{K}^n$ and L is a q-dimensional linear subspace of $\mathbb{R}^n$ with $q \in \{0, \ldots, n-1\}$. Let B_L denote the unit ball in L.

(a) Show that if $\alpha \geq 0$, then

$$V(K + \alpha B_L) = \sum_{j=0}^{q} \alpha^{q-j} \kappa_{q-j} \int_{L^\perp} V_j(K \cap (L + x)) \lambda_{n-q}(dx).$$

(b) The $(n - q)$-dimensional volume of the projection $K \mid L^\perp$ satisfies

$$V_{n-q}(K|L^\perp) = \frac{\binom{n}{q}}{\kappa_q} V(\underbrace{K, \ldots, K}_{n-q}, \underbrace{B_L, \ldots, B_L}_{q}).$$

Hint for (a): Use Fubini's theorem in $\mathbb{R}^n = L \times L^\perp$ for the left-hand side and apply Exercise 3.3.5.

7. Let $K \in \mathcal{K}^n$ and $\alpha \geq 0$. Prove the following Steiner formula for the intrinsic volumes:

$$V_k(K + B^n(\alpha)) = \sum_{j=0}^{k} \alpha^{k-j} \binom{n-j}{n-k} \frac{\kappa_{n-j}}{\kappa_{n-k}} V_j(K), \quad 0 \leq k \leq n-1.$$

8. Prove the following Theorem of Hadwiger (in dimension two, for a start):
   Let $f : \mathcal{K}^n \to \mathbb{R}$ be additive, motion invariant and continuous (resp. monotone). Then there are constants $\beta_j \in \mathbb{R}$ (resp. $\beta_j \geq 0$) such that

$$f = \sum_{j=0}^{n} \beta_j V_j.$$

9.* Let $P \in \mathcal{P}^n$ be a polytope in $\mathbb{R}^n$ Let $\mathcal{F}_k(P)$ denote the set of all k-dimensional faces of P, in particular, $\mathcal{F}_n(P) = \{P\}$, if $\dim P = n$. For $x \in \mathrm{bd}\, P$, recall that $N(P, x) = \{u \in \mathbb{R}^n : \langle x, u \rangle = h_P(u)\}$ and $N(P, z) := \{0\}$ for $z \in \mathrm{int}\, P$. Prove:

   (a) For $x \in P$, $N(P, x) = \{y - x : y \in \mathbb{R}^n, p(P, y) = x\}$ is a convex cone.
   (b) $N(P, x) = N(P, y) =: N(P, F) \subset F^{\perp}$ for $F \in \mathcal{F}_k(P)$, $k \in \{0, \ldots, n\}$ and $x, y \in \mathrm{relint}\, F$, where $F^{\perp}$ is the orthogonal complement of the linear subspace parallel to F.
   (c) The following disjoint decomposition holds:

$$P = \bigcup_{k=0}^{n} \bigcup_{F \in \mathcal{F}_k(P)} \mathrm{relint}\, F.$$

   (d) For $\epsilon > 0$ the following disjoint decomposition holds:

$$P + B^n(\epsilon) = \bigcup_{k=0}^{n} \bigcup_{F \in \mathcal{F}_k(P)} (\mathrm{relint}\, F + (N(P, F) \cap B^n(\epsilon))).$$

   (e) Let λ_F the denote the Lebesgue measure in the affine hull of F. We define

$$\gamma(P, F) := \frac{\lambda_{F^{\perp}}(N(P, F) \cap B^n)}{\kappa_{n-k}}.$$

   Then the kth intrinsic volume of P satisfies

$$V_k(P) = \sum_{F \in \mathcal{F}_k(P)} \gamma(P, F) \lambda_F(F), \quad k = 0, \ldots, n-1.$$

10. (a) Consider the rectangles $K := [0, a] \times [0, b]$ and $L := [0, c] \times [0, d]$ in $\mathbb{R}^2$. Determine the mixed area $V(K, L)$.
    (b) Consider the simplex $S := \mathrm{conv}\{0, e_1, e_2, e_3\}$ and the segments $I_1 := [0, e_1]$ and $I_2 := [0, e_2]$ in $\mathbb{R}^3$, where e_i denotes the ith unit vector of the standard basis. Determine the mixed volumes $V(S, I_1, I_2)$ and $V(S, I_1, I_1)$.
11. Let $P_1 = [0, 4] \times [0, 2] \subset \mathbb{R}^2$ and let P_2 be a rectangle with the vertices $(0, 0), (1, -1), (3, 1), (2, 2)$. Determine $V(P_1, P_2)$.
12. Let $K, L, M \in \mathcal{K}^n$ with $K = M + L$ and $R_1, \ldots, R_{n-j} \in \mathcal{K}^n$. Let $j \in \{0, \ldots, n\}$. Show that

$$V(K[j], R_1, \ldots, R_{n-j}) = \sum_{i=0}^{j} (-1)^{j-i} \binom{j}{i} V(K[i], L[j-i], R_1, \ldots, R_{n-j}).$$

13. Let $K \in \mathcal{K}^n$ and $d(K, x) := \min\{\|x - z\| : z \in K\}$ for $x \in \mathbb{R}^n$. Show that

$$\int_{\mathbb{R}^n} \exp\left(-\pi d(K, x)^2\right) dx = \sum_{i=0}^{n} V_i(K).$$

The functional of K defined by either side of this equation is called the *Wills functional*.

3.4 The Brunn–Minkowski Theorem

The Brunn–Minkowski Theorem is one of the first main results for convex bodies and was first proved around 1890. The theorem states that, for convex bodies $K, L \in \mathcal{K}^n$, the function

$$t \mapsto \sqrt[n]{V(tK + (1-t)L)}, \quad t \in [0, 1],$$

is concave. As consequences of the Brunn–Minkowski theorem, we shall derive several inequalities for mixed volumes, and in particular we thus obtain the celebrated isoperimetric inequality.

We first provide a purely analytic auxiliary result.

Lemma 3.3 *For $\alpha \in (0, 1)$ and $r, s, t > 0$,*

$$\left(\frac{\alpha}{r} + \frac{1-\alpha}{s}\right)\left[\alpha r^t + (1-\alpha)s^t\right]^{\frac{1}{t}} \geq 1$$

with equality if and only if $r = s$.

Proof The function $x \mapsto \ln x$ is strictly concave. Therefore we have

$$\ln \left\{ \left(\frac{\alpha}{r} + \frac{1-\alpha}{s} \right) [\alpha r^t + (1-\alpha)s^t]^{\frac{1}{t}} \right\}$$

$$= \frac{1}{t} \ln \left(\alpha r^t + (1-\alpha)s^t \right) + \ln \left(\frac{\alpha}{r} + \frac{1-\alpha}{s} \right)$$

$$\geq \frac{1}{t} \left(\alpha \ln r^t + (1-\alpha) \ln s^t \right) + \alpha \ln \frac{1}{r} + (1-\alpha) \ln \frac{1}{s}$$

$$= 0$$

with equality if and only if $r = s$ (the use of the logarithm is possible since its argument is always positive). Since the logarithm is a strictly increasing function, the result follows. $\qquad \square$

The following important inequality is known as the Brunn–Minkowski inequality. It has numerous applications to and connections with geometry, analysis and probability theory.

Theorem 3.13 (Brunn–Minkowski Inequality) *Let $K, L \in \mathcal{K}^n$ be convex bodies and $\alpha \in (0, 1)$. Then*

$$\sqrt[n]{V(\alpha K + (1-\alpha)L)} \geq \alpha \sqrt[n]{V(K)} + (1-\alpha) \sqrt[n]{V(L)}$$

with equality if and only if K and L lie in parallel hyperplanes or K and L are homothetic.

Remark 3.24 By definition, $K, L \in \mathcal{K}^n$ are homothetic if $K = \alpha L + x$ or $L = \alpha K + x$, for some $x \in \mathbb{R}^n$ and some $\alpha \geq 0$. Hence, if K or L is a point, then K and L are homothetic. The Brunn–Minkowski inequality trivially also holds for $\alpha \in \{0, 1\}$, but in view of the discussion of the equality cases, this is not included in the statement of the theorem.

Proof of Theorem 3.13 We distinguish four cases. The fourth case concerns n-dimensional convex bodies K, L. This main case can be reduced to the consideration of bodies with unit volume. Then we shall proceed by induction on the dimension. In this induction step, Lemma 3.3 will be used.

Case 1: K and L lie in parallel hyperplanes. Then $\alpha K + (1-\alpha)L$ also lies in a hyperplane, and hence $V(K) = V(L) = 0$ and $V(\alpha K + (1-\alpha)L) = 0$.

Case 2: $\dim K \leq n-1$ and $\dim L \leq n-1$, but K and L do not lie in parallel hyperplanes, i.e., $\dim(K+L) = n$. Then $\dim(\alpha K + (1-\alpha)L) = n$, for $\alpha \in (0, 1)$, hence

$$\sqrt[n]{V(\alpha K + (1-\alpha)L)} > 0 = \alpha \sqrt[n]{V(K)} + (1-\alpha) \sqrt[n]{V(L)},$$

for $\alpha \in (0, 1)$. Since K and L are not contained in parallel hyperplanes, they are not homothetic.

Case 3: $\dim K \leq n - 1$ and $\dim L = n$ (or vice versa). Then, for $x \in K$, we obtain

$$\alpha x + (1 - \alpha)L \subset \alpha K + (1 - \alpha)L,$$

and thus

$$(1 - \alpha)^n V(L) = V(\alpha x + (1 - \alpha)L) \leq V(\alpha K + (1 - \alpha)L)$$

with equality if and only if $K = \{x\}$.

Case 4: $\dim K = \dim L = n$. Define

$$\overline{K} := \frac{1}{\sqrt[n]{V(K)}}K, \quad \overline{L} := \frac{1}{\sqrt[n]{V(L)}}L$$

and

$$\overline{\alpha} := \frac{\alpha \sqrt[n]{V(K)}}{\alpha \sqrt[n]{V(K)} + (1 - \alpha)\sqrt[n]{V(L)}}.$$

Then $V(\overline{K}) = V(\overline{L}) = 1$ and $\overline{\alpha} \in (0, 1)$. If the Brunn–Minkowski theorem is already established for convex bodies of volume 1, then

$$\sqrt[n]{V(\overline{\alpha}\overline{K} + (1 - \overline{\alpha})\overline{L})} \geq 1,$$

which yields the Brunn–Minkowski inequality for arbitrary n-dimensional convex bodies, and hence the inequality is proved. Moreover, K and L are homothetic if and only if $\overline{K}$ and $\overline{L}$ are translates of each other.

Thus, in the following we may assume that $V(K) = V(L) = 1$ and have to show that

$$V(\alpha K + (1 - \alpha)L) \geq 1$$

with equality if and only if K, L are translates of each other. Because the volume is translation invariant, we can make the additional assumption that K and L have their center of gravity at 0. The center of gravity of an n-dimensional convex body M is the point $c \in \mathbb{R}^n$ for which

$$\langle c, u \rangle = \frac{1}{V(M)} \int_M \langle x, u \rangle \, dx$$

holds for $u \in \mathbb{S}^{n-1}$. Since K, L have volume one, assuming that K and L have their center of mass at the origin is equivalent to

$$\int_K \langle x, u \rangle \, dx = \int_L \langle x, u \rangle \, dx = 0$$

for $u \in \mathbb{S}^{n-1}$. The equality case then reduces to the claim that $K = L$.

We now prove the Brunn–Minkowski theorem by induction on n. For $n = 1$, the Brunn–Minkowski inequality follows from the linearity of the 1-dimensional volume. In particular, we always have equality, which corresponds to the fact that in $\mathbb{R}^1$ any two convex bodies (compact intervals) are homothetic. Now assume that $n \geq 2$ and the assertion of the Brunn–Minkowski theorem is true in dimension $n - 1$. We fix an arbitrary unit vector $u \in \mathbb{S}^{n-1}$ and denote by

$$E_\eta := H(u, \eta), \quad \eta \in \mathbb{R},$$

the hyperplane in direction u with (signed) distance η from the origin. The function

$$f : [-h_K(-u), h_K(u)] \to [0, 1], \quad \beta \mapsto V(K \cap H^-(u, \beta)),$$

is strictly increasing, onto, and continuous. In fact, this follows since

$$V(K \cap H^-(u, \beta)) = \int_{-h_K(-u)}^{\beta} v(K \cap E_\eta) \, d\eta$$

by Fubini's theorem and since $\eta \mapsto v(K \cap E_\eta)$ is positive and continuous on $[-h_K(-u), h_K(u)]$ (in fact, this map is continuous up to the boundary by Exercise 3.4.7). Moreover, the function f is differentiable on $[-h_K(-u), h_K(u)]$ and $f'(\beta) = v(K \cap E_\beta)$. Since f is invertible, the inverse function $\beta : [0, 1] \to [-h_K(-u), h_K(u)]$, which is also strictly increasing and continuous, satisfies $\beta(0) = -h_K(-u), \beta(1) = h_K(u)$, and

$$\beta'(\tau) = \frac{1}{f'(\beta(\tau))} = \frac{1}{v(K \cap E_{\beta(\tau)})}, \quad \tau \in (0, 1).$$

Analogously, for the convex body L we obtain a function $\gamma : [0, 1] \to [-h_L(-u), h_L(u)]$ with

$$\gamma'(\tau) = \frac{1}{v(L \cap E_{\gamma(\tau)})}, \quad \tau \in (0, 1).$$

Because of

$$\alpha(K \cap E_{\beta(\tau)}) + (1 - \alpha)(L \cap E_{\gamma(\tau)}) \subset (\alpha K + (1 - \alpha)L) \cap E_{\alpha\beta(\tau)+(1-\alpha)\gamma(\tau)},$$

for $\alpha, \tau \in [0, 1]$ and using a substitution by the map

$$[0, 1] \to [\alpha(-h_K(-u)) + (1 - \alpha)(-h_L(-u)), \alpha h_K(u) + (1 - \alpha)h_L(u)],$$
$$\tau \mapsto \alpha\beta(\tau) + (1 - \alpha)\gamma(\tau),$$

we obtain from the induction assumption

$$V(\alpha K + (1 - \alpha)L)$$

$$= \int_{-\infty}^{\infty} v((\alpha K + (1 - \alpha)L) \cap E_\eta)\, d\eta$$

$$= \int_0^1 v((\alpha K + (1 - \alpha)L) \cap E_{\alpha\beta(\tau)+(1-\alpha)\gamma(\tau)})(\alpha\beta'(\tau) + (1 - \alpha)\gamma'(\tau))\, d\tau$$

$$\geq \int_0^1 v\left(\alpha(K \cap E_{\beta(\tau)}) + (1 - \alpha)(L \cap E_{\gamma(\tau)})\right)$$

$$\times \left(\frac{\alpha}{v(K \cap E_{\beta(\tau)})} + \frac{1 - \alpha}{v(L \cap E_{\gamma(\tau)})}\right) d\tau$$

$$\geq \int_0^1 \left[\alpha \sqrt[n-1]{v(K \cap E_{\beta(\tau)})} + (1 - \alpha) \sqrt[n-1]{v(L \cap E_{\gamma(\tau)})}\right]^{n-1}$$

$$\times \left(\frac{\alpha}{v(K \cap E_{\beta(\tau)})} + \frac{1 - \alpha}{v(L \cap E_{\gamma(\tau)})}\right) d\tau.$$

Choosing $r := v(K \cap E_{\beta(\tau)})$, $s := v(L \cap E_{\gamma(\tau)})$, and $t := \frac{1}{n-1}$, we obtain from Lemma 3.3 that the integrand is ≥ 1, which yields the required inequality.

Now assume

$$V(\alpha K + (1 - \alpha)L) = 1.$$

Then we must have equality in our last estimation, which implies that the integrand equals 1, for all τ. Again by Lemma 3.3, this yields that

$$v(K \cap E_{\beta(\tau)}) = v(L \cap E_{\gamma(\tau)}), \quad \text{for } \tau \in [0, 1].$$

Therefore $\beta' = \gamma'$ on $(0, 1)$, hence the function $\beta - \gamma$ is a constant on $[0, 1]$. Because the center of gravity of K is at the origin, we obtain

$$0 = \int_K \langle x, u \rangle\, dx = \int_{\beta(0)}^{\beta(1)} \eta v(K \cap E_\eta)\, d\eta = \int_{\beta(0)}^{\beta(1)} \eta f'(\eta)\, d\eta = \int_0^1 \beta(\tau)\, d\tau,$$

where the change of variables $\eta = \beta(\tau)$ was used. In an analogous way,

$$0 = \int_0^1 \gamma(\tau)\,d\tau.$$

Consequently,

$$\int_0^1 (\beta(\tau) - \gamma(\tau))\,d\tau = 0$$

and therefore $\beta = \gamma$. In particular, we obtain

$$h_K(u) = \beta(1) = \gamma(1) = h_L(u).$$

Since $u \in \mathbb{S}^{n-1}$ was arbitrary, $V(\alpha K + (1-\alpha)L) = 1$ implies that $h_K = h_L$, and hence $K = L$.

Conversely, it is clear that $K = L$ implies that $V(\alpha K + (1-\alpha)L) = 1$.

$\square$

Remark 3.25 Theorem 3.13 implies that the function

$$f(t) := \sqrt[n]{V(tK + (1-t)L)}$$

is concave on $[0, 1]$. If $x, y, \alpha \in [0, 1]$, then

$$
\begin{aligned}
f(\alpha x + (1-\alpha)y) &= \sqrt[n]{V([\alpha x + (1-\alpha)y]K + [1 - \alpha x - (1-\alpha)y]L)} \\
&= \sqrt[n]{V(\alpha[xK + (1-x)L] + (1-\alpha)[yK + (1-y)L])} \\
&\geq \alpha\sqrt[n]{V(xK + (1-x)L)} + (1-\alpha)\sqrt[n]{V(yK + (1-y)L)} \\
&= \alpha f(x) + (1-\alpha)f(y).
\end{aligned}
$$

As a consequence of Theorem 3.13, we obtain an inequality for mixed volumes which was first proved by Hermann Minkowski.

Theorem 3.14 (Minkowski's Inequality) *Let $K, L \in \mathcal{K}^n$. Then*

$$V(K, \ldots, K, L)^n \geq V(K)^{n-1}V(L)$$

with equality if and only if $\dim K \leq n - 2$ *or K and L lie in parallel hyperplanes or K and L are homothetic.*

Proof For $\dim K \leq n - 1$, the inequality holds since the right-hand side is zero. Moreover, we then have equality, if and only if either $\dim K \leq n - 2$ or K and L lie in parallel hyperplanes (compare Exercise 3.1.1). Hence, we now assume $\dim K = n$.

By Theorem 3.13 (similarly to Remark 3.25), it follows that the function

$$f(t) := V(K + tL)^{\frac{1}{n}}, \quad t \in [0, 1],$$

is concave. Therefore

$$f^+(0) \geq f(1) - f(0) = V(K + L)^{\frac{1}{n}} - V(K)^{\frac{1}{n}}.$$

Since

$$f^+(0) = \frac{1}{n} V(K)^{\frac{1}{n}-1} n V(K, \ldots, K, L),$$

we arrive at

$$V(K)^{\frac{1}{n}-1} V(K, \ldots, K, L) \geq V(K + L)^{\frac{1}{n}} - V(K)^{\frac{1}{n}} \geq V(L)^{\frac{1}{n}},$$

where we used the Brunn–Minkowski inequality for the second inequality (with $t = \frac{1}{2}$). This yields the inequality. If equality holds, then equality holds in the Brunn–Minkowski inequality, which implies that K and L are homothetic. Conversely, if K and L are homothetic, then equality holds (as can be easily checked). $\qquad\quad\sqcup$

The *isoperimetric inequality* is undoubtedly one of the fundamental classical results in mathematics. In the present framework, it states that among all convex bodies of given volume, precisely the Euclidean balls minimize the surface area. Alternatively, if the surface area is fixed, then Euclidean balls maximize the volume functional. The following special case of the Minkowski inequality expresses this geometric fact in an analytic form.

Corollary 3.1 (Isoperimetric Inequality) *If $K \in \mathcal{K}^n$ is n-dimensional, then*

$$\left(\frac{F(K)}{F(B^n)} \right)^n \geq \left(\frac{V(K)}{V(B^n)} \right)^{n-1}.$$

Equality holds if and only if K is a ball.

Proof We put $L := B^n$ in Theorem 3.14 and get

$$V(K, \ldots, K, B^n)^n \geq V(K)^{n-1} V(B^n)$$

or, equivalently,

$$\frac{n^n V(K, \ldots, K, B^n)^n}{n^n V(B^n, \ldots, B^n, B^n)^n} \geq \frac{V(K)^{n-1}}{V(B^n)^{n-1}},$$

which is precisely what we had to show. $\qquad\qquad\qquad\qquad\qquad\qquad\quad\square$

Note that the inequality is scaling and motion invariant. It can also be expressed by saying that the *isoperimetric ratio* $F(K)^n/V(K)^{n-1}$ is minimized precisely by Euclidean balls.

Using $V(B^n) = \kappa_n$ and $F(B^n) = n\kappa_n$, we can re-write the inequality in the form

$$V(K)^{n-1} \leq \frac{1}{n^n \kappa_n} F(K)^n.$$

For $n = 2$ and using the common terminology $A(K)$ for the area (the "volume" in $\mathbb{R}^2$) and $L(K)$ for the boundary length (the "surface area" in $\mathbb{R}^2$), we obtain

$$A(K) \leq \frac{1}{4\pi} L(K)^2,$$

and, for $n = 3$,

$$V(K)^2 \leq \frac{1}{36\pi} F(K)^3.$$

An exchange of K and B^n in the proof above leads to a similar inequality for the mixed volume $V(B^n, \ldots B^n, K)$, whence we obtain the following corollary for the mean width $\overline{B}(K)$.

Corollary 3.2 *Let* $K \in \mathcal{K}^n$ *be a convex body. Then,*

$$\left(\frac{\overline{B}(K)}{\overline{B}(B^n)} \right)^n \geq \frac{V(K)}{V(B^n)}.$$

Equality holds if and only if K is a ball.

Remark 3.26 Since $\overline{B}(K)$ is not greater than the diameter of K, the corollary yields an inequality for the diameter. The resulting inequality is known as the isodiametric inequality.

Using Theorem 3.13 and second derivatives, we obtain in a similar manner inequalities of quadratic type.

Theorem 3.15 (Minkowski's Second Inequality) *For $K, L \in \mathcal{K}^n$,*

$$V(K, \ldots, K, L)^2 \geq V(K, \ldots, K, L, L)V(K). \tag{3.19}$$

The proof is left as an exercise (see Exercise 3.3.2). Equality holds if $\dim(K) \leq n-2$ or if K and L are homothetic, but there are also non-homothetic pairs of bodies (with interior points) for which equality holds. More precisely, the characterization of the case of equality involves the $(n-2)$-tangential body of L. We refer to [81, Theorem 7.6.19] for the proof. The solution of the long standing problem of characterizing the equality case for the more general quadratic Minkowski inequality

$$V(M[n-2], K, L)^2 \geq V(M[n-2], K, K)V(M[n-2], L, L), \tag{3.20}$$

for convex bodies $K, L, M \in \mathcal{K}^n$, has been announced in [86]. The even more general Alexandrov–Fenchel inequality is obtained if the $(n-2)$ bodies $M, \ldots, M$ in (3.20) are replaced by arbitrary convex bodies $M_1, \ldots, M_{n-2}$, so that

$$V(M_1, \ldots, M_{n-2}, K, L)^2$$
$$\geq V(M_1, \ldots, M_{n-2}, K, K)V(M_1, \ldots, M_{n-2}, L, L) \tag{3.21}$$

is obtained. The Alexandrov–Fenchel inequality is connected to various fields in mathematics and has many and surprising applications. A proof of this deep result and more detailed references to the literature can be found in [81, Chapter 7.3], a discussion of consequences and improvements are provided in [81, Chapter 7.4]. Recently, new approaches and variants of proofs of the Alexandrov–Fenchel inequality have been developed (see, for instance, [27, 85]). Despite these new insights and improvements, it is still not known for which convex bodies equality holds in (3.21).

Replacing K or L in (3.19) by the unit ball, we obtain more special inequalities, for example (in $\mathbb{R}^3$)

$$\pi \overline{B}(K)^2 \geq F(K)$$

or

$$F(K)^2 \geq 6\pi \overline{B}(K)V(K).$$

Exercises and Supplements for Sect. 3.4

1. Let $K, L \in \mathcal{K}_0^n$ and $K \subset L$. Show that $K = L$ if and only if $V(K) = V(L)$.
2.[*] Give a proof of Theorem 3.15.
3. The *diameter* diam(K) of a convex body $K \in \mathcal{K}^n$ is defined as

$$\mathrm{diam}(K) := \sup\{\|x - y\| : x, y \in K\}.$$

(a) Prove that

$$\overline{B}(K) \leq \mathrm{diam}(K) \leq \frac{n\kappa_n}{2\kappa_{n-1}} \cdot \overline{B}(K).$$

(b) If there is equality in one of the two inequalities, what can be said about K?

4. Let $K \in \mathcal{K}^n$ be an n-dimensional convex body. The *difference body* $D(K)$ of K is defined as the centrally symmetric convex body $D(K) := \frac{1}{2}(K + (-K))$. Show that

   (a) $D(K)$ has the same width as K in every direction.
   (b) $V(D(K)) \geq V(K)$ with equality if and only if K is centrally symmetric.

5. Let $K, L \in \mathcal{K}^n$, and let $u \in \mathbb{R}^n \setminus \{0\}$. Let K_u, L_u denote the orthogonal projection of K, L on $u^\perp$. Suppose that $K_u = L_u$. Show that

$$V((1-\alpha)K + \alpha L) \geq (1-\alpha)V(K) + \alpha V(L), \quad \alpha \in [0, 1].$$

   Show by an example that the assumption $K_u = L_u$ cannot be dropped.
   Hint: Use Fubini's theorem and check the inclusion

$$(1-\alpha)(K \cap G_x) + \alpha(L \cap G_x) \subset ((1-\alpha)K + \alpha L) \cap G_x,$$

   where $G_x := x + \mathbb{R}u$ for $x \in \mathbb{R}^n$ and the given vector $u \in \mathbb{R}^n \setminus \{0\}$.
6. We proved the Minkowski inequality (Theorem 3.14) with the help of the Brunn–Minkowski inequality (Theorem 3.13). Show that conversely the Brunn–Minkowski inequality can be deduced with the help of the Minkowski inequality.
   Hint: Use $V(K + L) = V((K + L)[n - 1], K + L)$.
7. Let $K \in \mathcal{K}^n$ with dim $K = n$, $u \in \mathbb{S}^{n-1}$ and $I_u = [-h(K, -u), h(K, u)]$. Show that the map

$$I_u \to \mathbb{R}, \quad t \mapsto V_{n-1}(K \cap H(u, t)),$$

   is continuous.
8. Let $K \in \mathcal{K}^n$ be an n-dimensional convex body and $u \in \mathbb{S}^{n-1}$. Let $I_u := [-h(K, -u), h(K, u)]$. Consider the function defined by

$$r(t) := [V_{n-1}(K \cap H(u, t))/\kappa_{n-1}]^{\frac{1}{n-1}}, \quad t \in I_u,$$

   and the set (*Schwarz symmetrization/rounding* of K)

$$S_u K := \bigcup_{t \in I_u} \left[B^n(tu, r(t)) \cap H(u, t) \right].$$

   Prove that the function r is concave, $S_u K$ is a convex body and $V(S_u K) = V(K)$.
9. Let $K \in \mathcal{K}^n$ with dim $K > 0$. Suppose that for every $(n - 1)$-dimensional linear subspace of $\mathbb{R}^n$ there is a parallel hyperplane with respect to which K is symmetric. Show that K is a Euclidean ball.

10. Let $K \in \mathcal{K}^n$ be an n-dimensional convex body ($n \geq 2$) with $h(K, e_n) = 1$ and $h(K, -e_n) = 0$ (e_n denotes the nth unit vector). Let $f(t) := V_{n-1}(K \cap H(e_n, t))$ for $t \in [0, 1]$. The nth coordinate c_n of the center of mass of K is given by

$$c_n = V(K)^{-1} \int_0^1 tf(t)dt.$$

Show the estimate

$$\frac{1}{n+1} \leq c_n \leq \frac{n}{n+1}.$$

11.* (a) Let $A, B \subset \mathbb{R}^n$ be boxes with parallel axes. Prove the Brunn–Minkowski inequality for A, B, based on the inequality between the arithmetic mean and the geometric mean.
   (b) Let A, and also B, be a finite union of boxes with parallel axes with pairwise disjoint interiors. Use induction over the total number of boxes and (a) to deduce the Brunn–Minkowski inequality for A, B.
   (c) Use approximation from outside to conclude the Brunn–Minkowski inequality for general compact sets $A, B \subset \mathbb{R}^n$.

12. Let $A, B \subset \mathbb{R}^n$ be compact sets and $\lambda \in [0, 1]$. By Exercise 3.4.11,

$$V((1 - \lambda)A + \lambda B) \geq \left[(1 - \lambda)V(A)^{\frac{1}{n}} + \lambda V(B)^{\frac{1}{n}}\right]^n$$

$$\geq \left[V(A)^{\frac{1-\lambda}{n}} V(B)^{\frac{\lambda}{n}}\right]^n,$$

by the inequality of the arithmetic mean and the geometric mean. Hence,

$$V((1 - \lambda)A + \lambda B) \geq V(A)^{1-\lambda}V(B)^{\lambda}$$

$$\geq \min\{V(A), V(B)\}. \tag{3.22}$$

Show that the seemingly weaker inequality (3.22) (for all sets A, B and parameters λ) is in fact equivalent to the Brunn–Minkowski inequality.

13.* Let $\lambda \in (0, 1)$ and let $f, g, h : \mathbb{R}^n \to [0, \infty)$ be measurable functions such that

$$h((1 - \lambda)x + \lambda y) \geq f(x)^{1-\lambda}g(y)^{\lambda}, \quad \text{for } x, y \in \mathbb{R}^n.$$

Show that

$$\int_{\mathbb{R}^n} h(x) \, dx \geq \left(\int_{\mathbb{R}^n} f(x) \, dx\right)^{1-\lambda} \left(\int_{\mathbb{R}^n} g(x) \, dx\right)^{\lambda}.$$

Recover the Brunn–Minkowski inequality for compact sets by specializing f, g, h.

3.5   The Alexandrov–Fenchel Inequality

In this section, we prove the famous Alexandrov–Fenchel inequality. In our presentation we essentially follow [27]. A special feature of the current approach is that from the very beginning it emphasizes the close connection to a classical Brunn–Minkowski type inequality for mixed volumes, and thus to a general form of Minkowski's inequality for mixed volumes. The following proof is similar to Alexandrov's first proof (which is described in detail in [81, Chapter 7.3]) in that it uses the special representation of mixed volumes available for strongly isomorphic polytopes, induction over the dimension and approximation, but in addition it involves analytic properties of convex functions (which are of independent interest) and a special case of the Perron–Frobenius theorem from linear algebra.

We use the following short notation. For integers $1 \le i \le j \le n$ and convex bodies $K_i, \ldots, K_j \in \mathcal{K}^n$, we write $\mathcal{K}_{i..j} = (K_i, \ldots, K_j)$ (with or without brackets) for a given finite sequence of $j - i + 1$ convex bodies. The sequence is empty (and omitted) if $i > j$. Moreover, we write $\mathcal{K}_0^n$ for the set of convex bodies with nonempty interiors.

Theorem 3.16 (Alexandrov–Fenchel Inequality) *Let $K_1, \ldots, K_n \in \mathcal{K}^n$. Then*

$$V(K_1, K_2, \mathcal{K}_{3..n})^2 \ge V(K_1[2], \mathcal{K}_{3..n})\, V(K_2[2], \mathcal{K}_{3..n}). \tag{AFI}$$

Note that by the symmetry properties of mixed volumes this is equivalent to (3.21). For $n = 2$ the Alexandrov–Fenchel inequality boils down to Minkowski's inequality. For this reason, we focus on dimension $n \ge 3$ in the following.

For $m \in \{2, \ldots, n\}$ and convex bodies $K_1, K_2, K_{m+1}, \ldots, K_n \in \mathcal{K}^n$, we consider the function defined by

$$f_m(t) := V(K_1 + tK_2\,[m], \mathcal{K}_{m+1..n})^{\frac{1}{m}}, \qquad t \ge 0.$$

We will see now that (AFI) is closely related to the fact that f_m is a concave function on $[0, \infty)$.

Lemma 3.4 *Let $m \in \{2, \ldots, n\}$, $K_1, K_2, K_{m+1}, \ldots, K_n \in \mathcal{K}_0^n$ and $t \ge 0$. Then*

$$f_m''(t) = -(m-1)\, f_m(t)^{1-2m}$$
$$\times \left(V(K_1, K_2, \overline{\mathcal{K}}_{3..n})^2 - V(K_1[2], \overline{\mathcal{K}}_{3..n})V(K_2[2], \overline{\mathcal{K}}_{3..n}) \right),$$

where $\overline{K}_t := K_1 + tK_2$ and $\overline{\mathcal{K}}_{3..n} := (\overline{K}_t[m-2], \mathcal{K}_{m+1..n})$, for $t \ge 0$.

Proof We define $h_m(t) := f_m(t)^m$ for $t \ge 0$. Then f_m and h_m are of class $\mathbf{C}^2$ and

$$f_m''(t) = \frac{1}{m} f_m(t)^{1-2m} \left(h_m''(t) h_m(t) + \frac{1-m}{m} h_m'(t)^2 \right), \qquad t \ge 0.$$

Using the Minkowski linearity of mixed volumes, we obtain

$$h_m(t) = V(\overline{K}_t[m], \mathcal{K}_{m+1..n}),$$

$$h'_m(t) = mV(\overline{K}_t[m-1], K_2, \mathcal{K}_{m+1..n}),$$

$$h''_m(t) = m(m-1)V(\overline{K}_t[m-2], K_2[2], \mathcal{K}_{m+1..n}).$$

Hence we get

$$h''_m(t)h_m(t) + \frac{1-m}{m}h'_m(t)^2$$

$$= -m(m-1)\left(V(\overline{K}_t, K_2, \overline{K}_t[m-2], \mathcal{K}_{m+1..n})^2\right.$$

$$\left. -V(\overline{K}_t[2], \overline{K}_t[m-2], \mathcal{K}_{m+1..n})V(K_2[2], \overline{K}_t[m-2], \mathcal{K}_{m+1..n})\right)$$

$$= -m(m-1)\left(V(K_1, K_2, \overline{K}_{3..n})^2 - V(K_1[2], \overline{K}_{3..n})V(K_2[2], \overline{K}_{3..n})\right),$$

from which the assertion follows. □

The following proposition is an immediate consequence of Lemma 3.4.

Proposition 3.3

(a) If (AFI) holds for all convex bodies, then f_m is a concave function for all convex bodies.

(b) For fixed convex bodies $K_1, \ldots, K_n \in \mathcal{K}_0^n$, (AFI) holds if and only if f_2 is concave. In particular, (AFI) holds for all convex bodies if and only if f_2 is always concave.

The following lemma shows that a converse of Proposition 3.3 (a) holds not only for $m = 2$, but also for $m = 3$. This observation will be a key fact in the inductive proof of the Alexandrov–Fenchel inequality. We start with another lemma.

We write $\mathbb{R}_+ = [0, \infty)$, $\mathbb{R}_+^3 = [0, \infty)^3$ and $x = (x_1, x_2, x_3)^\top \in \mathbb{R}^3$. For $K_1, \ldots, K_n \in \mathcal{K}_0^n$ and $x, y, z \in \mathbb{R}^3$, we define the functions

$$\widetilde{F}(x, y, z) := \sum_{i_1=1}^{3}\sum_{i_2=1}^{3}\sum_{i_3=1}^{3} x_{i_1} y_{i_2} z_{i_3} V\left(K_{i_1}, K_{i_2}, K_{i_3}, \mathcal{K}_{4..n}\right)$$

and $F(x) := \widetilde{F}(x, x, x)$. Note that for $x, y, z \in \mathbb{R}_+^3 \setminus \{0\}$, we have

$$\widetilde{F}(x, y, z) = V\left(\sum_{i=1}^{3} x_i K_i, \sum_{i=1}^{3} y_i K_i, \sum_{i=1}^{3} z_i K_i, \mathcal{K}_{4..n}\right) > 0.$$

In the following lemma, we use this notation for fixed $K_1, \ldots, K_n \in \mathcal{K}_0^n$.

Lemma 3.5 *If* $t \mapsto F(x + ty)^{\frac{1}{3}}$, $t \in \mathbb{R}_+$, *is concave for all* $x, y \in \mathbb{R}_+^3$, *then*

$$\widetilde{F}(x, y, z)^2 \geq \widetilde{F}(x, x, z)\widetilde{F}(y, y, z) \tag{3.23}$$

for all $x, y, z \in \mathbb{R}_+^3$.

Proof For the proof, we can assume that $\widetilde{F}(x, x, z) > 0$ and that the components of the vectors x, y, z are not zero. The proof is given in three steps.

(1) Let $x, y \in (0, \infty)^3$ be fixed for the moment. For $h(t) := F(x + ty)$, $t \geq 0$, we obtain $h'(t) = 3\widetilde{F}(y, x + ty, x + ty)$ and $h''(t) = 6\widetilde{F}(y, y, x + ty)$ and therefore

$$\frac{d^2}{dt^2} h^{\frac{1}{3}}(t) \mid_{t=0} = 2F(x)^{-\frac{5}{3}} \left(-\widetilde{F}(y, x, x)^2 + F(x)\widetilde{F}(y, y, x) \right).$$

Since by assumption $h^{\frac{1}{3}}$ is concave, it follows that

$$\widetilde{F}(y, x, x)^2 \geq \widetilde{F}(x, x, x)\widetilde{F}(y, y, x), \qquad x, y \in (0, \infty)^3. \tag{3.24}$$

(2) Now we extend (3.24) to $x \in (0, \infty)^3$ and all $y \in \mathbb{R}^3$. For this, consider

$$P_x(y) := \widetilde{F}(y, x, x)^2 - \widetilde{F}(x, x, x)\widetilde{F}(y, y, x), \qquad y \in \mathbb{R}^3.$$

If $t > 0$ is sufficiently large, then $y + tx \in (0, \infty)^3$ and hence $P_x(y + tx) \geq 0$ by (3.24). Moreover, the multilinearity of $\widetilde{F}$ (and cancellation) implies that $P_x(y + tx) = P_x(y)$, from which the assertion follows.

(3) Finally, for $x, y, z \in (0, \infty)^3$ and $t \in \mathbb{R}$ we consider

$$G(t) := \widetilde{F}(y + tz, y + tz, x) = \widetilde{F}(y, y, x) + 2t\widetilde{F}(y, z, x) + t^2\widetilde{F}(z, z, x).$$

There is a $t_0 \leq 0$ such that $\widetilde{F}(y + t_0z, x, x) = \widetilde{F}(y, x, x) + t_0\widetilde{F}(z, x, x,) = 0$ (here we use $\widetilde{F}(z, x, x) > 0$). An application of the result of (2) to $y + t_0z$ then yields that $G(t_0) = \widetilde{F}(y + t_0z, y + t_0z, x) \leq 0$, since $F(x) > 0$. On the other hand, $G(0) = \widetilde{F}(y, y, x) \geq 0$. Therefore, the discriminant of the quadratic function $t \mapsto G(t)$ is nonnegative, which is equivalent to the assertion. $\square$

Lemma 3.6 *If* f_3 *is a concave function for all convex bodies, then* (AFI) *holds for all convex bodies.*

Proof Let $K_1, \ldots, K_n \in \mathcal{K}_0^n$ be given. We continue to use the notation from Lemma 3.5. For $x, y \in \mathbb{R}_+^3$ and $t \geq 0$, consider

$$F(x + ty) = V\left(\sum_{i=1}^{3}(x_i + ty_i)K_i\,[3], \mathcal{K}_{4..n} \right) = V\left(\overline{K}_1 + t\overline{K}_2\,[3], \mathcal{K}_{4..n} \right),$$

where $\overline{K}_1 = \sum_{i=1}^3 x_i K_i$ and $\overline{K}_2 = \sum_{i=1}^3 y_i K_i$. By the assumption of the lemma, the function $t \mapsto F(x + ty)^{\frac{1}{3}}$, $t \geq 0$, is concave for each choice of $x, y \in \mathbb{R}_+^3$. An application of Lemma 3.5 with the standard unit vectors $x = e_1$, $y = e_2$, $z = e_3$ now yields that (AFI) holds for the given convex bodies.   □

The final part of the proof of the Alexandrov–Fenchel inequality requires further preparations.

Recall that we denote by $\mathcal{P}^n$ the set of polytopes in $\mathbb{R}^n$. We write $\mathcal{P}_0^n$ for the subset of n-dimensional polytopes of $\mathcal{P}^n$. For vectors $u_1, \ldots, u_N \in \mathbb{S}^{n-1}$ and $h = (h_1, \ldots, h_N)^\top \in \mathbb{R}^N$, we consider polyhedral sets of the form

$$P_{[h]} := \bigcap_{i=1}^N H^-(u_i, h_i).$$

Clearly, if $h \in \mathbb{R}_+^N$, then $0 \in P_{[h]}$ and $P_{[h]}$ is a polytope if and only if the vectors $u_1, \ldots, u_N \in \mathbb{S}^{n-1}$ are not contained in any hemisphere. Further, $0 \in \operatorname{int}(P_{[h]})$ if and only if $h_1, \ldots, h_N > 0$. The vector h is called the vector of support numbers of $P_{[h]}$ if $u_1, \ldots, u_N$ are the exterior unit facet normals of $P_{[h]}$, that is, if $\dim(P_{[h]}(u_i)) = n - 1$ for $i = 1, \ldots, N$. In this case, the support numbers are uniquely determined by $P_{[h]}$, since $h(P_{[h]}, u_i) = h_i$.

Strongly Isomorphic Polytopes

Definition 3.7

(a) A polytope $P \in \mathcal{P}_0^n$ is *simple* if each vertex of P is contained in precisely n facets of P.

(b) Two polytopes $P_1, P_2 \in \mathcal{P}_0^n$ are *strongly isomorphic* if $\dim(P_1(u)) = \dim(P_2(u))$ for all $u \in \mathbb{S}^{n-1}$.

Clearly, strong isomorphism of n-polytopes is an equivalence relation, the equivalence class of a polytope $P \in \mathcal{P}_0^n$ is called the *a-type* of P. For $Q \in \mathcal{P}_0^n$ we write $Q \in a(P)$ if Q and P belong to the same class. The following lemma collects several geometric facts that will be used in the following.

Lemma 3.7

(a) If $P_1, P_2 \in \mathcal{P}_0^n$ are strongly isomorphic, then $P_1(u)$ and $P_2(u)$ are also strongly isomorphic for each $u \in \mathbb{S}^{n-1}$.

(b) If $P_1, \ldots, P_m \in \mathcal{P}_0^n$ are strongly isomorphic, then all polytopes $\alpha_1 P_1 + \cdots + \alpha_m P_m$ with $\alpha_1, \ldots, \alpha_m \geq 0$ and $\alpha_1 + \cdots + \alpha_m > 0$ are strongly isomorphic.

(c) If $P = P_{[h]} \in \mathcal{P}_0^n$ is simple and has exterior facet normals $u_1, \ldots, u_N \in \mathbb{S}^{n-1}$, then there is some $\beta > 0$ such that any two of the polytopes $P_{[h+\alpha]}$ with $\alpha = (\alpha_1, \ldots, \alpha_N)^\top$ and $|\alpha_i| \leq \beta$ are strongly isomorphic.

(d) *For any* $(K_1, \ldots, K_n) \in (\mathcal{K}^n)^n$ *there is a sequence* $(P_1(m), \ldots, P_n(m)) \in (\mathcal{P}_0^n)^n$, $m \in \mathbb{N}$, *such that* $P_j(m) \to K_j$ *as* $m \to \infty$ *(in the Hausdorff metric), for* $j = 1, \ldots, n$, *and* $P_1(m), \ldots, P_n(m)$ *are simple and strongly isomorphic for each* $m \in \mathbb{N}$.

Proof (a) is [81, Lemma 2.4.10], **(b)** is [81, Lemma 2.4.12], **(c)** is [81, Lemma 2.4.13], **(d)** is [81, Theorem 2.4.15].   □

Mixed Volumes of Strongly Isomorphic Polytopes

In comparison to general convex polytopes, the representation of mixed volumes of strongly isomorphic polytopes slightly simplifies, which will be convenient for the subsequent analysis.

Let $P \in \mathcal{P}_0^n$ be a simple polytope with facet normals $u_1, \ldots, u_N \in \mathbb{S}^{n-1}$. Then

$$\mathcal{C}(P) := \left\{ h \in (0, \infty)^N : P_{[h]} \in a(P) \right\}$$

is an open convex cone. In fact, if $h, h' \in \mathcal{C}(P)$, then $P_{[h]} + P_{[h']} \in a(P)$ by Lemma 3.7 (b), and thus

$$(P_{[h]} + P_{[h']})(u_i) = P_{[h]}(u_i) + P_{[h']}(u_i) = h_i + h_i' = P_{[h+h']}(u_i)$$

implies that $P_{[h]} + P_{[h']} = P_{[h+h']}$. The fact that $\mathcal{C}(P)$ is open follows from Lemma 3.7 (c).

Lemma 3.8 *Let* $P \in \mathcal{P}_0^n$ *be a simple polytope with exterior facet normals* $u_1, \ldots, u_N \in \mathbb{S}^{n-1}$. *For* $i = 1, \ldots, n$, *let* $P_i = P_{[h^{(i)}]} \in a(P)$ *with* $h_j^{(i)} = h(P_{[h^{(i)}]}, u_j)$ *for* $j = 1, \ldots, N$. *Then there are real numbers* $a_{j_1 \cdots j_n}$, *for* $j_1, \ldots, j_n \in \{1, \ldots, N\}$, *depending only on* $a(P)$ *(and independent of the support numbers of the polytopes) and symmetric in the lower indices, such that*

$$V(P_1, \ldots, P_n) = \sum_{j_1, \ldots, j_n = 1}^{N} a_{j_1 \cdots j_n} h_{j_1}^{(1)} \cdots h_{j_n}^{(n)}.$$

In particular, the map $\mathcal{C}(P)^N \ni (h^{(1)}, \ldots, h^{(n)}) \mapsto V(P_{[h^{(1)}]}, \ldots, P_{[h^{(n)}]})$ *is of class* $\mathbf{C}^\infty$.

Proof We proceed by induction. For $n = 1$, we have $h = (h_1, h_2)^\top$ and $P = [h_1 e_1, h_2(-e_1)]$ with $h_1 > -h_2$, and thus $V(P) = h_1 + h_2$. For the induction step, we use that $N(P_1 + \cdots + P_{n-1}) = \{u_1, \ldots, u_N\}$, and hence by definition

$$V(P_{[h^{(1)}]}, \ldots, P_{[h^{(n)}]}) = \frac{1}{n} \sum_{i=1}^{N} h_i^{(n)} V^{(n-1)}(P_{[h^{(1)}]}(u_i), \ldots, P_{[h^{(n-1)}]}(u_i)).$$

$$(3.25)$$

By Lemma 3.7 (a), for $i = 1, \ldots, N$ the $(n-1)$-polytopes $P_{[h^{(l)}]}(u_i), l = 1, \ldots, n-1$, are strongly isomorphic in $u_i^\perp$ (say) and their a-type is determined by $a(P)$ and i. Hence, using $J := \{(r, s) \in \{1, \ldots, N\}^2 : \dim(P(u_r) \cap P(u_s)) = n - 2\}$, we obtain from the induction hypothesis that there are numbers $a^i_{j_1 \cdots j_{n-1}}$ such that

$$V^{(n-1)}(P_{[h^{(1)}]}(u_i), \ldots, P_{[h^{(n-1)}]}(u_i)) = \sum_{(i,j_r) \in J} a^i_{j_1 \cdots j_{n-1}} h^{(1)}_{ij_1} \cdots h^{(n-1)}_{ij_{n-1}}, \qquad (3.26)$$

where $j_1, \ldots, j_{n-1} \in \{1, \ldots, N\}$ and the summation is extended over all pairs $(i, j_r) \in J$. It follows from relation (3.9) that

$$h^{(r)}_{ij_r} = h(P_{[h^{(r)}]}(u_i), \tilde{u}_{j_r}) = \frac{1}{\sin \angle(u_i, u_{j_r})} h^{(r)}_{j_r} - \frac{1}{\tan \angle(u_i, u_{j_r})} h^{(r)}_i. \qquad (3.27)$$

Note that the angles $\angle(u_i, u_{j_r})$ depend only on $a(P)$. Inserting (3.26) and (3.27) into (3.25), the assertion follows if additional zero coefficients are inserted (if needed) and a symmetrization is carried out to obtain coefficients with symmetric indices.   □

Recall from Sect. 2.2 that for a (twice continuously) differentiable function $\mathcal{F}$ on an open subset of $\mathbb{R}^N$ we write $\nabla \mathcal{F}(h) = (\mathcal{F}_1(h), \ldots, \mathcal{F}_N(h))^\top \in \mathbb{R}^N$ for the gradient and $\nabla^2 \mathcal{F}(h) := \partial^2 \mathcal{F}(h) = (\mathcal{F}_{ij}(h))^N_{i,j=1} \in \mathbb{R}^{N,N}$ for the Hessian matrix of $\mathcal{F}$ at h, with respect to the standard basis of $\mathbb{R}^N$.

Now we are prepared for the final step of the proof of the Alexandrov–Fenchel inequality.

Proof of Theorem 3.16 We proceed by induction. The theorem has already been proved in the case $n = 2$ (and also the case $n = 3$). Hence let $n \geq 3$ (or $n \geq 4$) and assume the theorem holds in smaller dimensions.

Let $P \in \mathcal{P}^n_0$ be a simple polytope and fix $\mathcal{K}_{n-3} := \mathcal{K}_{4..n} := (P_4, \ldots, P_n)$ with n-polytopes $P_r \in a(P)$ for $r = 4, \ldots, n$. Then the function $\mathcal{F} : \mathcal{C}(P) \to (0, \infty)$ defined by

$$\mathcal{F}(h) := V(P_{[h]}[3], \mathcal{K}_{n-3})$$

has the following properties.

(i) $\mathcal{F}$ is $\mathbf{C}^\infty$ (by Lemma 3.8) and positively homogeneous of degree 3 on the open convex cone $\mathcal{C}(P)$.

(ii) For $h \in \mathcal{C}(P)$, we have

$$\mathcal{F}_i(h) = \frac{3}{n} V^{(n-1)}\left(P_{[h]}(u_i)[2], \mathcal{K}_{n-3}(u_i)\right) > 0,$$

where $\mathcal{K}_{n-3}(u_i) = (P_4(u_i), \ldots, P_n(u_i))$. The derivative follows from (3.25) and the symmetry of mixed volumes. Also note that $\dim(P_{[h]}(u_i)) = n-1 \geq 2$ and $\dim(P_r(u_i)) = n - 1$ for $r \geq 4$.

(iii) By the induction hypothesis, the Alexandrov–Fenchel inequality holds for the mixed volume $V^{(n-1)}$ in $u_i^\perp$. Hence, since for $\lambda \in [0, 1]$ we have

$$P_{[(1-\lambda)h+\lambda h']}(u_i) = \left((1-\lambda)P_{[h]} + \lambda P_{[h']}\right)(u_i) = (1-\lambda)P_{[h]}(u_i) + \lambda P_{[h']}(u_i),$$

the map $C(P) \ni h \mapsto \mathcal{F}_i(h)^{\frac{1}{2}}$ is concave (see Exercise 3.5.1).

(iv) For $(i, j) \in J$ and $h \in C(P)$, it follows from (3.8) and (3.27) that

$$\mathcal{F}_{ij}(h) = \frac{3}{n}\frac{2}{n-1}\frac{1}{\sin \angle(u_i, u_j)} V^{(n-2)}\left(P_{[h]}(u_i) \cap P_{[h]}(u_j), \mathcal{K}_{n-3}(u_i, u_j)\right)$$

$$> 0,$$

where $\mathcal{K}_{n-3}(u_i, u_j) = (P_4(u_i) \cap P_4(u_j), \ldots, P_n(u_i) \cap P_n(u_j))$. Note that all these intersections are $(n-2)$-dimensional since $(i, j) \in J$. Further, we have $\mathcal{F}_{ii}(h) < 0$ and $\mathcal{F}_{ij}(h) = 0$ if $(i, j) \in \{(r, s) \in \{1, \ldots, N\}^2 \setminus J : r \neq s\}$. Moreover, since any two facets are connected by a sequence of facets such that the intersection of successive facets has dimension $n-2$, it follows that the matrix $\nabla^2 \mathcal{F}(h)$ is irreducible, that is, for any $i, j \in \{1, \ldots, N\}$ with $i \neq j$ there is a sequence $i_0 := i, i_1, \ldots, i_m := j \in \{1, \ldots, N\}$ with $i_r \neq i_{r+1}$ such that $\mathcal{F}_{i_r i_{r+1}}(h) > 0$.

The following Lemma 3.9 shows these properties imply that $C(P) \ni h \mapsto \mathcal{F}(h)^{\frac{1}{3}}$ is concave. Now from Lemma 3.7 (d) and the continuity of mixed volumes it follows that

$$L \mapsto V(L[3], K_3, \ldots, K_n)^{\frac{1}{3}}, \qquad L \in \mathcal{K}^n,$$

is a concave map, and therefore f_3 (as in Lemma 3.6) is concave for all convex bodies (see also Exercise 3.5.1). Hence, by Lemma 3.6 it follows that (AFI) holds in n-dimensional Euclidean space, which completes the induction step.   □

For $v \in \mathbb{R}^N$, we write $v \otimes v$ for the square matrix $v \cdot v^\top \in \mathbb{R}^{N,N}$. If $A, B \in \mathbb{R}^{N,N}$ are symmetric matrices, we write $A \geq 0$ to express that A is positive semi-definite and $A \geq B$ to say that $A - B \geq 0$. A vector is said to be positive if its coordinates with respect to the standard basis are all positive.

Lemma 3.9 *Let $C \subset (0, \infty)^N$ be an open, convex cone. Let $\mathcal{F} : C \to \mathbb{R}$ be a function of class $\mathbf{C}^3$ such that the following conditions are satisfied:*

(a) *$\mathcal{F}$ is positively 3-homogeneous.*
(b) *$C \ni h \mapsto \mathcal{F}_i(h)^{\frac{1}{2}}$ is concave and $\mathcal{F}_i(h) > 0$ for $h \in C$.*
(c) *$\nabla^2 \mathcal{F}(h)$ is irreducible and $\mathcal{F}_{ij}(h) \geq 0$ for $i \neq j$ and $h \in C$.*

Then $C \ni h \mapsto \mathcal{F}(h)^{\frac{1}{3}}$ is a concave function.

Proof Since the map $C \ni h \mapsto \mathcal{F}_i(h)^{\frac{1}{2}}$ is concave and of class $\mathbf{C}^2$, the Hessian matrix $\nabla^2(\mathcal{F}_i)^{\frac{1}{2}}(h)$ is negative semi-definite, which can be expressed by

$$\nabla^2 \mathcal{F}_i(h) \leq \frac{1}{2} \frac{\nabla \mathcal{F}_i(h) \otimes \nabla \mathcal{F}_i(h)}{\mathcal{F}_i(h)}, \quad h \in C. \tag{3.28}$$

Since $\nabla^2 \mathcal{F}$ is positively 1-homogeneous, we get

$$\sum_{i=1}^{N} h_i \nabla^2 \mathcal{F}_i(h) = \sum_{i=1}^{N} h_i \left(\nabla^2 \mathcal{F}\right)_i(h) = \nabla^2 \mathcal{F}(h).$$

Therefore, (3.28) yields

$$\nabla^2 \mathcal{F}(h) = \sum_{i=1}^{N} h_i \nabla^2 \mathcal{F}_i(h) \leq \frac{1}{2} \sum_{i=1}^{N} h_i \frac{\nabla \mathcal{F}_i(h) \otimes \nabla \mathcal{F}_i(h)}{\mathcal{F}_i(h)}, \quad h \in C. \tag{3.29}$$

With the diagonal matrix $\mathcal{D}(h) = \operatorname{diag}\left(\mathcal{F}_i(h)^{-1} h_i : i = 1 \ldots, N\right)$, the symmetric matrix $M(h) := \mathcal{D}(h)^{\frac{1}{2}} \nabla^2 \mathcal{F}(h) \mathcal{D}(h)^{\frac{1}{2}}$ and using

$$\sum_{i=1}^{N} h_i \frac{\nabla \mathcal{F}_i(h) \otimes \nabla \mathcal{F}_i(h)}{\mathcal{F}_i(h)} = \nabla^2 \mathcal{F}(h) \mathcal{D}(h) \nabla^2 \mathcal{F}(h),$$

we can rewrite (3.29) in the form

$$M(h) \leq \frac{1}{2} M(h)^2, \quad h \in C. \tag{3.30}$$

Clearly, $M(h) \in \mathbb{R}^{N,N}$ is symmetric, irreducible and $M(h)_{ij} \geq 0$ for $i \neq j$. If $v \in \mathbb{R}^N \setminus \{0\}$ is an eigenvector of $M(h)$ with corresponding eigenvalue λ, then (3.30) implies that $\lambda \geq 2$ or $\lambda \leq 0$. Moreover, $\mathcal{D}(h)^{-\frac{1}{2}} h$ is an eigenvector of $M(h)$ with positive coordinates $\sqrt{h_i \mathcal{F}_i(h)} > 0$ and corresponding eigenvalue 2. This can be seen from

$$M(h) \mathcal{D}(h)^{-\frac{1}{2}} h = \mathcal{D}(h)^{\frac{1}{2}} \nabla^2 \mathcal{F}(h) h = \mathcal{D}(h)^{\frac{1}{2}} \left(\sum_{i=1}^{N} h_i (\mathcal{F}_j)_i(h)\right)_{j=1}^{N}$$

$$= \mathcal{D}(h)^{\frac{1}{2}} \left(2\mathcal{F}_j(h)\right)_{j=1}^{N} = 2\mathcal{D}(h)^{\frac{1}{2}} \nabla \mathcal{F}(h) = 2\mathcal{D}(h)^{-\frac{1}{2}} h,$$

where for the third equality we used that $\mathcal{F}_j$ is positively 2-homogeneous.

Note that

$$\|\mathcal{D}(h)^{-\frac{1}{2}}h\|^2 = \langle h, \mathcal{D}(h)^{-1}h \rangle = \langle h, \nabla \mathcal{F}(h) \rangle = 3\mathcal{F}(h),$$

since $\mathcal{F}$ is positively 3-homogeneous. An application of Lemma 3.10 below now shows that

$$M(h) \leq 2 \, \frac{\mathcal{D}(h)^{-\frac{1}{2}}h \otimes \mathcal{D}(h)^{-\frac{1}{2}}h}{3\mathcal{F}(h)}.$$

Thus we arrive at

$$\nabla^2 \mathcal{F}(h) \leq \frac{2}{3} \frac{\mathcal{D}(h)^{-1}h \otimes \mathcal{D}(h)^{-1}h}{\mathcal{F}(h)} = \frac{2}{3} \frac{\nabla \mathcal{F}(h) \otimes \nabla \mathcal{F}(h)}{\mathcal{F}(h)}, \quad h \in \mathcal{C},$$

which shows that $\nabla^2 \left(\mathcal{F}^{\frac{1}{3}} \right)$ is negative semi-definite.

This finally proves that $\mathcal{F}^{\frac{1}{3}}$ is concave.   □

It remains to establish the following lemma from linear algebra.

Lemma 3.10 *Let $M = (m_{ij}) \in \mathbb{R}^{N,N}$ be symmetric, irreducible and such that $m_{ij} \geq 0$ for $i \neq j$. Suppose that $v_1 \in \mathbb{S}^{N-1}$ is a positive eigenvector of M and corresponding eigenvalue $\lambda_1 > 0$. If M does not have any eigenvalues in $(0, \lambda_1)$, then*

$$M \leq \lambda_1 v_1 \otimes v_1.$$

Proof Let $v_1, \ldots, v_N$ be an orthonormal basis of eigenvectors of M with corresponding real eigenvalues $\lambda_1, \ldots, \lambda_N$. Choosing $a := \max\{|m_{ii}| : i = 1, \ldots, N\}$, all components of the matrix $M + aI$ are nonnegative (I denotes the identity matrix). Then $v_1, \ldots, v_N$ is an orthonormal basis of eigenvectors of $M + aI$ with corresponding eigenvalues $\lambda_1 + a, \ldots, \lambda_N + a$. An application of (a special case of) the Perron–Frobenius theorem to the symmetric, irreducible matrix $M + aI$, which has v_1 as a positive eigenvector with positive eigenvalue $\lambda_1 + a$, shows that $\lambda_i + a \leq |\lambda_i + a| \leq \lambda_1 + a$ and $\lambda_i + a \neq \lambda_1 + a$ for $i \geq 2$. Hence $\lambda_i < \lambda_1$ for $i \geq 2$, and therefore (by assumption) $\lambda_i \leq 0$ for $i = 2, \ldots, N$.

For $x = \sum_{i=1}^{N} x_i v_i \in \mathbb{R}^N$ with $x_i \in \mathbb{R}$ we get

$$\langle (\lambda_1 v_1 \otimes v_1 - M) x, x \rangle = \sum_{j=2}^{N} (-\lambda_j) x_j^2 \geq 0,$$

which proves the assertion.   □

Remark 3.27 The general Perron–Frobenius theorem is for instance covered in [47, Satz 6.3.3], [34, Section 8.2, Theorem 2], [46, Theorem 8.4.4], [72, Chapter 8]; see also Exercise 3.5.5.

The following Theorem 3.17 states that the function f_m is concave for arbitrary convex bodies.

Theorem 3.17 *For* $m \in \{2, \ldots, n\}$ *and* $K_{m+1}, \ldots, K_n \in \mathcal{K}^n$, *let* $\mathcal{K} = (K_{m+1}, \ldots, K_n)$. *Then the map*

$$\mathcal{K}^n \ni L \mapsto V(L[m], \mathcal{K})^{\frac{1}{m}}$$

is concave.

From the inequalities derived up to this point, a variety of strong geometric inequalities can be deduced. For example, we obtain a general version of Minkowski's inequality (see Exercise 3.5.2).

Theorem 3.18 *For* $m \in \{2, \ldots, n\}$ *and* $K_1, K_2, K_{m+1}, \ldots, K_n \in \mathcal{K}^n$, *let* $\mathcal{K} = (K_{m+1}, \ldots, K_n)$. *Then*

$$V(K_1[m-1], K_2, \mathcal{K})^m \geq V(K_1[m], \mathcal{K})^{m-1} V(K_2[m], \mathcal{K}).$$

Exercises and Supplements for Sect. 3.5

1. For a map $f : \mathcal{K}^n \to \mathbb{R}$, we consider the following properties.

   (a) f is concave, that is,

   $$f((1-\lambda)K + \lambda L) \geq (1-\lambda)f(K) + \lambda f(L), \quad K, L \in \mathcal{K}^n, \lambda \in [0, 1].$$

   (b) $[0, 1] \ni s \mapsto f((1-s)K + sL)$ is concave for $K, L \in \mathcal{K}^n$.
   (c) $\mathbb{R}_+ \ni t \mapsto f(K + tL)$ is concave for $K, L \in \mathcal{K}^n$.

   Show that $(a) \iff (b) \implies (c)$.
   If f is positively homogeneous of degree 1 and continuous, then all three properties are equivalent, that is, we also have $(c) \implies (b)$.
2. For the proof of Theorem 3.18 proceed as in the proof of Minkowski's inequality.
3. Deduce Theorem 3.17 from Theorem 3.18.
   Note that this exercise and Exercise 3.5.2 together show that for convex bodies Theorems 3.17 and 3.18 are essentially equivalent.
4. The argument for the crucial Lemma 3.9 essentially follows the proof of Proposition 3 in [27] in the special case $p = 3$ which is needed here.
5. Provide a direct (and elementary) argument for the Perron–Frobenius theorem as needed in the proof of Lemma 3.10 for symmetric matrices with the additional information which is available in the proof (the existence of a positive eigenvector with corresponding positive eigenvalue is already available).

3.6 Steiner Symmetrization

The isoperimetric inequality for a convex body $K \in \mathcal{K}^n$ with nonempty interior (or satisfying $F(K) > 0$) can be written in the form

$$\frac{V(K)^{n-1}}{F(K)^n} \leq \frac{V(B^n)^{n-1}}{F(B^n)^n}$$

and expresses an extremal property of Euclidean balls. Euclidean balls are distinguished by their perfect symmetry. For instance, if B^n is a Euclidean ball and $u \in \mathbb{S}^{n-1}$ is a unit vector, then there is a hyperplane H orthogonal to u such that all secants of B^n with direction u have their midpoints in H, in other words, B^n is symmetric with respect to H.

There is a simple procedure which transforms a given convex body K into a convex body $K_1 := S_{H_1}(K)$ which is symmetric with respect to a hyperplane H_1. Clearly, $S_{H_1}(K)$ need not be a ball after just one symmetrization step. Therefore the symmetrization is now repeated with respect to another hyperplane H_2. This leads to a convex body $K_2 := S_{H_2}(S_{H_1}(K))$. After m steps, we finally arrive at a convex body K_m. In general, K_2 will no longer be symmetric with respect to H_1 and K_m will not be a ball. Still, we shall see that there is a sequence $(K_i)_{i \in \mathbb{N}}$ of convex bodies converging to a ball and such that each K_i is obtained from K by successive symmetrization. Since this type of symmetrization does not change the volume of a convex body, that is, $V(S_H(K)) = V(K)$, and since it does not increase the surface area, that is, $F(S_H(K)) \leq F(K)$, we finally obtain

$$\frac{V(K)^{n-1}}{F(K)^n} \leq \frac{V(K_1)^{n-1}}{F(K_1)^n} \leq \cdots \leq \frac{V(K_i)^{n-1}}{F(K_i)^n} \to \frac{V(B^n)^{n-1}}{F(B^n)^n} \quad \text{as } i \to \infty.$$

A refinement of the argument also yields the uniqueness assertion.

In the following, we shall provide an introduction to what is called *Steiner symmetrization* and establish some of its basic properties. As applications, we shall derive some geometric inequalities. Although Steiner symmetrization can be applied to compact sets in general, in the following we consider only the case of convex bodies.

We start with an intuitive description. Let H be a hyperplane and K a convex body in $\mathbb{R}^n$. Let G be a line orthogonal to H intersecting K in a segment S. Then S is translated within G to $\overline{S}$ so that the midpoint of $\overline{S}$ is in H. The union of all segments $\overline{S}$, which are obtained in this way, yields a new set $S_H(K) \in \mathcal{K}^n$. The map $S_H : \mathcal{K}^n \to \mathcal{K}^n$ is called *Steiner symmetrization* with respect to the hyperplane H.

In a more formal way, this can be expressed as follows. Let $H \subset \mathbb{R}^n$ be a hyperplane, which is fixed for the moment. Let $u \in \mathbb{S}^{n-1}$ be a unit vector orthogonal to H (which is uniquely determined up to the orientation). Then each point $x \in \mathbb{R}^n$

has a unique representation

$$x = x_0 + z u, \quad x_0 \in H, z \in \mathbb{R}.$$

Since u and H are fixed for the moment, the dependence on u and H is not indicated in the notation. Thus we simply write

$$G_x := \{x + \lambda u : \lambda \in \mathbb{R}\}$$

for the line through x orthogonal to H. Let $K|H$ be the orthogonal projection of K to H, that is,

$$K|H = \{x \in H : G_x \cap K \neq \emptyset\}.$$

For $x \in K|H$, the intersection $G_x \cap K$ is a (possibly degenerate) segment of the form

$$G_x \cap K = \{x + z u : \underline{z}(x) \leq z \leq \overline{z}(x)\}.$$

Thus, two functions $\underline{z} : K|H \to \mathbb{R}$ and $\overline{z} : K|H \to \mathbb{R}$ are implicitly defined. Explicitly, they are given by

$$\underline{z}(x) = \min\{z \in \mathbb{R} : x + zu \in K\} \quad \text{and} \quad \overline{z}(x) = \max\{z \in \mathbb{R} : x + zu \in K\}.$$

Clearly, $\underline{z}$ is convex and $\overline{z}$ is concave, and hence $\underline{z}$ and $\overline{z}$ are continuous on $\mathrm{relint}(K|H)$. However, for $n \geq 3$ these functions need not be continuous on $K|H$, as the example $K = \mathrm{conv}([-e_3, e_3] \cup B)$ with $B = e_1 + B^3 \cap \mathrm{lin}\{e_1, e_2\} \subset \mathbb{R}^3$ and $H = e_3^\perp$ shows. Still, $\underline{z}$ is lower semi-continuous and $\overline{z}$ is upper semi-continuous. The former means that for $x \in K|H$ and $\varepsilon > 0$, there is a neighbourhood U of x such that

$$\underline{z}(y) \geq \underline{z}(x) - \varepsilon \quad \text{for } y \in U \cap (K|H).$$

To verify this, we assume that this is not the case. Then there is some $\varepsilon > 0$ and there is some $x \in K|H$ such that there is a sequence $(y_i)_{i \in \mathbb{N}}$ in $K|H$ satisfying $y_i \to x$ and $\underline{z}(y_i) < \underline{z}(x) - \varepsilon$. We may assume (passing to a subsequence) that $\underline{z}(y_i) \to a \in \mathbb{R}$, hence $a \leq \underline{z}(x) - \varepsilon$. Then $K \ni y_i + \underline{z}(y_i)u \to x + au$, and hence $x + au \in K$, and therefore $\underline{z}(x) \leq a \leq \underline{z}(x) - \varepsilon$, a contradiction. The upper semi-continuity of $\overline{z}$ can be proved in the same way.

The difference $\overline{z}(x) - \underline{z}(x)$ equals $\lambda_1(G_x \cap K)$. For $x \in K|H$, we put

$$s(x) := \left\{ x + zu : -\frac{1}{2}(\overline{z}(x) - \underline{z}(x)) \leq z \leq \frac{1}{2}(\overline{z}(x) - \underline{z}(x)) \right\}$$

and then define

$$S_H(K) := \bigcup_{x \in K|H} s(x).$$

Clearly, the definition is independent of the orientation of $u \perp H$. In the following lemma, we collect some basic properties of $S_H(K)$.

Lemma 3.11 *If $K \in \mathcal{K}^n$ and $H =$ is a hyperplane, then*

(a) $S_H(K) \in \mathcal{K}^n$,
(b) $V(S_H(K)) = V(K)$, *and*
(c) $S_H(K)$ *is symmetric with respect to H.*

Proof The symmetry assertion is clear by definition, and the volume does not change by Fubini's theorem. Furthermore, $S_H(K)$ is convex, since $\overline{z} - \underline{z}$ is concave and bounded. It remains to be shown that $S_H(K)$ is closed. Let $(y_i)_{i \in \mathbb{N}} \subset S_H(K)$ with $y_i \to y$ as $i \to \infty$. There are $x_i \in K|H$ and $z_i \in \mathbb{R}$, $i \in \mathbb{N}$, such that $y_i = x_i + z_i u$. Let $x_0 \in H$ be an arbitrary fixed point. Since $z_i = \langle y_i - x_0, u \rangle \to \langle y - x_0, u \rangle =: z$, we get $x_i = y_i - z_i u \to y - zu =: x \in H$, that is, $y = x + zu$. Since $x_i \in K|H$ and $K|H$ is closed, we have $x \in K|H$. Furthermore,

$$2|z_i| \le \overline{z}(x_i) - \underline{z}(x_i),$$

hence

$$2|z| = \lim_{i \to \infty} 2|z_i| \le \limsup_{i \to \infty} \overline{z}(x_i) - \liminf_{i \to \infty} \underline{z}(x_i) \le \overline{z}(x) - \underline{z}(x),$$

which shows that $y = x + zu \in S_H(K)$.   □

Definition 3.8 For $K \in \mathcal{K}^n$ and a hyperplane $H \subset \mathbb{R}^n$, the convex body $S_H(K)$ is called the *Steiner symmetral* of K with respect to H. The map $S_H: \mathcal{K}^n \to \mathcal{K}^n$, $K \mapsto S_H(K)$, is called *Steiner symmetrization*.

The map $S_H: \mathcal{K}^n \to \mathcal{K}^n$ is *not* continuous in general (see the exercises). However, the following weaker property is true. In the proof, we write $\sigma_H: \mathbb{R}^n \to \mathbb{R}^n$ for the orthogonal reflection at H.

Lemma 3.12 *Let $(K_j)_{j \in \mathbb{N}}$ be a sequence in $\mathcal{K}^n$ with $K_j \to K$ and such that $S_H(K_j) \to \widetilde{K}$ as $j \to \infty$. Then $\widetilde{K} \subset S_H(K)$.*

Proof Let $x \in \widetilde{K}$. There are $x_j \in S_H(K_j)$ with $x_j \to x$ as $j \to \infty$. This implies that there are $\underline{y}_j, \overline{y}_j \in G_{x_j} \cap K_j$ such that

$$\|x_j - \sigma_H(x_j)\| \le \|\overline{y}_j - \underline{y}_j\|.$$

Passing to a subsequence, we can assume that $\underline{y}_j \to \underline{y}$ and $\overline{y}_j \to \overline{y}$ as $j \to \infty$. Then $\underline{y}, \overline{y} \in G_x \cap K$ and

$$\|x - \sigma_H(x)\| \le \|\overline{y} - \underline{y}\|,$$

which shows that $x \in S_H(K)$. $\qquad\square$

Let $H_1, \ldots, H_k$ be hyperplanes passing through the origin. Then the composition map $S_{H_m} \circ \cdots \circ S_{H_1}$ is referred to as a *repeated* (or *iterated*) Steiner symmetrization. For $K \in \mathcal{K}^n$, let $\mathscr{S}(K)$ denote the set of all convex bodies obtained from K by repeated Steiner symmetrization.

Theorem 3.19 *For $K \in \mathcal{K}^n$, there is a sequence in $\mathscr{S}(K)$ converging to a ball.*

Proof Let

$$R(L) := \min\{r \ge 0 : L \subset rB^n\}, \quad L \in \mathcal{K}^n,$$

and

$$R_0 := \inf\{R(K') : K' \in \mathscr{S}(K)\}.$$

Since $\mathscr{S}(K)$ is bounded, it follows from Blaschke's selection theorem that there is a sequence $(K_j)_{j \in \mathbb{N}}$ in $\mathscr{S}(K)$ with

$$\lim_{j \to \infty} R(K_j) = R_0, \quad K_j \to K_0 \in \mathcal{K}^n.$$

Since R is continuous, we have $R_0 = R(K_0)$. Assume K_0 is not the ball B_0 centered at 0 of radius R_0. Then there is a point $z \in \mathrm{bd}(B_0) \setminus K_0$. Hence there is a ball C with center z and $C \cap K_0 = \emptyset$. For any hyperplane H through 0,

$$\mathrm{bd}\, B_0 \cap C \cap S_H(K_0) = \emptyset = \mathrm{bd}\, B_0 \cap \sigma_H(C) \cap S_H(K_0).$$

Let $C_1, \ldots, C_m$ be balls congruent to C with centers in $\mathrm{bd}\, B_0$ which cover $\mathrm{bd}\, B_0$. Let H_i be a hyperplane through 0 with respect to which C and C_i are symmetric. Then

$$\mathrm{bd}\, B_0 \cap C \cap S_{H_1}(K_0) = \emptyset = \mathrm{bd}\, B_0 \cap C_1 \cap S_{H_1}(K_0).$$

Then also

$$\mathrm{bd}\, B_0 \cap C \cap S_{H_2} \circ S_{H_1}(K_0) = \emptyset = \mathrm{bd}\, B_0 \cap C_1 \cap S_{H_2} \circ S_{H_1}(K_0)$$
$$= \mathrm{bd}\, B_0 \cap \underbrace{\sigma_{H_2}(C)}_{=C_2} \cap S_{H_2} \circ S_{H_1}(K_0).$$

For $S^* := S_{H_m} \circ \cdots \circ S_{H_1}$, iteration yields

$$\mathrm{bd}\, B_0 \cap C_i \cap S^*(K_0) = \emptyset \qquad \text{for } i = 1, \ldots, m,$$

that is,

$$\mathrm{bd}\, B_0 \cap S^*(K_0) = \emptyset,$$

since $C_1 \cup \cdots \cup C_m \supset \mathrm{bd}\, B_0$. Since $S^*(K_0)$ is compact, it follows that $R(S^*(K_0)) < R_0$. We can assume that $S_{H_r} \circ \cdots \circ S_{H_1}(K_j) \to \widetilde{K}_r$ for $r = 1, \ldots, m$. Then we deduce from Lemma 3.12 that

$$S_{H_1}(K_j) \to \widetilde{K}_1 \Rightarrow \widetilde{K}_1 \subset S_{H_1}(K_0)$$

$$\underbrace{S_{H_2} \circ S_{H_1}(K_j)}_{=S_{H_2}(S_{H_1}(K_j))} \to \widetilde{K}_2 \Rightarrow \widetilde{K}_2 \subset S_{H_2}(\widetilde{K}_1) \subset S_{H_2} \circ S_{H_1}(K_0)$$

$$\vdots$$

$$\underbrace{S_{H_m} \circ \cdots \circ S_{H_1}(K_j)}_{=S_{H_m}(S_{H_{m-1}} \circ \cdots \circ S_{H_1}(K_j))} \to \widetilde{K}_m \Rightarrow \widetilde{K}_m \subset S_{H_m}(\widetilde{K}_{m-1})$$

and

$$S_{H_m}(\widetilde{K}_{m-1}) \subset S_{H_m}(S_{H_{m-1}} \circ \cdots \circ S_{H_1}(K_0)) = S^*(K_0).$$

Thus we get $R(\widetilde{K}_m) \leq R(S^*(K_0)) < R_0$. Since R is continuous and since we have $S_{H_m} \circ \cdots \circ S_{H_1}(K_j) \to \widetilde{K}_m$, we have $R(S_{H_m} \circ \cdots \circ S_{H_1}(K_j)) < R_0$ if j is sufficiently large. This is a contradiction, since $S_{H_m} \circ \cdots \circ S_{H_1}(K_j) \in \mathscr{S}(K)$, which proves the theorem. $\qquad\qquad\square$

Before continuing with some examples of typical applications of Steiner symmetrization, we will collect some of its further useful properties. The obvious properties (a) and (b) have already been used in the preceding argument.

Lemma 3.13 *Let $K, L \in \mathcal{K}^n$ and let H be a hyperplane. Then the following assertions hold.*

(a) *If $K \subset L$, then $S_H(K) \subset S_H(L)$.*
(b) *$S_H(S_H(K)) = S_H(K)$.*
(c) *$\mathrm{diam}(S_H(K)) \leq \mathrm{diam}(K)$.*
(d) *If $0 \in H$ and $\lambda \in \mathbb{R}$, then $S_H(\lambda K) = \lambda S_H(K)$.*
(e) *If $0 \in H$, then $S_H(K) + S_H(L) \subset S_H(K + L)$.*
(f) *$F(S_H(K)) \leq S_H(K)$.*

Proof The assertions (a), (b), (d) are easy to see.

(c) Let $y_i = x_i + z_i u \in S_H(K)$ for $i = 1, 2$. Then $\underline{y}_i = x_i + \underline{z}_i u$ and $\overline{y}_i = x_i + \overline{z}_i u$ satisfy $\overline{z}_i - \underline{z}_i \geq 2|z_i|$ for $i = 1, 2$. Then, we get $\|y_2 - y_1\|^2 = \|x_2 - x_1\|^2 + (z_2 - z_1)^2$ and

$$\|\overline{y}_2 - \underline{y}_1\|^2 = \|x_2 - x_1\|^2 + (\overline{z}_2 - \underline{z}_1)^2, \quad \|\overline{y}_1 - \underline{y}_2\|^2 = \|x_2 - x_1\|^2 + (\overline{z}_1 - \underline{z}_2)^2.$$

Since

$$(z_2 - z_1)^2 \leq |z_1| + |z_2| \leq \frac{1}{2}(\overline{z}_1 - \underline{z}_1) + \frac{1}{2}(\overline{z}_2 - \underline{z}_2) \leq \frac{1}{2}|\overline{z}_2 - \underline{z}_1| + \frac{1}{2}|\overline{z}_1 - \underline{z}_2|$$

$$\leq \max\{|\overline{z}_2 - \underline{z}_1|, |\overline{z}_1 - \underline{z}_2|\},$$

we get

$$\|y_2 - y_1\| \leq \max\{\|\overline{y}_2 - \underline{y}_1\|, \|\overline{y}_1 - \underline{y}_2\|\}.$$

Hence, it follows that $\operatorname{diam}(S_H(K)) \leq \operatorname{diam}(K)$.

(e) Let $H = H(u, 0)$ with $u \in \mathbb{S}^{n-1}$. We have to show that if $y_1 \in S_H(K)$ and $y_2 \in S_H(L)$, then $y_1 + y_2 \in S_H(K + L)$. We have $y_i = x_i + z_i u$ with $|z_1| \leq \frac{1}{2}\lambda_1(G_{x_1} \cap K)$ and $|z_2| \leq \frac{1}{2}\lambda_1(G_{x_2} \cap L)$. Since $(G_{x_1} \cap K) + (G_{x_2} \cap L) \subset G_{x_1+x_2} \cap (K + L)$, we obtain

$$|z_1 + z_2| \leq |z_1| + |z_2| \leq \frac{1}{2}(\lambda_1(G_{x_1} \cap K) + \lambda_1(G_{x_2} \cap L))$$

$$= \frac{1}{2}\lambda_1((G_{x_1} \cap K) + (G_{x_2} \cap L)) \leq \frac{1}{2}\lambda_1(G_{x_1+x_2} \cap (K + L)),$$

which yields the assertion.

(f) Using (e), we get

$$F(S_H(K)) = \lim_{t \downarrow 0} \frac{1}{t}(V(S_H(K) + tB^n) - V(S_H(K)))$$

$$= \lim_{t \downarrow 0} \frac{1}{t}(V(S_H(K) + S_H(tB^n)) - V(S_H(K)))$$

$$\leq \lim_{t \downarrow 0} \frac{1}{t}(V(S_H(K + tB^n)) - V(S_H(K)))$$

$$= \lim_{t \downarrow 0} \frac{1}{t}(VK + tB^n) - V(K)) = F(K),$$

as asserted. $\qquad\square$

Recall that $\mathcal{K}_0^n$ denotes the set of convex bodies having nonempty interiors.

Example 3.2 (Isoperimetric Inequality) Let $K \in \mathcal{K}_0^n$. Then the *isoperimetric inequality* is obtained as described before. The crucial fact that surface area does not increase under Steiner symmetrization is established in Lemma 3.13 (f). However, note that the sequence $K_m \to rB^n$, where K_m is an iterated Steiner symmetral of K, need not be of the form

$$K_m = S_{H_m} \circ \cdots \circ S_{H_1}(K)$$

for a single sequence $H_1, H_2, \ldots$ of hyperplanes through 0. (Although it can be shown that such a single sequence does indeed exist.) The argument shows that the isoperimetric problem has a solution (in the class of convex bodies) and the balls are extremal bodies, but it does not imply that balls are the unique maximizers.

Example 3.3 (Brunn–Minkowski Inequality) Let $K, L \in \mathcal{K}^n$. By Lemma 3.13 (e) we get

$$V(K + L) = V(S_H(K + L)) \geq V(S_H(K) + S_H(L)). \tag{3.31}$$

Choose a sequence $K_i \to r(K)B^n$ of iterated symmetrals of K. Then $V(K) = V(K_i) = V(r(K)B^n)$. Passing to a subsequence we can also assume that the corresponding sequence of iterated symmetrals L_i of L converges, say to $\widetilde{L} \in \mathcal{K}^n$. Then $V(L) = V(L_i) = V(\widetilde{L})$. Hence we get from (3.31)

$$V(K + L) \geq V(K_i + L_i) \to V(r(K)B^n + \widetilde{L}).$$

Now we choose a sequence of Steiner symmetrals $\widetilde{L}_i$ of $\widetilde{L}$ with $\widetilde{L}_i \to r(L)B^n$ and clearly $V(\widetilde{L}_i) = V(\widetilde{L}) = V(L) = V(r(L)B^n)$. Hence

$$\begin{aligned}
V(r(K)B^n + \widetilde{L}) &= V(S_H(r(K)B^n + \widetilde{L})) \geq V(S_H(r(K)B^n)) + S_H(\widetilde{L}) \\
&= V(r(K)B^n + \widetilde{L}_1) \geq \cdots \geq V(r(K)B^n + \widetilde{L}_i) \\
&\to V(r(K)B^n + r(L)B^n).
\end{aligned}$$

This finally yields that

$$\begin{aligned}
V(K + L)^{\frac{1}{n}} &\geq (r(K) + r(L))V(B^n)^{\frac{1}{n}} \\
&= V(r(K)B^n)^{\frac{1}{n}} + V(r(L)B^n)^{\frac{1}{n}} \\
&= V(K)^{\frac{1}{n}} + V(L)^{\frac{1}{n}}.
\end{aligned}$$

Example 3.4 (Isodiametric Inequality) By similar arguments as above and using Lemma 3.13 (c), we deduce that

$$\frac{V(K)}{(\operatorname{diam}(K))^n} \le \frac{V(S_H(K))}{(\operatorname{diam}(S_H(K)))^n} \le \cdots$$

$$\cdots \le \frac{V(K_i)}{(\operatorname{diam}(K_i))^n} \to \frac{V(r(K)B^n)}{(\operatorname{diam}(r(K)B^n))^n} = \frac{V(B^n)}{(\operatorname{diam}(B^n))},$$

and therefore

$$\frac{V(K)}{(\operatorname{diam}(K))^n} \le \frac{\kappa_n}{2^n}.$$

Although it is appealing that a single method (the Steiner symmetrization) is successful in these three examples, this is not surprising in view of the fact that the isoperimetric inequality and the isodiametric inequality both follow from Minkowski's inequality, which in turn is a consequence (in fact, it is equivalent in our setting) to the Brunn–Minkowski inequality. The following application of Steiner symmetrization is even more convincing, since the inequality is new for us and symmetrization is indeed a highly efficient tool for establishing it.

Definition 3.9 A real-valued functional $f \colon \mathcal{K}_0^n \to \mathbb{R}$ is called *affine invariant* if $f(\alpha K) = f(K)$ for $K \in \mathcal{K}^n$ and all affinities (regular affine transformations) $\alpha \colon \mathbb{R}^n \to \mathbb{R}^n$.

Theorem 3.20 *Let $f \colon \mathcal{K}_0^n \to \mathbb{R}$ be continuous and affine invariant. Then f has a maximum and a minimum.*

Proof See Exercise 3.2.5. □

Let $\mathcal{K}_c^n$ denote the set of all centrally symmetric convex bodies with center at the origin, that is,

$$\mathcal{K}_c^n := \{K \in \mathcal{K}_0^n : K = -K\}.$$

The volume product of $K \in \mathcal{K}_c^n$ is defined by

$$\operatorname{vp}(K) := V_n(K) V_n(K^\circ),$$

where $K^\circ = \{x \in \mathbb{R}^n : \langle x, z \rangle \le 1 \text{ for } z \in K\}$ is the polar body of K (recall Exercises 1.1.14 and 2.3.2). For a regular linear map $\alpha \in \mathrm{GL}(n)$ and $K \in \mathcal{K}_c^n$, we have $(\alpha K)^\circ = \alpha^{-\top}(K^\circ)$, where $\alpha^{-\top}$ is the inverse of the adjoint map of α. Thus it follows that

$$\mathrm{vp}(\alpha K) = V(\alpha K)V((\alpha K)^\circ) = |\det\alpha|V(K)\cdot V_n(\alpha^{-\top}(K^\circ))$$

$$= |\det\alpha|V(K)\frac{1}{|\det\alpha|}V(K^\circ) = V(K)V(K^\circ)$$

$$= \mathrm{vp}(K).$$

The map $K \mapsto \mathrm{vp}(K)$ is continuous on $\mathcal{K}_c^n$. In the following theorem, the maximum of vp is determined. The inequality thus obtained is known as the Blaschke–Santaló inequality (for centrally symmetric convex bodies). There is also a version of this inequality for not necessarily centrally symmetric bodies. The determination of the minimum is a notoriously difficult open problem (also in the nonsymmetric case) known as K. Mahler's problem (conjecture). It is conjectured that the minimum is attained for the cube (and the cross-polytope), but not only for these.

Theorem 3.21 (Blaschke–Santaló Inequality) *If $K \in \mathcal{K}_c^n$ is an origin-centred convex body, then $\mathrm{vp}(K) \leq \mathrm{vp}(B^n)$.*

Proof Let $H = H(u, 0)$ be a hyperplane with $u \in \mathbb{S}^{n-1}$. We show that $V(K^\circ) \leq V(S_H(K)^\circ)$. Once this is shown, it follows that $\mathrm{vp}(K) \leq \mathrm{vp}(S_H(K))$. Choosing a sequence $(K_j) \subset \mathscr{S}(K)$ with $K_j \to r(K)B^n$, the assertion follows.

Each $x \in \mathbb{R}^n$ can be written in the form $x = y + tu$ with $y \in H$ and $t \in \mathbb{R}$, in which case we write $x = (y, t)$. Then

$$(y, s) \in S_H(K) \iff s = \frac{s_1 - s_2}{2} \text{ for some } (y, s_1), (y, s_2) \in K.$$

Therefore, we have

$$(z, t) \in S_H(K)^\circ \iff \langle (z, t), (y, s) \rangle \leq 1 \quad \text{for } (y, s) \in S_H(K)$$

$$\iff \langle z, y \rangle + ts \leq 1 \quad \text{for } (y, s_j) \in K, s = \frac{s_1 - s_2}{2}, j = 1, 2,$$

$$\iff \langle z, y \rangle + t\frac{s_1 - s_2}{2} \leq 1 \quad \text{for } (y, s_j) \in K, j = 1, 2.$$

For $E \subset \mathbb{R}^n$ and $t \in \mathbb{R}$, we define the t-section of E as the subset of H which is given by $E_t := \{y \in H : (y, t) \in E\}$. Then $K^\circ = -K^\circ$ implies that $-(K^\circ)_t = (K^\circ)_{-t}$. Next we show that

$$\frac{1}{2}(K^\circ)_t + \frac{1}{2}(-(K^\circ)_t) \subset (S_H(K)^\circ)_t.$$

To verify this, first observe that

$$z \in \frac{1}{2}(K^{\circ})_t + \frac{1}{2}(-(K^{\circ})_t) = \frac{1}{2}(K^0)_t + \frac{1}{2}(K^0)_{-t}$$

can be written in the form $z = \frac{1}{2}p + \frac{1}{2}q$ with $p \in (K^0)_t, q \in (K^0)_{-t}$. Thus,

$$\langle z, y \rangle + t \frac{s_1 - s_2}{2} = \left\langle \frac{p+q}{2}, y \right\rangle + t \frac{s_1 - s_2}{2}$$

$$= \frac{1}{2}(\langle p, y \rangle + t s_1) + \frac{1}{2}(\langle q, y \rangle + (-t)s_2)$$

$$= \frac{1}{2}\langle(p, t), (y, s_1)\rangle + \frac{1}{2}\langle(q, -t), (y, s_2)\rangle \le 1$$

since $(p, t) \in K^{\circ}$, $(y, s_1) \in K$ and $(q, -t) \in K^{\circ}$, $(y, s_2) \in K$, respectively. This shows that $(z, t) \in S_H(K)^{\circ}$, that is, $z \in (S_H(K)^{\circ})_t$.

By the Brunn–Minkowski inequality, we then obtain

$$V_{n-1}((K^{\circ})_t) \le V_{n-1}\left(\frac{1}{2}(K^{\circ})_t + \frac{1}{2}(-(K^{\circ})_t))\right) \le V_{n-1}((S_H(K)^{\circ})_t).$$

Integration with respect to $t \in \mathbb{R}$ and Fubini's theorem yield

$$V_n(K^{\circ}) \le V_n(S_H(K)^{\circ}),$$

as requested.   □

Exercises and Supplements for Sect. 3.6

1. Show by an example that the Steiner symmetrization is not a continuous transformation in general.
2. Let $K_i, K \in \mathcal{K}^n$, $i \in \mathbb{N}$, with int $K \ne \emptyset$. Let H be a hyperplane. Show that $K_i \to K$ as $i \to \infty$ implies that $S_H(K_i) \to S_H(K)$.
3. Hints to the literature: Classical textbooks dealing with various aspects of convexity are [11, 15, 18, 25, 30, 32, 42, 50, 57, 64, 92]. More recent expositions are provided in [10, 24, 38, 58, 59, 81, 88, 91, 93]. For connections to algebraic aspects of convexity, including algebraic geometry, see [31, 49].

Chapter 4
From Area Measures to Valuations

In Chap. 3, we studied the mixed volume as a real-valued functional on n-tuples of convex bodies in $\mathbb{R}^n$. Already the recursive definition of the mixed volume of polytopes $P_1, \ldots, P_n$ involves the support function of one of the bodies, say P_n, and mixed functionals of the facets with the same exterior unit normal vector of the remaining bodies $P_1, \ldots, P_{n-1}$. This defining relation will lead to a fundamental and general relation between the mixed volume of convex bodies $K_1, \ldots, K_n$, the support function of one of the bodies, say K_n, and (what we shall call) the mixed area measure of the remaining convex bodies $K_1, \ldots, K_{n-1}$. Specializing these mixed area measures, we shall obtain the area measures $S_j(K, \cdot)$, $j \in \{0, \ldots, n-1\}$, of a single convex body K.

The area measures can also be introduced independently as coefficients of a local Steiner formula. This leads to explicit integral representations of these measures as integrals of elementary symmetric functions of principal radii of curvature. A particular role is played by the top order measure $S_{n-1}(K, \cdot)$, which is just the $(n-1)$-dimensional Hausdorff measure of the set of all boundary points of K having an exterior unit normal in a given Borel subset of the unit sphere. Minkowski's existence and uniqueness theorem provides a deep and extremely useful characterization of top order area measures. The top order area measure of a convex body K can also be used to provide an analytic expression for the volume of the projection of K to a subspace orthogonal to a given direction $u \in \mathbb{S}^{n-1}$. The function which assigns to a given direction the projection volume of a given convex body turns out to be the support function of a convex body ΠK. The class of convex bodies thus obtained are precisely the zonoids, which show up in various contexts and are explored in the current chapter as well.

Finally, we close this chapter with an introductory study of valuations. In fact, most of the functionals (such as the intrinsic volumes or the support function) we have seen so far enjoy an important finite additivity property, which is the defining property of a valuation. We have also seen the diameter, which does not enjoy the additivity property. A particular highlight in the context of valuation theory is

© Springer Nature Switzerland AG 2020
D. Hug, W. Weil, *Lectures on Convex Geometry*, Graduate Texts
in Mathematics 286, https://doi.org/10.1007/978-3-030-50180-8_4

Hadwiger's characterization theorem. Applications of this fundamental result to the proof of integral-geometric formulas will be given in the following chapter.

4.1 Mixed Area Measures

In Sect. 3.3, we have shown that for polytopes $P_1, \ldots, P_n \in \mathcal{P}^n$ the mixed volume satisfies the formula

$$V(P_1, \ldots, P_{n-1}, P_n) = \frac{1}{n} \sum_{u \in \mathbb{S}^{n-1}} h_{P_n}(u) v(P_1(u), \ldots, P_{n-1}(u)).$$

On the right-hand side, the summation extends over all unit vectors u for which $v(P_1(u), \ldots, P_{n-1}(u)) > 0$ or, alternatively, over all facet normals of the polytope $P_1 + \cdots + P_{n-1}$. In any case, the summation is independent of P_n. By approximation and using the continuity of mixed volumes and support functions, we therefore get the same formula for an arbitrary convex body $K_n \in \mathcal{K}^n$, that is,

$$V(P_1, \ldots, P_{n-1}, K_n) = \frac{1}{n} \sum_{u \in \mathbb{S}^{n-1}} h_{K_n}(u) v(P_1(u), \ldots, P_{n-1}(u)). \tag{4.1}$$

We define

$$S(P_1, \ldots, P_{n-1}, \cdot) := \sum_{u \in \mathbb{S}^{n-1}} v(P_1(u), \ldots, P_{n-1}(u)) \delta_u, \tag{4.2}$$

where δ_u denotes the Dirac measure with unit point mass in $u \in \mathbb{S}^{n-1}$, that is,

$$\delta_u(A) := \begin{cases} 1, & \text{if } u \in A, \\ 0, & \text{if } u \notin A, \end{cases}$$

for Borel sets $A \subset \mathbb{S}^{n-1}$. Then, $S(P_1, \ldots, P_{n-1}, \cdot)$ is a finite Borel measure on the unit sphere $\mathbb{S}^{n-1}$, which is called the *mixed surface area measure* or simply mixed area measure of the polytopes $P_1, \ldots, P_{n-1}$. Equation (4.1) is then equivalent to

$$V(P_1, \ldots, P_{n-1}, K_n) = \frac{1}{n} \int_{\mathbb{S}^{n-1}} h_{K_n}(u) \, S(P_1, \ldots, P_{n-1}, du). \tag{4.3}$$

Our next goal is to extend this integral representation to arbitrary convex bodies $K_1, \ldots, K_{n-1}$ and thus to define mixed (surface) area measures for general convex bodies.

We first need an auxiliary result.

Lemma 4.1 *If $K_1, \ldots, K_{n-1}, K_n, K_n' \in \mathcal{K}^n$ are arbitrary convex bodies, then*

$$|V(K_1, \ldots, K_{n-1}, K_n) - V(K_1, \ldots, K_{n-1}, K_n')|$$
$$\leq \|h_{K_n} - h_{K_n'}\| V(K_1, \ldots, K_{n-1}, B^n).$$

Proof First, let $K_1, \ldots, K_{n-1}$ be polytopes. Since $h_{B^n} \equiv 1$ on $\mathbb{S}^{n-1}$, we obtain from (4.1) that

$$|V(K_1, \ldots, K_{n-1}, K_n) - V(K_1, \ldots, K_{n-1}, K_n')|$$

$$= \frac{1}{n} \left| \sum_{u \in \mathbb{S}^{n-1}} (h_{K_n}(u) - h_{K_n'}(u)) v(K_1(u), \ldots, K_{n-1}(u)) \right|$$

$$\leq \frac{1}{n} \sum_{u \in \mathbb{S}^{n-1}} |h_{K_n}(u) - h_{K_n'}(u)| v(K_1(u), \ldots, K_{n-1}(u))$$

$$\leq \frac{1}{n} \sup_{v \in \mathbb{S}^{n-1}} |h_{K_n}(v) - h_{K_n'}(v)| \sum_{u \in \mathbb{S}^{n-1}} v(K_1(u), \ldots, K_{n-1}(u))$$

$$= \frac{1}{n} \|h_{K_n} - h_{K_n'}\| \sum_{u \in \mathbb{S}^{n-1}} h_{B^n}(u) v(K_1(u), \ldots, K_{n-1}(u))$$

$$= \|h_{K_n} - h_{K_n'}\| V(K_1, \ldots, K_{n-1}, B^n).$$

By Theorem 3.9, which provides the continuity of the mixed volume, the inequality extends to arbitrary convex bodies. $\qquad\qquad\qquad\qquad\qquad\qquad\qquad\qquad\square$

Now we can extend (4.3) to arbitrary convex bodies.

Theorem 4.1 *For $K_1, \ldots, K_{n-1} \in \mathcal{K}^n$, there exists a uniquely determined finite Borel measure $S(K_1, \ldots, K_{n-1}, \cdot)$ on $\mathbb{S}^{n-1}$ such that*

$$V(K_1, \ldots, K_{n-1}, K) = \frac{1}{n} \int_{\mathbb{S}^{n-1}} h_K(u)\, S(K_1, \ldots, K_{n-1}, du)$$

for $K \in \mathcal{K}^n$.

Proof We consider the Banach space $\mathbf{C}(\mathbb{S}^{n-1})$ and the linear subspace $\mathbf{C}^2(\mathbb{S}^{n-1})$ of twice continuously differentiable functions. Here, a function f on $\mathbb{S}^{n-1}$ is called twice continuously differentiable if the homogeneous extension $\tilde{f}$ of f, which is given by

$$\tilde{f}(x) := \begin{cases} \|x\| f\left(\frac{x}{\|x\|}\right), & x \in \mathbb{R}^n \setminus \{0\}, \\ 0, & x = 0, \end{cases}$$

is twice continuously differentiable on $\mathbb{R}^n \setminus \{0\}$. From analysis we use the fact that the subspace $\mathbf{C}^2(\mathbb{S}^{n-1})$ is dense in $\mathbf{C}(\mathbb{S}^{n-1})$, that is, for each $f \in \mathbf{C}(\mathbb{S}^{n-1})$ there is a sequence of functions $f_i \in \mathbf{C}^2(\mathbb{S}^{n-1})$ with $f_i \to f$ as $i \to \infty$, in the maximum norm (this can be proved either by a convolution argument or by using a result of Stone–Weierstrass type; see Exercise 4.1.1).

Further, we consider the set $\mathcal{L}^n$ of all functions $f \in \mathbf{C}(\mathbb{S}^{n-1})$ which have a representation $f = h_K - h_{K'}$ with convex bodies $K, K' \in \mathcal{K}^n$. Obviously, $\mathcal{L}^n$ is also a linear subspace. Exercise 2.3.1 shows that $\mathbf{C}^2(\mathbb{S}^{n-1}) \subset \mathcal{L}^n$, therefore $\mathcal{L}^n$ is dense in $\mathbf{C}(\mathbb{S}^{n-1})$.

We now define a functional $T_{K_1,\dots,K_{n-1}}$ on $\mathcal{L}^n$ by

$$T_{K_1,\dots,K_{n-1}}(f) := nV(K_1, \dots, K_{n-1}, K) - nV(K_1, \dots, K_{n-1}, K'),$$

where $f = h_K - h_{K'}$. This definition is actually independent of the particular representation of f. Namely, if $f = h_K - h_{K'} = h_L - h_{L'}$, then $K + L' = K' + L$ and hence

$$V(K_1, \dots, K_{n-1}, K) + V(K_1, \dots, K_{n-1}, L')$$
$$= V(K_1, \dots, K_{n-1}, K') + V(K_1, \dots, K_{n-1}, L),$$

by the multilinearity of mixed volumes. This yields

$$nV(K_1, \dots, K_{n-1}, K) - nV(K_1, \dots, K_{n-1}, K')$$
$$= nV(K_1, \dots, K_{n-1}, L) - nV(K_1, \dots, K_{n-1}, L').$$

The argument just given also implies that $T_{K_1,\dots,K_{n-1}}$ is linear. Moreover, $T_{K_1,\dots,K_{n-1}}$ is a positive functional since $f = h_K - h_{K'} \geq 0$ implies $K \supset K'$. Hence

$$V(K_1, \dots, K_{n-1}, K) \geq V(K_1, \dots, K_{n-1}, K')$$

and therefore $T_{K_1,\dots,K_{n-1}}(f) \geq 0$. Finally, $T_{K_1,\dots,K_{n-1}}$ is continuous (with respect to the maximum norm), since Lemma 4.1 shows that

$$|T_{K_1,\dots,K_{n-1}}(f)| \leq c(K_1, \dots, K_{n-1})\|f\|$$

with $c(K_1, \dots, K_{n-1}) := nV(K_1, \dots, K_{n-1}, B^n)$.

Since $\mathcal{L}^n$ is dense in $\mathbf{C}(\mathbb{S}^{n-1})$, the inequality just proven (or alternatively, the Hahn–Banach theorem) implies that there is a unique continuous extension of $T_{K_1,\dots,K_{n-1}}$ to a positive linear functional on $\mathbf{C}(\mathbb{S}^{n-1})$ (see Exercise 4.1.2). To verify that $T_{K_1,\dots,K_{n-1}}$ is indeed positive, let $f \in \mathbf{C}(\mathbb{S}^{n-1})$ be nonnegative. There is a sequence $f_i = h_{K_i} - h_{L_i} \in \mathcal{L}^n$, $i \in \mathbb{N}$, such that $\|f_i - f\| \leq 1/i$ for $i \in \mathbb{N}$. Then $\bar{f}_i := f_i + \frac{1}{i} \geq 0$ and $\bar{f}_i = h_{K_i + \frac{1}{i}B^n} - h_{L_i} \in \mathcal{L}^n$. Moreover, $\|\bar{f}_i - f\| \to 0$ as $i \to \infty$. Thus we conclude that $T_{K_1,\dots,K_{n-1}}(f) = \lim_{i \to \infty} T_{K_1,\dots,K_{n-1}}(\bar{f}_i) \geq 0$.

The Riesz representation theorem (see [79, Theorem 2.14]) then shows that there is a finite (nonnegative) Borel measure $S(K_1, \ldots, K_{n-1}, \cdot)$ on $\mathbb{S}^{n-1}$ such that

$$T_{K_1, \ldots, K_{n-1}}(f) = \int_{\mathbb{S}^{n-1}} f(u)\, S(K_1, \ldots, K_{n-1}, du)$$

for $f \in \mathbf{C}(\mathbb{S}^{n-1})$. The existence assertion of the theorem now follows, if we put $f = h_K$.

For the uniqueness part, let $K_1, \ldots, K_{n-1} \in \mathcal{K}^n$ be given and let μ, μ' be two Borel measures on $\mathbb{S}^{n-1}$, depending on $K_1, \ldots, K_{n-1}$, such that

$$\int_{\mathbb{S}^{n-1}} h_K(u)\, \mu(du) = \int_{\mathbb{S}^{n-1}} h_K(u)\, \mu'(du)$$

for $K \in \mathcal{K}^n$. By linearity, we get

$$\int_{\mathbb{S}^{n-1}} f(u)\, \mu(du) = \int_{\mathbb{S}^{n-1}} f(u)\, \mu'(du),$$

first for $f \in \mathcal{L}^n$, and then for $f \in \mathbf{C}(\mathbb{S}^{n-1})$. The uniqueness assertion in the Riesz representation theorem then implies that $\mu = \mu'$. $\qquad\square$

Definition 4.1 The measure $S(K_1, \ldots, K_{n-1}, \cdot)$ is called the *mixed surface area measure* or simply *mixed area measure* of the bodies $K_1, \ldots, K_{n-1}$. In particular,

$$S_j(K, \cdot) := S(\underbrace{K, \ldots, K}_{j}, \underbrace{B^n, \ldots, B^n}_{n-1-j}, \cdot)$$

is called the jth order *surface area measure* or simply jth *area measure* of K for $j \in \{0, \ldots, n-1\}$.

Remark 4.1 For polytopes $K_1, \ldots, K_{n-1} \in \mathcal{P}^n$, the mixed area measure $S(K_1, \ldots, K_{n-1}, \cdot)$ equals the measure defined in (4.2).

Remark 4.2 All area measures have centroid 0. In fact, since

$$V(K_1, \ldots, K_{n-1}, \{x\}) = 0,$$

we have

$$\int_{\mathbb{S}^{n-1}} \langle x, u \rangle\, S(K_1, \ldots, K_{n-1}, du) = 0$$

for $x \in \mathbb{R}^n$. It is instructive to interpret this centredness condition for $S_{n-1}(P, \cdot)$ and, in particular, for $n = 2$.

Remark 4.3 For the total mass of the mixed area measure, we get

$$S(K_1, \ldots, K_{n-1}, \mathbb{S}^{n-1}) = nV(K_1, \ldots, K_{n-1}, B^n);$$

in particular,

$$S_j(K, \mathbb{S}^{n-1}) = nV(\underbrace{K, \ldots, K}_{j}, \underbrace{B^n, \ldots, B^n}_{n-j})$$

$$= \frac{n\kappa_{n-j}}{\binom{n}{j}} V_j(K),$$

and

$$S_{n-1}(K, \mathbb{S}^{n-1}) = 2V_{n-1}(K) = F(K),$$

which explains the name surface area measure.

Remark 4.4 For $j \in \{0, \ldots, n-1\}$ and $K \in \mathcal{K}^n$, the measure

$$S_0(K, \cdot) = S(B^n, \ldots, B^n, \cdot) = S_j(B^n, \cdot)$$

equals the spherical Lebesgue measure $\sigma = \mathcal{H}^{n-1}$ on $\mathbb{S}^{n-1}$. This follows from part (d) of the following theorem. Hence we obtain the equation

$$V(K, B^n, \ldots, B^n) = \frac{1}{n} \int_{\mathbb{S}^{n-1}} h_K(u) \, \sigma(du),$$

which we used already at the end of Sect. 3.3.

Further properties of mixed area measures follow, if we combine Theorem 4.1 with Theorem 3.9. In order to formulate a continuity result, we make use of the weak convergence of measures on $\mathbb{S}^{n-1}$ (since $\mathbb{S}^{n-1}$ is compact, weak and vague convergence are the same). A sequence of finite Borel measures μ_i, $i \in \mathbb{N}$, on $\mathbb{S}^{n-1}$ is said to *converge weakly* to a finite Borel measure μ on $\mathbb{S}^{n-1}$ if and only if

$$\int_{\mathbb{S}^{n-1}} f(u) \, \mu_i(du) \to \int_{\mathbb{S}^{n-1}} f(u) \, \mu(du) \quad \text{as } i \to \infty$$

for $f \in \mathbf{C}(\mathbb{S}^{n-1})$.

Theorem 4.2 *The measure-valued mapping*

$$S : (K_1, \ldots, K_{n-1}) \mapsto S(K_1, \ldots, K_{n-1}, \cdot)$$

on $(\mathcal{K}^n)^{n-1}$ has the following properties:

(a) *S is symmetric, that is,*

$$S(K_1, \ldots, K_{n-1}, \cdot) = S(K_{\pi(1)}, \ldots, K_{\pi(n-1)}, \cdot)$$

for $K_1, \ldots, K_{n-1} \in \mathcal{K}^n$ and permutations π of $\{1, \ldots, n-1\}$.

(b) *S is multilinear, that is,*

$$S(\alpha K + \beta L, K_2, \ldots, K_{n-1}, \cdot)$$
$$= \alpha S(K, K_2, \ldots, K_{n-1}, \cdot) + \beta S(L, K_2, \ldots, K_{n-1}, \cdot)$$

for $\alpha, \beta \geq 0$ and $K, L, K_2, \ldots, K_{n-1} \in \mathcal{K}^n$.

(c) *S is translation invariant, that is,*

$$S(K_1 + x_1, \ldots, K_{n-1} + x_{n-1}, \cdot) = S(K_1, \ldots, K_{n-1}, \cdot)$$

for $K_1, \ldots, K_{n-1} \in \mathcal{K}^n$ and $x_1, \ldots, x_{n-1} \in \mathbb{R}^n$.

(d) *S is rotation-covariant, that is,*

$$S(\vartheta K_1, \ldots, \vartheta K_{n-1}, \vartheta A) = S(K_1, \ldots, K_{n-1}, A)$$

for $K_1, \ldots, K_{n-1} \in \mathcal{K}^n$, Borel sets $A \subset \mathbb{S}^{n-1}$, and rotations $\vartheta \in O(n)$.

(e) *S is continuous, that is,*

$$S(K_1^{(m)}, \ldots, K_{n-1}^{(m)}, \cdot) \to S(K_1, \ldots, K_{n-1}, \cdot)$$

weakly, as $m \to \infty$, provided that $K_i^{(m)} \to K_i$ for $i = 1, \ldots, n-1$.

Proof (a)–(c) follow directly from the integral representation and the uniqueness in Theorem 4.1 together with the corresponding properties of mixed volumes in Theorem 3.9.

(d) If $\rho \circ \mu$ denotes the image of a measure μ on $\mathbb{S}^{n-1}$ under the rotation ρ, then

$$\int_{\mathbb{S}^{n-1}} h_{K_n}(u) \, [\vartheta^{-1} \circ S(\vartheta K_1, \ldots, \vartheta K_{n-1}, \cdot)](du)$$

$$= \int_{\mathbb{S}^{n-1}} h_{K_n}(\vartheta^{-1} u) \, S(\vartheta K_1, \ldots, \vartheta K_{n-1}, du)$$

$$= \int_{\mathbb{S}^{n-1}} h_{\vartheta K_n}(u) \, S(\vartheta K_1, \ldots, \vartheta K_{n-1}, du)$$

$$= n V(\vartheta K_1, \ldots, \vartheta K_{n-1}, \vartheta K_n)$$

$$= n V(K_1, \ldots, K_{n-1}, K_n)$$

$$= \int_{\mathbb{S}^{n-1}} h_{K_n}(u) \, S(K_1, \ldots, K_{n-1}, du),$$

where $K_n \in \mathcal{K}^n$ is arbitrary. The assertion now follows from the uniqueness part of Theorem 4.1.

(e) For $\varepsilon > 0$ and $f \in C(\mathbb{S}^{n-1})$, choose $K, L \in \mathcal{K}^n$ with

$$\|f - (h_K - h_L)\| \le \varepsilon.$$

Further, choose $m_0 \in \mathbb{N}$ such that for $m \ge m_0$ we have $K_i^{(m)} \subset K_i + B^n$, for $i = 1, \ldots, n-1$, and

$$|V(K_1^{(m)}, \ldots, K_{n-1}^{(m)}, M) - V(K_1, \ldots, K_{n-1}, M)| \le \varepsilon$$

for $M \in \{K, L\}$ and $m \ge m_0$. Then, by the triangle inequality, we get

$$\left| \int_{\mathbb{S}^{n-1}} f(u)\, S(K_1^{(m)}, \ldots, K_{n-1}^{(m)}, du) - \int_{\mathbb{S}^{n-1}} f(u)\, S(K_1, \ldots, K_{n-1}, du) \right|$$

$$\le \left| \int_{\mathbb{S}^{n-1}} (f - (h_K - h_L))(u)\, S(K_1^{(m)}, \ldots, K_{n-1}^{(m)}, du) \right|$$

$$+ \left| \int_{\mathbb{S}^{n-1}} (h_K - h_L)(u)\, S(K_1^{(m)}, \ldots, K_{n-1}^{(m)}, du) \right.$$

$$\left. - \int_{\mathbb{S}^{n-1}} (h_K - h_L)(u)\, S(K_1, \ldots, K_{n-1}, du) \right|$$

$$+ \left| \int_{\mathbb{S}^{n-1}} (f - (h_K - h_L))(u)\, S(K_1, \ldots, K_{n-1}, du) \right|$$

$$\le \|f - (h_K - h_L)\| n\, V(K_1 + B^n, \ldots, K_{n-1} + B^n, B^n)$$

$$+ n |V(K_1^{(m)}, \ldots, K_{n-1}^{(m)}, K) - V(K_1, \ldots, K_{n-1}, K)|$$

$$+ n |V(K_1^{(m)}, \ldots, K_{n-1}^{(m)}, L) - V(K_1, \ldots, K_{n-1}, L)|$$

$$+ \|f - (h_K - h_L)\| n\, V(K_1, \ldots, K_{n-1}, B^n)$$

$$\le c(K_1, \ldots, K_{n-1})\varepsilon$$

for $m \ge m_0$. $\square$

Recall that $S_{n-1}(K, \cdot) = S(K, \ldots, K, \cdot)$ for $K \in \mathcal{K}^n$. Let $K_1, \ldots, K_m \in \mathcal{K}^n$ and $\alpha_1, \ldots, \alpha_m \ge 0$. Then the multilinearity of mixed area measures and the invariance with respect to permutations of the bodies involved implies that

$$S_{n-1}(\alpha_1 K_1 + \cdots + \alpha_m K_m, \cdot)$$

$$= \sum_{i_1=1}^{m} \cdots \sum_{i_{n-1}=1}^{m} \alpha_{i_1} \cdots \alpha_{i_{n-1}} S(K_{i_1}, \ldots, K_{i_{n-1}}, \cdot) \tag{4.4}$$

$$= \sum_{r_1=0}^{n-1} \cdots \sum_{r_m=0}^{n-1} \binom{n-1}{r_1, \ldots, r_m} \alpha_1^{r_1} \cdots \alpha_m^{r_m} S(K_1[r_1], \ldots, K_m[r_m], \cdot). \tag{4.5}$$

Moreover, this formula can be inverted so that we obtain

$$S(K_1, \ldots, K_{n-1}, \cdot)$$

$$= \frac{1}{(n-1)!} \sum_{k=1}^{n-1} (-1)^{n-1+k} \sum_{1 \le r_1 < \cdots < r_k \le n-1} S_{n-1}(K_{r_1} + \cdots + K_{r_k}, \cdot). \tag{4.6}$$

As a special case of (4.5) and the preceding results, we summarize some of the properties of area measures.

Corollary 4.1 *For* $j = 0, \ldots, n - 1$, *the mapping* $K \mapsto S_j(K, \cdot)$ *on* $\mathcal{K}^n$ *is translation invariant, rotation-covariant and continuous.*
Moreover,

$$S_{n-1}(K + B^n(\alpha), \cdot) = \sum_{j=0}^{n-1} \alpha^{n-1-j} \binom{n-1}{j} S_j(K, \cdot)$$

for $\alpha \ge 0$ *(local Steiner formula).*

Proof We only have to prove the local Steiner formula. The latter follows from Theorem 4.2 (a) and (b). □

Exercises and Supplements for Sect. 4.1

1. State the Stone–Weierstrass approximation theorem and provide a reference. Show that $\mathbf{C}^2(\mathbb{S}^{n-1})$ is a dense subset of $\mathbf{C}(\mathbb{S}^{n-1})$ with respect to the maximum norm.
2. Prove that $T_{K_1,\ldots,K_{n-1}}$ can be continuously extended as a linear functional from $\mathcal{L}^n$ to $\mathbf{C}(\mathbb{S}^{n-1})$.
3. State the Riesz representation theorem in a form which is suitable for the application in the proof of Theorem 4.1.

4. Let $K, M, L \in \mathcal{K}^n$ be such that $K = M + L$. Show that

$$S_j(M, \cdot) = \sum_{i=0}^{j} (-1)^{j-i} \binom{j}{i} S(\underbrace{K, \ldots, K}_{i}, \underbrace{L, \ldots, L}_{j-i}, \underbrace{B^n, \ldots, B^n}_{n-1-j}, \cdot),$$

for $j = 0, \ldots, n - 1$.

5. Prove the inversion formula (polarization formula) (4.6) for the mixed area measures.

6. Let $K \in \mathcal{K}^n, \alpha \geq 0$, and let $r \in \{0, \ldots, n - 1\}$. Then

$$S_r(K + B^n(\alpha), \cdot) = \sum_{l=0}^{r} \alpha^r \binom{n}{l} S_{r-l}(K, \cdot).$$

4.2 An Existence and Uniqueness Result

The interpretation of the surface area measure $S_{n-1}(P, \cdot)$ for a convex polytope $P \subset \mathbb{R}^n$ is quite simple. For a Borel set $A \subset \mathbb{S}^{n-1}$, the value of $S_{n-1}(P, A)$ gives the total surface area of the set of all boundary points of P which have an outer normal in A (since this set is a union of facets, the surface area is defined). In an appropriate way (and using approximation by polytopes), this interpretation carries over to arbitrary bodies K. It states that $S_{n-1}(K, A)$ measures the total surface area of the set of all boundary points of K which have an outer normal in A. Furthermore, we have $S_{n-1}(K, \cdot) = 0$ if and only if $\dim K \leq n-2$, and $S_{n-1}(K, \cdot) = V_{n-1}(K)(\delta_u + \delta_{-u})$, if $\dim K = n - 1$ and K is contained in an affine subspace parallel to $u^\perp$, for some $u \in \mathbb{S}^{n-1}$.

Now we investigate to what extent a convex body $K \subset \mathbb{R}^n$ is determined by its top order area measure $S_{n-1}(K, \cdot)$. The following result provides a strong positive answer to the corresponding uniqueness problem.

Theorem 4.3 *Let $K, L \in \mathcal{K}^n$ with $\dim K = \dim L = n$. Then*

$$S_{n-1}(K, \cdot) = S_{n-1}(L, \cdot)$$

if and only if K and L are translates.

Proof If K, L are translates of each other, the equality of the area measures follows from Corollary 4.1.

Assume now $S_{n-1}(K, \cdot) = S_{n-1}(L, \cdot)$. Then Theorem 4.1 implies that

$$V(K, \ldots, K, L) = \frac{1}{n} \int_{\mathbb{S}^{n-1}} h_L(u) \, S_{n-1}(K, du)$$

$$= \frac{1}{n} \int_{\mathbb{S}^{n-1}} h_L(u) \, S_{n-1}(L, du)$$

$$= V(L).$$

In the same way, we obtain $V(L, \ldots, L, K) = V(K)$. The Minkowski inequality (Theorem 3.14) therefore yields that

$$V(L)^n \geq V(K)^{n-1} V(L)$$

and

$$V(K)^n \geq V(L)^{n-1} V(K),$$

which implies that $V(K) = V(L)$. But then we have equality in both inequalities, and hence K and L are homothetic. Since K and L have the same volume, they must be translates of each other.   □

More generally, a corresponding uniqueness result holds for the jth order area measure, for $j \in \{1, \ldots, n-1\}$, if the underlying convex body has dimension at least $j + 1$ (and for $j = 1$ even without a dimensional restriction). The proof is based on a deep generalization of the Minkowski inequalities (the Alexandrov–Fenchel inequality) and a successive reduction to a situation where spherical harmonics can be used (as in the case $j = 1$).

Theorem 4.3 can be applied to express certain properties of convex bodies in terms of their area measures. We mention only one application of this type, other results can be found in the exercises. We recall that a convex body $K \in \mathcal{K}^n$ is *centrally symmetric* if there is a point $x \in \mathbb{R}^n$ such that $K - x = -(K - x)$. The uniquely determined point x is contained in K and is called the center of symmetry of K. Also, a Borel measure μ on $\mathbb{S}^{n-1}$ is called *even* if μ is invariant under reflection, that is, $\mu(A) = \mu(-A)$ for all Borel sets $A \subset \mathbb{S}^{n-1}$.

Corollary 4.2 *Let $K \in \mathcal{K}^n$ with $\dim K = n$. Then, K is centrally symmetric if and only if $S_{n-1}(K, \cdot)$ is an even measure.*

Proof If $S_{n-1}(K, \cdot)$ is an even measure, then $S_{n-1}(K, B) = S_{n-1}(K, -B) = S_{n-1}(-K, B)$ for all Borel sets $B \subset \mathbb{S}^{n-1}$. Then K and $-K$ are translates of each other. Hence there is some $t \in \mathbb{R}^n$ such that $K = -K + t$. This shows that $c = t/2$.

If $K - x = -(K - x)$, for some $x \in \mathbb{R}^n$, then $S_{n-1}(K - x, \cdot) = S_{n-1}(-K + x, \cdot)$. By the translation invariance of the area measures, we obtain that $S_{n-1}(K, \cdot) = S_{n-1}(-K, \cdot)$.   □

In the following, we explore which measures μ on $\mathbb{S}^{n-1}$ arise as area measures $S_{n-1}(K, \cdot)$ of convex bodies $K \subset \mathbb{R}^n$. This topic is well known as Minkowski's *existence problem*. A necessary condition is

$$\int_{\mathbb{S}^{n-1}} u\,\mu(du) = 0.$$

In this case, we say that μ is *centred* or that μ has centroid 0.

Another condition arises from a dimensional restriction. Namely, if $\dim K \leq n - 2$, then $S_{n-1}(K, \cdot) = 0$, whereas for $\dim K = n - 1$, $K \subset u^\perp$ with $u \in \mathbb{S}^{n-1}$, we have $S_{n-1}(K, \cdot) = V_{n-1}(K)(\delta_u + \delta_{-u})$ (both results follow from Theorem 4.1). Therefore, we now concentrate on bodies $K \in \mathcal{K}^n$ with $\dim K = n$.

A finite Borel measure μ on the unit sphere $\mathbb{S}^{n-1}$ is said to be *n-dimensional* if $\mu(\mathbb{S}^{n-1} \setminus E) > 0$ for all $(n-1)$-dimensional linear subspaces E in $\mathbb{R}^n$. In this case, we write $\dim(\mu) = n$. This condition is equivalent to

$$\int_{S^{n-1}} |\langle x, u \rangle|\,\mu(du) > 0$$

for $x \in \mathbb{R}^n \setminus \{0\}$. Moreover, if μ is centred, then

$$\int_{\mathbb{S}^{n-1}} |\langle x, u \rangle|\,\mu(du) = 2 \int_{\mathbb{S}^{n-1}} \langle x, u \rangle_+\,\mu(du)$$

for $x \in \mathbb{R}^n$, where $t_+ := \max\{t, 0\}$ for $t \in \mathbb{R}$.

With this terminology, a given convex body $K \in \mathcal{K}^n$ satisfies $\dim K = n$ if and only if $\dim S_{n-1}(K, \cdot) = n$. In fact, $\dim K = n$ holds if and only if

$$V(K[n-1], [-x, x]) > 0 \quad \text{for } x \in \mathbb{R}^n \setminus \{0\},$$

and by Theorem 4.1 it follows that this is true if and only if

$$\int_{\mathbb{S}^{n-1}} |\langle x, u \rangle|\,S_{n-1}(K, du) > 0 \quad \text{for } x \in \mathbb{R}^n \setminus \{0\},$$

that is $\dim S_{n-1}(K, \cdot) = n$.

As we shall show now, these two conditions (the centredness condition and the dimensional condition) characterize area measures of order $n - 1$. We first consider the case of polytopes and discrete measures which admits an elementary approach. The general case will then be deduced by an approximation argument. Observe that if $P \subset \mathbb{R}^n$ is a polytope, then it follows from (4.2) in the special case $P_1 = \cdots = P_{n-1} = P$ that

$$S_{n-1}(P, \cdot) = \sum_{i=1}^{k} V_{n-1}(P(u_i))\,\delta_{u_i},$$

where $u_1, \ldots, u_k \in \mathbb{S}^{n-1}$ are the exterior facet normals of the polytope P and $V_{n-1}(P(u_i)) = v(P(u_i))$ is the $(n-1)$-dimensional volume of $P(u_i)$. Hence, the area measure of a polytope is a discrete measure. The following theorem describes precisely which discrete measures arise in this way.

Theorem 4.4 *Let $u_1, \ldots, u_k \in \mathbb{S}^{n-1}$ be pairwise distinct unit vectors which span $\mathbb{R}^n$, and let $v^{(1)}, \ldots, v^{(k)} > 0$ be numbers such that*

$$\sum_{i=1}^{k} v^{(i)} u_i = 0.$$

Then there exists a polytope $P \in \mathcal{P}^n$ with $\dim P = n$ (uniquely determined up to a translation) such that

$$S_{n-1}(P, \cdot) = \sum_{i=1}^{k} v^{(i)} \delta_{u_i},$$

that is, $u_1, \ldots, u_k$ are the facet normals of P and $v^{(1)}, \ldots, v^{(k)}$ are the corresponding facet contents.

Proof The uniqueness assertion follows from Theorem 4.3.

For the existence, we denote by $\mathbb{R}_+^k$ the set of all $y = (y^{(1)}, \ldots, y^{(k)})^\top \in \mathbb{R}^k$ with $y^{(i)} \geq 0$ for $i = 1, \ldots, k$. For $y \in \mathbb{R}_+^k$, let

$$P_{[y]} := \bigcap_{i=1}^{k} H^-(u_i, y^{(i)}).$$

Since $0 \in P_{[y]}$, this set is nonempty and polyhedral. Moreover, $P_{[y]}$ is bounded and hence a convex polytope in $\mathbb{R}^n$. To see this, we proceed by contradiction and assume that $\alpha x \in P_{[y]}$ for some $x \in \mathbb{S}^{n-1}$ and all $\alpha \geq 0$, hence $\langle x, u_i \rangle \leq 0$ for $i = 1, \ldots, k$. By the centredness condition,

$$\sum_{i=1}^{k} v^{(i)} \langle x, u_i \rangle = 0$$

with $v^{(i)} > 0$ and $\langle x, u_i \rangle \leq 0$, and hence $\langle x, u_1 \rangle = \cdots = \langle x, u_k \rangle = 0$. But then $u_1, \ldots, u_k$ do not span $\mathbb{R}^n$, a contradiction.

We next show that the mapping $y \mapsto P_{[y]}$ is concave, that is,

$$\gamma P_{[y]} + (1 - \gamma) P_{[z]} \subset P_{[\gamma y + (1-\gamma)z]} \tag{4.7}$$

for $y, z \in \mathbb{R}_+^k$ and $\gamma \in [0, 1]$. This follows since a point $x \in \gamma P_{[y]} + (1 - \gamma) P_{[z]}$ satisfies $x = \gamma a + (1 - \gamma) b$ with some $a \in P_{[y]}$, $b \in P_{[z]}$, and hence

$$\langle x, u_i \rangle = \gamma \langle a, u_i \rangle + (1 - \gamma) \langle b, u_i \rangle \leq \gamma y^{(i)} + (1 - \gamma) z^{(i)},$$

which shows that $x \in P_{[\gamma y + (1 - \gamma) z]}$.

Since the normal vectors u_i of the halfspaces $H^-(u_i, y^{(i)})$ are fixed and only their distances $y^{(i)}$ from the origin vary, the mapping $y \mapsto P_{[y]}$ is continuous with respect to the Hausdorff metric (see Exercise 4.2.3). Therefore, $y \mapsto V(P_{[y]})$ is continuous, which implies that the set

$$\mathcal{M} := \{ y \in \mathbb{R}_+^k : V(P_{[y]}) = 1 \}$$

is nonempty and closed (note that $P_{[0]} = \{0\}$ and $r B^n \subset P_{[y]}$ if $y^{(i)} \geq r$ for $i = 1, \ldots, k$). The linear function

$$\varphi := \frac{1}{n} \langle \cdot, v \rangle, \quad v := (v^{(1)}, \ldots, v^{(k)})^\top,$$

is nonnegative on $\mathcal{M}$ (and continuous). Since $v^{(i)} > 0$ for $i = 1, \ldots, k$, there is a vector $y_0 \in \mathcal{M}$ such that $\varphi(y_0) =: \alpha \geq 0$ is the minimum of φ on $\mathcal{M}$. Since $y_0 \in \mathcal{M}$ implies that $y_0^{(i)} > 0$ for some $i \in \{1, \ldots, k\}$, we get $\alpha > 0$.

We consider the polytope $Q := P_{[y_0]}$. Since $V(Q) = 1$, it follows that Q has interior points. Moreover, we have $0 \in Q$. We may assume that $0 \in \mathrm{int}\, Q$. To see this, we argue as follows. If $0 \in \mathrm{bd}\, Q$, we can choose a translation vector $t \in \mathbb{R}^n$ such that $0 \in \mathrm{int}(Q + t)$. Then

$$Q + t = \bigcap_{i=1}^{k} H^-(u_i, \tilde{y}_0^{(i)})$$

with $\tilde{y}_0^{(i)} := y_0^{(i)} + \langle t, u_i \rangle$ for $i = 1, \ldots, k$. In particular, $Q + t = P_{[\tilde{y}_0]}$. Since $0 \in \mathrm{int}(Q + t)$, we have $\tilde{y}_0^{(i)} > 0$. Moreover, $V(Q + t) = V(Q) = 1$ and

$$\varphi(\tilde{y}_0) = \frac{1}{n} \langle y_0, v \rangle + \frac{1}{n} \sum_{i=1}^{k} \langle t, u_i \rangle v^{(i)} = \varphi(y_0) + \frac{1}{n} \left\langle t, \sum_{i=1}^{k} v^{(i)} u_i \right\rangle = \alpha.$$

Hence, we can assume that $0 \in \mathrm{int}\, Q$, which gives us $y_0^{(i)} > 0$ for $i = 1, \ldots, k$.

We define a vector $w = (w^{(1)}, \ldots, w^{(k)})$, where $w^{(i)} := V_{n-1}(Q(u_i))$ is the $(n-1)$-dimensional volume of the support set of Q in direction u_i, for $i = 1, \ldots, k$. Then,

$$\frac{1}{\alpha n} \langle y_0, v \rangle = \frac{1}{\alpha} \varphi(y_0) = 1 = V(Q) = \frac{1}{n} \sum_{i=1}^{k} y_0^{(i)} w^{(i)} = \frac{1}{n} \langle y_0, w \rangle,$$

and hence,

$$\langle y_0, w \rangle = \left\langle y_0, \alpha^{-1} v \right\rangle = n.$$

Next, we define the hyperplanes

$$E := H(w, n) \qquad \text{and} \qquad F := H(\alpha^{-1} v, n)$$

in $\mathbb{R}^k$. We want to show that $E = F$. First, note that $y_0 \in E \cap F$.

Since y_0 has positive components, we can find a convex neighbourhood U of y_0 such that any $y \in U$ has the following two properties. First, $y^{(i)} > 0$ for $i = 1, \ldots, k$, and second, every facet normal of $Q = P_{[y_0]}$ is also a facet normal of $P_{[y]}$. We now claim that $V(P_{[y]}) \leq 1$ for $y \in F \cap U$. Aiming at a contradiction, assume that $V(P_{[y]}) > 1$ for some $y \in F \cap U$. Then there exists $0 < \beta < 1$ with

$$V(P_{[\beta y]}) = \beta^n V(P_{[y]}) = 1.$$

Since $y \in F$,

$$\varphi(\beta y) = \frac{1}{n} \langle \beta y, v \rangle = \beta \alpha < \alpha,$$

a contradiction, which proves the claim.

For $\vartheta \in [0, 1]$ and $y, y_0 \in F \cap U$, we also have $\vartheta y + (1 - \vartheta) y_0 \in F \cap U$. Therefore the volume inequality just proven applies and we get from (4.7)

$$V(\vartheta P_{[y]} + (1 - \vartheta) Q) \leq V(P_{[\vartheta y + (1-\vartheta) y_0]}) \leq 1.$$

This yields

$$V(Q, \ldots, Q, P_{[y]}) = \frac{1}{n} \lim_{\vartheta \to 0+} \frac{V(\vartheta P_{[y]} + (1 - \vartheta) Q) - (1 - \vartheta)^n}{\vartheta}$$

$$\leq \frac{1}{n} \lim_{\vartheta \to 0+} \frac{1 - (1 - \vartheta)^n}{\vartheta} = 1.$$

By our choice of U, each facet normal of Q is a facet normal of $P_{[y]}$ for $y \in F \cap U$, and thus we have $h_{P_{[y]}}(u_i) = y^{(i)}$ for all i for which $w^{(i)} > 0$. Hence

$$1 \geq V(Q, \ldots, Q, P_{[y]}) = \frac{1}{n} \sum_{i=1}^{k} h_{P_{[y]}}(u_i) w^{(i)} = \frac{1}{n} \langle y, w \rangle,$$

for $y \in F \cap U$. This shows that $F \cap U \subset E^-$. Since $y_0 \in E \cap F$, we conclude that $E = F$. Hence $w = \frac{1}{\alpha} v$, and thus the polytope $P := \sqrt[n-1]{\alpha}\, Q$ satisfies all assertions of the theorem.   □

We now extend this result to arbitrary bodies $K \in \mathcal{K}^n$. First, we provide an approximation result.

Lemma 4.2 *Let μ be a finite, n-dimensional, centred Borel measure on $\mathbb{S}^{n-1}$. Then there is a sequence μ_k, $k \in \mathbb{N}$, of finite, discrete, n-dimensional, centred Borel measures on $\mathbb{S}^{n-1}$ such that $\mu_k \to \mu$ weakly as $k \to \infty$ and*

$$\int_{\mathbb{S}^{n-1}} \langle v, u \rangle_+ \, \mu_k(du) \to \int_{\mathbb{S}^{n-1}} \langle v, u \rangle_+ \, \mu(du),$$

uniformly in $v \in \mathbb{S}^{n-1}$, as $k \to \infty$.

Proof For $x \in \mathbb{S}^{n-1}$ and $k \in \mathbb{N}$, let $B_k(x) \subset \mathbb{S}^{n-1}$ be a spherical cap with centre x and height $1/(2k^2)$. By elementary geometry it follows that the diameter of $B_k(x)$ is at most $2/k$ (with respect to the Euclidean metric of $\mathbb{R}^n$). The sphere is covered by finitely many such caps. From these caps we deduce a finite sequence $B_{k,1}, \ldots, B_{k,N_k}$ of mutually disjoint Borel sets, each of which is contained in one of the caps, which satisfy $\mu(B_{k,i}) > 0$, and which are such that the union of these Borel sets is $\mathbb{S}^{n-1}$ up to a set of μ-measure zero. Then

$$c_{k,i} := \frac{1}{\mu(B_{k,i})} \int_{B_{k,i}} u \, \mu(du)$$

satisfies

$$1 \geq \|c_{k,i}\| \geq 1 - \frac{1}{2k^2} > 0,$$

and therefore $f_{k,i} u_{k,i} := c_{k,i}$ with $f_{k,i} \in (0, 1]$ and $u_{k,i} \in \mathbb{S}^{n-1}$ are such that

$$1 \geq f_{k,i} \geq 1 - \frac{1}{2k^2} > 0.$$

The finite, discrete Borel measure

$$\mu_k := \sum_{i=1}^{N_k} \mu(B_{k,i}) f_{k,i} \, \delta_{u_{k,i}}$$

is centered, since

$$\int_{\mathbb{S}^{n-1}} u \, \mu_k(du) = \sum_{i=1}^{N_k} \mu(B_{k,i}) f_{k,i} u_{k,i} = \sum_{i=1}^{N_k} \mu(B_{k,i}) c_{k,i}$$

$$= \sum_{i=1}^{N_k} \int_{B_{k,i}} u \, \mu(du) = \int_{\mathbb{S}^{n-1}} u \, \mu(du) = 0.$$

Next we show that μ_k converges weakly to μ as $k \to \infty$. For this, let $g \in C(\mathbb{S}^{n-1})$ be given. Then

$$\int_{\mathbb{S}^{n-1}} g(u) \, \mu_k(du) - \int_{\mathbb{S}^{n-1}} g(u) \, \mu(du)$$

$$= \sum_{i=1}^{N_k} \left[\mu(B_{k,i}) f_{k,i} g(u_{k,i}) - \int_{B_{k,i}} g(u) \, \mu(du) \right]$$

$$= \sum_{i=1}^{N_k} \int_{B_{k,i}} \left(f_{k,i} g(u_{k,i}) - g(u) \right) \, \mu(du).$$

If $k \geq k_0$, then $\|u_{k,i} - u\| \leq 2/k \leq 2/k_0$ for $u \in B_{k,i}$. Since g is bounded and uniformly continuous, for given $\varepsilon > 0$, we get

$$|g(u_{k,i}) - g(u)| \leq \frac{\varepsilon}{2} \quad \text{and} \quad \frac{1}{2k^2} |g(u)| \leq \frac{\varepsilon}{2}$$

for $u \in B_{k,i}$ and $k \geq k_0 = k_0(\varepsilon)$. Thus, for $u \in B_{k,i}$ and $k \geq k_0 = k_0(\varepsilon)$, we obtain

$$|f_{k,i} g(u_{k,i}) - g(u)| = |f_{k,i} g(u_{k,i}) - f_{k,i} g(u)| + |f_{k,i} g(u) - g(u)|$$

$$= |g(u_{k,i}) - g(u)| + |1 - f_{k,i}| \, |g(u)|$$

$$\leq \varepsilon/2 + \varepsilon/2 = \varepsilon.$$

Hence, for $k \geq k_0$ we conclude that

$$\left| \int_{\mathbb{S}^{n-1}} g(u) \, \mu_k(du) - \int_{\mathbb{S}^{n-1}} g(u) \, \mu(du) \right| \leq \varepsilon \mu(\mathbb{S}^{n-1}),$$

which yields the weak convergence $\mu_k \to \mu$ as $k \to \infty$.

To obtain the uniform convergence result, we argue similarly for $g_v(u) := \langle v, u \rangle_+, v, u \in \mathbb{S}^{n-1}$. Then $|g_v(u)| \leq 1$ and

$$|g_v(u_{k,i}) - g_v(u)| = |\langle v, u_{k,i} \rangle_+ - \langle v, u \rangle_+| \leq \|u_{k,i} - u\| \leq 2/k,$$

independently of $v \in \mathbb{S}^{n-1}$. Now the argument can be completed as before.

This latter fact also implies that μ_k is n-dimensional if k is sufficiently large. In fact, since μ is centred and n-dimensional, the map

$$v \mapsto \int_{\mathbb{S}^{n-1}} \langle v, u \rangle_+ \, \mu(du), \quad v \in \mathbb{S}^{n-1},$$

is positive. Since it is also continuous, there is a positive constant $\rho_0 > 0$ such that

$$\int_{\mathbb{S}^{n-1}} \langle v, u \rangle_+ \, \mu(du) \geq \rho_0$$

for $v \in \mathbb{S}^{n-1}$. By uniform convergence, we obtain

$$\int_{\mathbb{S}^{n-1}} \langle v, u \rangle_+ \, \mu_j(du) \geq \rho_1 > 0, \tag{4.8}$$

for all $v \in \mathbb{S}^{n-1}$ and some constant $\rho_1 > 0$, if $j \in \mathbb{N}$ is large enough. This shows that μ_k is n-dimensional if k is large enough. $\square$

Now we can deduce Theorem 4.5 from Theorem 4.4 via Lemma 4.2 and Blaschke's selection theorem.

Theorem 4.5 *Let μ be a finite Borel measure on $\mathbb{S}^{n-1}$ with centroid 0 and $\dim \mu = n$. Then, there exists a (unique up to translation) convex body $K \in \mathcal{K}^n$ such that*

$$S_{n-1}(K, \cdot) = \mu.$$

Proof Again, we only need to show the existence of K.

Let μ_j, $j \in \mathbb{N}$, be a sequence of approximating measures for μ, as provided by Lemma 4.2. From Theorem 4.4, we then obtain polytopes P_j with $0 \in P_j$ and such that

$$\mu_j = S_{n-1}(P_j, \cdot), \quad j \in \mathbb{N}.$$

We show that the sequence $(P_j)_{j \in \mathbb{N}}$ is uniformly bounded. First, $F(P_j) = \mu_j(\mathbb{S}^{n-1}) \to \mu(\mathbb{S}^{n-1})$ as $j \to \infty$ implies that

$$F(P_j) \leq c, \quad j \in \mathbb{N},$$

for some $c > 0$. The isoperimetric inequality shows that then

$$V(P_j) \le C, \quad j \in \mathbb{N},$$

with a constant $C > 0$. Now let $x \in \mathbb{S}^{n-1}$ and $\alpha \ge 0$ be such that $\alpha x \in P_j$, hence $[0, \alpha x] \subset P_j$. Since $h_{[0,\alpha x]} = \alpha \langle x, \cdot \rangle_+$, we get

$$V(P_j) = \frac{1}{n} \int_{\mathbb{S}^{n-1}} h_{P_j}(u)\, S_{n-1}(P_j, du)$$

$$\ge \frac{1}{n} \int_{\mathbb{S}^{n-1}} \alpha \langle x, u \rangle_+\, S_{n-1}(P_j, du)$$

$$= \frac{\alpha}{n} \int_{\mathbb{S}^{n-1}} \langle x, u \rangle_+\, \mu_j(du).$$

Since μ is centred and n-dimensional, by the uniform convergence result stated in Lemma 4.2, and arguing as in the derivation of (4.8) we get

$$\int_{\mathbb{S}^{n-1}} \langle x, u \rangle_+\, \mu_j(du) \ge \rho_2$$

for all $x \in \mathbb{S}^{n-1}$, $j \in \mathbb{N}$ and some constant $\rho_2 > 0$. Hence,

$$\rho_2 \frac{\alpha}{n} \le V(P_j) \le C,$$

which implies that $\alpha \le (nC)/\rho_2$. This shows that the sequence $(P_j)_{j \in \mathbb{N}}$ is contained in $B^n(0, (nC)/\rho_2)$.

By Blaschke's selection theorem, we can choose a convergent subsequence $P_{j_r} \to K$, $r \in \mathbb{N}$, with $K \in \mathcal{K}^n$. Then

$$S_{n-1}(P_{j_r}, \cdot) \to S_{n-1}(K, \cdot),$$

but also

$$S_{n-1}(P_{j_r}, \cdot) \to \mu.$$

Therefore, we obtain $S_{n-1}(K, \cdot) = \mu$. □

We conclude this section by establishing a sharp upper bound for the mixed volume $V(K, M[n-1])$. To prepare this, we first derive an inequality between circumradius and mean width (the first intrinsic volume) of a convex body. The proof is a revision of an argument due to J. Linhart [62].

Theorem 4.6 *Let* $K \in \mathcal{K}^n$ *with* $\mathrm{diam}(K) > 0$, *and let* $R(K)$ *denote the circumradius of* K. *Then* $V_1 K) \geq 2R(K)$, *with equality if and only if* K *is a segment.*

Proof By homogeneity and translation invariance, we can assume that $R(K) = 1$ and B^n is the circumball of K. If K is a nondegenerate segment, the assertion is clear. Hence we can assume that $\dim(K) \geq 2$. By a separation argument and by Carathéodory's theorem, there is a k-simplex $S \subset K$ which is inscribed to B^n and such that $0 \in \mathrm{relint}\, S$. If $k = 1$, then S is a diameter and $V_1(S) = 2R(K)$. Since $K \neq S$, we have $V_1(K) > V_1(S)$. Hence, we can assume that $k \geq 2$. In this case, we prove that $V_1(K) > 2$.

Let $x_0, \ldots, x_k \in \mathbb{S}^{n-1}$ denote the vertices of S, and let $N_j = N(S, x_j) \cap \mathbb{S}^{n-1}$ be the normal cone at x_j intersected with the unit sphere. We claim that $N_j \subset x_j^+ := \{z \in \mathbb{R}^n : \langle z, x_j \rangle \geq 0\}$. To see this, suppose to the contrary that there is some $u \in \mathbb{S}^{n-1}$ such that $\langle u, x_i - x_j \rangle \leq 0$ for $i \in \{1, \ldots, k\}$ and $\langle u, x_j \rangle < 0$. But then $\langle u, x_i \rangle \leq \langle u, x_j \rangle < 0$, hence $0 \notin \mathrm{relint}\, S$, a contradiction.

We write $D_j := \mathbb{S}^{n-1} \cap x_j^+$ for $j = 1, \ldots, k$. Then

$$\int_{\mathbb{S}^{n-1}} h_K(u) \, \mathcal{H}^{n-1}(du) = \sum_{j=0}^{k} \int_{N_j} \langle x_j, u \rangle \, \mathcal{H}^{n-1}(du),$$

and we assert that

$$\int_{N_j} \langle x_j, u \rangle \, \mathcal{H}^{n-1}(du) \geq \mathcal{H}^{n-1}(N_j) \frac{1}{\mathcal{H}^{n-1}(D_j)} \int_{D_j} \langle x_j, u \rangle \, \mathcal{H}^{n-1}(du) \qquad (4.9)$$

$$= \mathcal{H}^{n-1}(N_j) \frac{2}{n\kappa_n} \kappa_{n-1}$$

with equality if and only if $N_j = D_j$. Since $\dim S \geq 2$, the inequality is indeed always strict.

Thus we obtain

$$\int_{\mathbb{S}^{n-1}} h_K(u) \, \mathcal{H}^{n-1}(du) \geq \sum_{j=0}^{k} \int_{N_j} \mathcal{H}^{n-1}(N_j) \frac{2}{n\kappa_n} \kappa_{n-1} = 2\kappa_{n-1},$$

where the inequality is strict, since K is not a segment.

This proves the assertion of the theorem.

It remains to establish (4.9).

Since $N_j \subset x_j^+$, for each $v \in \mathbb{S}^{n-1} \cap x_j^\perp$ there is a unique $\varphi(v) \in [0, \pi/2]$ such that $\cos(\varphi(v))x_j + \sin(\varphi(v))v \in \mathrm{bd}\, N_j$. Let $\varphi'(v) \in (0, \pi/2)$ and $h(s) := (\sin s)^{n-2}$.

Using that cos is strictly decreasing on $[0, \pi/2]$, we obtain

$$\left(\int_{\varphi'(v)}^{\pi/2} h(s)\,ds\right)^{-1} \int_{\varphi'(v)}^{\pi/2} \cos(s)h(s)\,ds$$

$$< \cos(\varphi'(s)) < \left(\int_0^{\varphi'(v)} h(s)\,ds\right)^{-1} \int_0^{\varphi'(v)} \cos(s)h(s)\,ds.$$

From this we deduce that

$$\int_0^{\pi/2} \cos(s)h(s)\,ds$$

$$\leq \int_0^{\varphi'(v)} \cos(s)h(s)\,ds$$

$$+ \int_{\varphi'(v)}^{\pi/2} h(s)\,ds \left(\int_0^{\varphi'(v)} h(s)\,ds\right)^{-1} \int_0^{\varphi'(v)} \cos(s)h(s)\,ds$$

$$= \left(\int_0^{\varphi'(v)} h(s)\,ds\right)^{-1} \int_0^{\varphi'(v)} \cos(s)h(s)\,ds$$

$$\times \left(\int_0^{\varphi'(v)} h(s)\,ds + \int_{\varphi'(v)}^{\pi/2} h(s)\,ds\right),$$

and therefore

$$\int_0^{\varphi'(v)} h(s)\,ds \left(\int_0^{\pi/2} h(s)\,ds\right)^{-1} \int_0^{\pi/2} \cos(s)h(s)\,ds$$

$$\leq \int_0^{\varphi'(v)} \cos(s)h(s)\,ds.$$

This inequality is strict, except if $\varphi'(v) \in \{0, \pi/2\}$. Integration of this inequality, for $\varphi(v)$ instead of $\varphi'(v)$, over $v \in \mathbb{S}^{n-1} \cap x_j^{\perp}$ yields the required inequality (here we use spherical polar coordinates). The resulting inequality is strict, since $\varphi(v) \notin \{0, \pi/2\}$ on a set of $v \in \mathbb{S}^{n-1} \cap x_j^{+}$ of positive measure (as S is not a segment). $\qquad\square$

Theorem 4.6 is the main ingredient in the proof of the following inequality, which can be viewed as a reverse Minkowski-type inequality.

Theorem 4.7 *Let* $K, M \in \mathcal{K}^n$. *Then*

$$V(K, M[n-1]) \leq \frac{1}{n} V_1(K) V_{n-1}(M).$$

If $\dim(K) \geq 1$ *and* $\dim(M) \geq n - 1$, *then equality holds if and only if* K *is a segment and* M *is contained in a hyperplane orthogonal to* K.

Proof We can assume that $B^n(0, R(K))$ is the circumball of K. Then

$$V(K, M[n-1]) = \frac{1}{n} \int_{\mathbb{S}^{n-1}} h_K(u) \, S_{n-1}(M, du) \leq \frac{1}{n} R(K) F(M)$$

$$\leq \frac{1}{n} \frac{1}{2} V_1(K) 2 V_{n-1}(M) = \frac{1}{n} V_1(K) V_{n-1}(M),$$

where we used Theorem 4.6 for the second inequality. If equality holds, then equality holds in Theorem 4.6, since $V_{n-1}(M) > 0$, and therefore $K = [-Re, Re]$ for some $e \in \mathbb{S}^{n-1}$. Moreover, we then also have equality in the first inequality, which yields

$$\int_{\mathbb{S}^{n-1}} |\langle u, e \rangle| \, S_{n-1}(M, du) = S_{n-1}(M, \mathbb{S}^{n-1}).$$

This implies that the area measure of M is concentrated in $\{-e, e\}$, hence M is contained in a hyperplane orthogonal to e. $\square$

Improvements of Theorems 4.6 and 4.7, in terms of stability results, have recently been established in [20].

Exercises and Supplements for Sect. 4.2

1. Let $K \in \mathcal{K}^n$ and $R(K)$ be the circumradius of K. Show that $R(K) \leq 1$ if and only if $V(K, M, \ldots, M) \leq \frac{1}{n} F(M)$ for all $M \in \mathcal{K}^n$.
2. Let $\alpha \in (0, 1)$ and $M, L \in \mathcal{K}^n$ with $\dim M = \dim L = n$.

   (a) Show that there is a convex body $K_\alpha \in \mathcal{K}^n$ with $\dim K_\alpha = n$ and

   $$S_{n-1}(K_\alpha, \cdot) = \alpha S_{n-1}(M, \cdot) + (1 - \alpha) S_{n-1}(L, \cdot).$$

   (b) Show that

   $$V(K_\alpha)^{\frac{n-1}{n}} \geq \alpha V(M)^{\frac{n-1}{n}} + (1 - \alpha) V(L)^{\frac{n-1}{n}},$$

   with equality if and only if M and L are homothetic.

3. *Prove the following two facts which are used in the proof of Theorem 4.4.

   (a) Let $K \subset \mathbb{R}^n$ be an unbounded, closed, convex set with $0 \in K$. Show that there is a unit vector $u \in \mathbb{S}^{n-1}$ such that $\alpha u \in K$ for all $\alpha \geq 0$.
   (b) The map $\mathbb{R}_+^k \to \mathcal{P}^n$, $y \mapsto P_{[y]}$, is continuous.

4. Let $v, a, b \in \mathbb{R}^n$. Show that $|\langle v, a \rangle_+ - \langle v, b \rangle_+| \leq \|a - b\| \|v\|$.
5. Let $K \subset \mathbb{R}^2$ denote a triangle inscribed to the unit sphere which contains the origin. Show that $F(K) \geq 4$ with equality if and only if K is a segment (degenerate triangle).

4.3 A Local Steiner Formula

In Sect. 4.1, we introduced the area measure $S_j(K, \cdot)$ of a convex body $K \in \mathcal{K}^n$ as the special mixed measure $S(K[j], B^n[n-1-j], \cdot)$, for $j \in \{0, \ldots, n-1\}$. In this section, we show that the measures $S_j(K, \cdot)$ are also determined as coefficients of a local Steiner formula. Moreover, we obtain a description of the area measures for bodies with support functions of class $\mathbf{C}^2$ in terms of curvature integrals. This indicates that the area measures encode second-order information about the underlying convex set.

For $K \in \mathcal{K}^n$ and $\omega \subset \mathbb{S}^{n-1}$, we define the *reverse spherical image* of K at ω by

$$\tau(K, \omega) = \bigcup_{u \in \omega} K(u) = \{x \in \partial K : N(K, x) \cap \omega \neq \emptyset\}$$

$$= \{x \in \partial K : \langle x, u \rangle = h(K, u) \text{ for some } u \in \omega\}.$$

For a polytope $P \subset \mathbb{R}^n$, the set $\tau(P, \omega)$ is the union of all support sets $P(u)$ of P with $u \in \omega$. Since $\mathcal{H}^{n-1}(P(u)) = 0$ if $P(u)$ is not a facet of P and since P has only finitely many supports sets, we obtain

$$S_{n-1}(P, \cdot) = \sum_{u \in \mathbb{S}^{n-1}} v(P(u)) \delta_u = \mathcal{H}^{n-1}(\tau(P, \cdot)).$$

In order to prove that this relation extends to general convex bodies, we first show that $\tau(K, \omega)$ is $\mathcal{H}^{n-1}$-measurable if $\omega \subset \mathbb{S}^{n-1}$ is a Borel set.

Lemma 4.3 *Let $K \in \mathcal{K}^n$ and $\omega \in \mathcal{B}(\mathbb{S}^{n-1})$. Then $\tau(K, \omega)$ is $\mathcal{H}^{n-1}$-measurable.*

Proof If $\omega \subset \mathbb{S}^{n-1}$ is closed, then $\tau(K, \omega)$ is closed. For this, let $x_i \in \tau(K, \omega)$, $i \in \mathbb{N}$, and $x_i \to x$ as $i \to \infty$. Then $x \in K$ and $\langle x_i, u_i \rangle = h(K, u_i)$ for some $u_i \in \omega$, for $i \in \mathbb{N}$. Since $(u_i)_{i \in \mathbb{N}}$ is bounded and ω is closed, there are a subsequence u_{i_j}, $j \in \mathbb{N}$, and $u \in \omega$ such that $u_{i_j} \to u$ as $j \to \infty$. Hence, we get $\langle x, u \rangle = h(K, u)$, which shows that $x \in \tau(K, \omega)$.

Let $\mathcal{A}$ denote the class of all $\omega \in \mathcal{B}(\mathbb{S}^{n-1})$ for which $\tau(K, \omega)$ is $\mathcal{H}^{n-1}$-measurable. As we have just shown, $\mathcal{A}$ contains all closed subsets of $\mathbb{S}^{n-1}$. Next we show that $\mathcal{A}$ is a σ-algebra. This will complete the proof. It remains to be shown that $\mathcal{A}$ is closed under countable unions and complements. Since $\tau(K \cup_{i \in \mathbb{N}} \omega_i) = \cup_{i \in \mathbb{N}} \tau(K, \omega_i)$, the first assertion is obvious. Suppose that $\omega \in \mathcal{A}$, that is, $\tau(K, \omega)$ is $\mathcal{H}^{n-1}$-measurable. Let $\text{reg}(K)$ denote the set of regular boundary points of K, that is, the set of all $x \in \partial K$ such that $\dim N(K, x) = 1$. By Exercise 4.3.2 we have $\mathcal{H}^{n-1}(\partial K \setminus \operatorname{reg}(K)) = 0$. Since $\tau(K, \omega) \cap \tau(K, \omega^c) \subset \partial K \setminus \operatorname{reg}(K)$, it follows that $\tau(K, \omega^c) = (\partial K \setminus \tau(K, \omega)) \cup N$, where $\mathcal{H}^{n-1}(N) = 0$. Hence $\tau(K, \omega^c)$ is $\mathcal{H}^{n-1}$-measurable, that is, $\omega^c \in \mathcal{A}$. $\square$

We write $\mathcal{K}_0^n$ for the space of n-dimensional compact convex sets in $\mathbb{R}^n$. A map from $\mathcal{K}_0^n$ to $\mathcal{M}(\mathbb{S}^{n-1})$, the space of finite Borel measure on the unit sphere, is said to be weakly continuous if it is continuous with respect to the topology induced by the Hausdorff metric on $\mathcal{K}_0^n$ and the weak topology on $\mathcal{M}(\mathbb{S}^{n-1})$.

Lemma 4.4 *The map* $F : \mathcal{K}_0^n \to \mathcal{M}(\mathbb{S}^{n-1})$, $K \mapsto \mathcal{H}^{n-1}(\tau(K, \omega))$, *is weakly continuous.*

Proof We write $F(K, \cdot)$ instead of $F(K)(\cdot)$. The argument in the proof of Lemma 4.3 shows that $\mathcal{H}^{n-1}(\tau(K, \omega_i) \cap \tau(K, \omega_j)) = 0$ if $\omega_i \cap \omega_j = \emptyset$. Hence $F(K, \cdot)$ is indeed a finite measure.

Let $K_i, K \in \mathcal{K}_0^n, i \in \mathbb{N}$, and $K_i \to K$ as $i \to \infty$. Since F is translation invariant, we can assume that $o \in \operatorname{int}(K_i), \operatorname{int}(K)$ for $i \in \mathbb{N}$. Let $g : \mathbb{S}^{n-1} \to \mathbb{R}$ be continuous. Then

$$\int_{\mathbb{S}^{n-1}} g(u) \, F(K, du) = \int_{\partial K} g(\sigma_K(x)) \, \mathcal{H}^{n-1}(dx),$$

where $\sigma_K : \partial K \to \mathbb{S}^{n-1}$ is an exterior unit normal of K at $x \in \partial K$, which is uniquely determined for $\mathcal{H}^{n-1}$-almost all $x \in \partial K$. The restriction of σ_K to $\operatorname{reg}(K)$ is continuous, hence σ_K is measurable.

Let $\tilde{\pi} : \partial K \to \mathbb{S}^{n-1}$, $x \mapsto \|x\|^{-1} x =: \tilde{x}$, be the radial projection map with inverse $\tilde{\pi}^{-1}(u) = \rho_K(u) u$ for $u \in \mathbb{S}^{n-1}$. (See Exercise 2.3.3 for a definition and discussion of the radial function ρ_K.) Clearly, $\tilde{\pi}$ is a bi-Lipschitz homeomorphism. Let $x \in \operatorname{reg}(K)$, $u := \sigma_K(x)$, and let $u_1, \ldots, u_{n-1}$ be an orthonormal basis of $u^\perp$. Then

$$d\tilde{\pi}(x)(u_i) = \frac{1}{\|x\|} (u_i - \langle \tilde{x}, u_i \rangle \tilde{x}), \quad i = 1, \ldots, n-1,$$

and hence

$$
\begin{aligned}
J_{n-1}\tilde{\pi}(x) &= \left| \det \left(\langle d\tilde{\pi}(x)(u_i), d\tilde{\pi}(x)(u_j) \rangle \right)_{i,j=1}^{n-1} \right|^{\frac{1}{2}} \\
&= \frac{1}{\|x\|^{n-1}} \left| \det \left(\delta_{ij} - \langle \tilde{x}, u_i \rangle \langle \tilde{x}, u_j \rangle \right)_{i,j=1}^{n-1} \right|^{\frac{1}{2}} \\
&= \frac{1}{\|x\|^{n-1}} \left(1 - \sum_{i=1}^{n-1} \langle \tilde{x}, u_i \rangle^2 \right)^{\frac{1}{2}} \\
&= \frac{1}{\|x\|^{n-1}} \langle \tilde{x}, u \rangle.
\end{aligned}
$$

Using the coarea formula and the fact that $\mathcal{H}^{n-1}$-almost all boundary points of K are regular, we get

$$\int_{\partial K} g(\sigma_K(x))\, \mathcal{H}^{n-1}(dx) = \int_{\partial K} g(\sigma_K(x)) \frac{\|x\|^{n-1}}{\langle \tilde{x}, \sigma_K(x) \rangle} J_{n-1} \tilde{\pi}(x)\, \mathcal{H}^{n-1}(dx)$$

$$= \int_{\mathbb{S}^{n-1}} g(\sigma_K(\rho_K(u)u)) \frac{\rho_K(u)^{n-1}}{\langle u, \sigma_K(\rho_K(u)u) \rangle}\, \mathcal{H}^{n-1}(du).$$

Since

$$\mathcal{H}^{n-1}\left(\mathbb{S}^{n-1} \setminus \bigcap_{i \geq 1} \tilde{\pi}(\mathrm{reg}(K_i)) \cap \tilde{\pi}(\mathrm{reg}(K)) \right) = 0,$$

$\rho_{K_i}(u) \to \rho_K(u)$ as $i \to \infty$ for $u \in \mathbb{S}^{n-1}$, and $\sigma_{K_i}(\rho_{K_i}(u)u) \to \sigma_K(\rho_K(u)u)$ as $i \to \infty$ for $u \in \bigcap_{i \geq 1} \tilde{\pi}(\mathrm{reg}(K_i)) \cap \tilde{\pi}(\mathrm{reg}(K))$, it follows that

$$g(\sigma_{K_i}(\rho_{K_i}(u)u)) \frac{\rho_{K_i}(u)^{n-1}}{\langle u, \sigma_{K_i}(\rho_{K_i}(u)u) \rangle}$$

$$\to g(\sigma_K(\rho_K(u)u)) \frac{\rho_K(u)^{n-1}}{\langle u, \sigma_K(\rho_K(u)u) \rangle} \qquad (4.10)$$

as $i \to \infty$ for $\mathcal{H}^{n-1}$-almost all $u \in \mathbb{S}^{n-1}$ (see Exercise 4.3.1). Since all expressions in (4.10) are uniformly bounded (see Exercise 4.3.1), the assertion follows from the dominated convergence theorem. □

Theorem 4.8 *If $K \in \mathcal{K}_0^n$, then $S_{n-1}(K, \cdot) = \mathcal{H}^{n-1}(\tau(K, \cdot))$.*

Proof The result holds for n-dimensional polytopes. Since both sides of the asserted relation are measures which depend continuously on the underlying convex body, the general equality immediately follows by approximation with polytopes. □

Let $K \in \mathcal{K}_0^n$ be a convex body with support function h_K of class $\mathbf{C}^2$ on $\mathbb{R}^n \setminus \{0\}$. Since h_K is differentiable on $\mathbb{R}^n \setminus \{0\}$, K is strictly convex and the unique boundary point of K with exterior unit normal vector u is given by the gradient $\nabla h_K(u)$. Then the coarea formula yields

$$\int_{\mathbb{S}^{n-1}} g(u)\, F(K, du) = \int_{\mathrm{bd}\, K} g(\sigma_K(x))\, \mathcal{H}^{n-1}(dx)$$

$$= \int_{\mathbb{S}^{n-1}} g(u) J_{n-1} \nabla h_K(u)\, \mathcal{H}^{n-1}(du).$$

The map $\nabla h_K : \mathbb{R}^n \setminus \{0\} \to \mathbb{R}^n$ is differentiable and its differential at $u \in \mathbb{S}^{n-1}$ yields a symmetric positive semi-definite linear map $d^2 h_K(u) : u^\perp \to u^\perp$, since

$d^2 h_K(u)(u) = 0$, that is, u is an eigenvector with eigenvalue zero of $d^2 h_K(u)$. We can choose an orthonormal basis of $u^\perp$ consisting of eigenvectors $u_1, \ldots, u_{n-1}$ of $d^2 h_K(u)$, hence $d^2 h_K(u)(u_i) = r_i(K, u)u_i$ for $i = 1, \ldots, n-1$. The numbers $r_i(K, u)$ for $i = 1, \ldots, n-1$ are the *principal radii of curvature* of K in direction u. Then we obtain

$$J_{n-1} \nabla h_K(u) = \det(d^2 h_K(u)|u^\perp) = \prod_{i=1}^{n-1} r_i(K, u) =: R_{n-1}(K, u). \tag{4.11}$$

This finally yields

$$\int_{\mathbb{S}^{n-1}} g(u)\, F(K, du) = \int_{\mathbb{S}^{n-1}} g(u) R_{n-1}(K, u)\, \mathcal{H}^{n-1}(du). \tag{4.12}$$

In the following, more generally we write

$$R_j(K, u) := \binom{n-1}{j}^{-1} \sum_{1 \le i_1 < \cdots < i_j \le n-1} r_{i_1}(K, u) \cdots r_{i_j}(K, u),$$

where $R_0(K, u) = 1$, for the jth normalized elementary symmetric function of the principal radii of curvature of K in direction u, for $j = 0, \ldots, n-1$.

Theorem 4.9 *Let $K \in \mathcal{K}^n$ be a convex body with support function h_K of class $\mathbf{C}^2$. Then*

$$S_j(K, \cdot) = \int_{\mathbb{S}^{n-1}} \mathbf{1}\{u \in \cdot\} R_j(K, u)\, \mathcal{H}^{n-1}(du)$$

for $j = 0, \ldots, n-1$.

Proof The assertion is true for $j = n-1$ and the convex body $K + \alpha B^n$, $\alpha > 0$, by Theorem 4.8 and (4.12). To extend the result to general convex bodies with support function of class $\mathbf{C}^2$ and to all j, recall from Corollary 4.1 that

$$S_{n-1}(K + \alpha B^n, \cdot) = \sum_{j=0}^{n-1} \alpha^{n-1-j} \binom{n-1}{j} S_j(K, \cdot). \tag{4.13}$$

Moreover, we have

$$d^2 h_{K+\alpha B^n}(u)|u^\perp = d^2 h_K(u)|u^\perp + \alpha \operatorname{id}_{u^\perp},$$

for $u \in \mathbb{S}^{n-1}$, hence $r_i(K + \alpha B^n, u) = r_i(K, u) + \alpha$ for $i = 1, \ldots, n - 1$. This implies that

$$R_{n-1}(K + \alpha B^n, u) = \prod_{i=1}^{n-1} (r_i(K, u) + \alpha)$$

$$= \sum_{j=0}^{n-1} \alpha^{n-1-j} \binom{n-1}{j} R_j(K, u) \qquad (4.14)$$

for $u \in \mathbb{S}^{n-1}$. From (4.14) we deduce that

$$S_{n-1}(K + \alpha B^n, \cdot)$$

$$= \int_{\mathbb{S}^{n-1}} \mathbf{1}\{u \in \cdot\} R_{n-1}(K + \alpha B^n, u) \, \mathcal{H}^{n-1}(du)$$

$$= \sum_{j=0}^{n-1} \alpha^{n-1-j} \binom{n-1}{j} \int_{\mathbb{S}^{n-1}} \mathbf{1}\{u \in \cdot\} R_j(K, u) \, \mathcal{H}^{n-1}(du). \qquad (4.15)$$

A comparison of coefficients of (4.13) and (4.15) completes the proof. □

The intrinsic volumes of convex bodies are determined as coefficients of a Steiner formula. The area measures $S_j(K, \cdot)$ will now be obtained as coefficients of a local Steiner formula. For this, we consider $K \in \mathcal{K}^n$, a Borel set $\omega \subset \mathbb{S}^{n-1}$, and $\varepsilon > 0$. We define the local parallel set

$$B_\varepsilon(K, \omega) := \{x \in \mathbb{R}^n \setminus K : d(K, x) \leq \varepsilon, u(K, x) \in \omega\},$$

where $u(K, x) = d(K, x)^{-1}(x - p(K, x))$ for $x \in \mathbb{R}^n \setminus K$. Since $d(K, \cdot)$ and $p(K, \cdot)$ are continuous and hence Borel measurable, the set $B_\varepsilon(K, \omega) \subset \mathbb{R}^n$ is a Borel set.

Lemma 4.5 *The map* $\mathcal{K}^n \to \mathcal{M}(\mathbb{S}^{n-1})$, $K \mapsto \eta(K, \cdot) := \mathcal{H}^n(B_\varepsilon(K, \cdot))$ *is weakly continuous.*

Proof Clearly, $\eta(K, \cdot)$ is a measure and finite. Let $g : \mathbb{S}^{n-1} \to \mathbb{R}$ be continuous. Then

$$\int_{\mathbb{S}^{n-1}} g(u) \, \eta(K, du) = \int_{K_\varepsilon \setminus K} g(u(K, x)) \, \mathcal{H}^n(dx)$$

and we have to show that

$$K \mapsto \int_{K_\varepsilon \setminus K} g(u(K, x)) \, \mathcal{H}^n(dx)$$

is continuous. Let $K_i, K \in \mathcal{K}^n, i \in \mathbb{N}$, and $K_i \to K$ as $i \to \infty$. If $x \in \text{int}(K_\varepsilon) \setminus K$, then $0 < d(K, x) < \varepsilon$, and hence $0 < d(K_i, x) < \varepsilon$ for $i \geq i(x)$. Moreover, in this case we also have

$$u(K_i, x) = \frac{x - p(K_i, x)}{d(K_i, x)} \to \frac{x - p(K, x)}{d(K, x)} = u(K, x) \quad \text{as } i \to \infty.$$

If $x \notin K_\varepsilon$, then also $x \notin (K_i)_\varepsilon$ for almost all $i \in \mathbb{N}$, and if $x \in \text{int}(K)$, then also $x \in \text{int}(K_i)$ for almost all $i \in \mathbb{N}$. Since $\mathcal{H}^n(\partial K_\varepsilon \cup \partial K) = 0$, it follows that

$$\mathbf{1}\{x \in (K_i)_\varepsilon \setminus K_i\} g(u(K_i, x)) \to \mathbf{1}\{x \in K_\varepsilon \setminus K\} g(u(K, x))$$

for $\mathcal{H}^n$-almost all $x \in \mathbb{R}^n$. By the dominated convergence theorem, the assertion follows. $\square$

Theorem 4.10 *Let $K \in \mathcal{K}^n$ and $\varepsilon > 0$. Then*

$$\mathcal{H}^n(B_\varepsilon(K, \cdot)) = \frac{1}{n} \sum_{j=0}^{n-1} \varepsilon^{n-j} \binom{n}{j} S_j(K, \cdot).$$

Proof Since on both sides of the asserted equation we have Borel measures which depend continuously on the underlying convex body, it is sufficient to prove the result for convex bodies K with support function of class $\mathbf{C}^2$.

The map $T : \mathbb{S}^{n-1} \times (0, \infty) \to \mathbb{R}^n$, $(u, t) \mapsto \nabla h_K(u) + tu$, is differentiable and injective. To see this, let $u_1, u_2 \in \mathbb{S}^{n-1}$ and $t_1, t_2 > 0$ be such that $T(u_1, t_1) = T(u_2, t_2)$, that is, $\nabla h_K(u_1) + t_1 u_1 = \nabla h_K(u_2) + t_2 u_2$. Since $u_2 \in N(K, \nabla h_K(u_2))$, we have $\langle \nabla h_K(u_1) - \nabla h_K(u_2), u_2 \rangle \leq 0$, and hence we get $\langle t_2 u_2 - t_1 u_1, u_2 \rangle \leq 0$. This yields $0 < t_2 \leq t_1 \langle u_1, u_2 \rangle$. But then $1 \geq \langle u_1, u_2 \rangle > 0$ and $0 < t_2 \leq t_1$. By symmetry we also obtain $t_1 \leq t_2$, hence $t_1 = t_2$. But then $t_2 \leq t_1 \langle u_1, u_2 \rangle \leq t_1 = t_2$ shows that $\langle u_1, u_2 \rangle = 1$, which implies that $u_1 = u_2$.

Let $\omega \subset \mathbb{S}^{n-1}$ be a Borel set. Since $B_\varepsilon(K, \omega) = T(\omega \times (0, \varepsilon])$ and, for given $u \in \mathbb{S}^{n-1}$, choosing an orthonormal basis $u_1, \ldots, u_{n-1}$ of $u^\perp$ consisting of eigenvectors of $d^2 h_K(u)$ with eigenvalues $r_1(K, u), \ldots, r_{n-1}(K, u)$, the Jacobian of T can be expressed in the form

$$J_n T(u, t) = \left| \det \left(d^2 h_K(u)(u_1) + tu_1, \ldots, d^2 h_K(u)(u_{n-1}) + tu_{n-1}, u \right) \right|$$

$$= |\det ((r_1(K, u) + t)u_1, \ldots, (r_{n-1}(K, u) + t)u_{n-1}, u)|$$

$$= \prod_{i=1}^{n-1} (r_i(K, u) + t).$$

Then the transformation formula (coarea formula) yields that

$$\mathcal{H}^n(B_\varepsilon(K, \omega)) = \mathcal{H}^n(T(\omega \times (0, \varepsilon]))$$

$$= \int_{\omega \times (0, \varepsilon]} J_n T(u, t) \, \mathcal{H}^n(d(u, t))$$

$$= \int_\omega \int_0^\varepsilon \prod_{i=1}^{n-1} (r_i(K, u) + t) \, dt \, \mathcal{H}^{n-1}(du)$$

$$= \sum_{j=0}^{n-1} \binom{n-1}{j} \int_\omega \int_0^\varepsilon R_j(K, u) t^{n-1-j} \, dt \, \mathcal{H}^{n-1}(du)$$

$$= \frac{1}{n} \sum_{j=0}^{n-1} \varepsilon^{n-j} \binom{n}{j} \int_\omega R_j(K, u) \, \mathcal{H}^{n-1}(du)$$

$$= \frac{1}{n} \sum_{j=0}^{n-1} \varepsilon^{n-j} \binom{n}{j} S_j(K, \omega),$$

where we used Theorem 4.9 in the last step. $\qquad \square$

In order to extend Theorem 4.9 to mixed area measures, we introduce the notion of a *mixed discriminant*. Let $A(1), \ldots, A(n) \in \mathbb{R}^{n,n}$ be real (symmetric) $n \times n$ matrices. We shall write $a_i(j)$ for the ith column of $A(j)$, that is, $A(j) = (a_1(j), \ldots, a_n(j))$. We define

$$D(A(1), \ldots, A(n)) := \frac{1}{n!} \sum_{\sigma \in S(n)} \det(a_1(\sigma(1)), \ldots, a_n(\sigma(n))),$$

where $S(n)$ is the set of permutations of $\{1, \ldots, n\}$. By definition it is clear that D is symmetric in the n matrices and that $D(A, \ldots, A) = \det(A)$. The functional $D(\cdot)$ on n-tuples of real (symmetric) $n \times n$ matrices is called the mixed discriminant (of the matrices to which D is applied).

For $k \in \mathbb{N}$, real (symmetric) $n \times n$ matrices $A(1), \ldots, A(k) \in \mathbb{R}^{n,n}$, and $\alpha_1, \ldots, \alpha_k \in \mathbb{R}$, we get

$$\det\left(\sum_{i=1}^k \alpha_i A(i)\right) = \det\left(\sum_{r_1=1}^k \alpha_{r_1} a_1(r_1), \ldots, \sum_{r_n=1}^k \alpha_{r_n} a_n(r_n)\right)$$

$$= \sum_{r_1=1}^k \cdots \sum_{r_n=1}^k \alpha_{r_1} \cdots \alpha_{r_n} \det(a_1(r_1), \ldots, a_n(r_n))$$

$$= \sum_{r_1=1}^{k} \cdots \sum_{r_n=1}^{k} \alpha_{r_1} \cdots \alpha_{r_n} \frac{1}{n!} \sum_{\sigma \in S(n)} \det \left(a_1 (r_{\sigma(1)}), \ldots, a_n (r_{\sigma(n)}) \right)$$

$$= \sum_{r_1=1}^{k} \cdots \sum_{r_n=1}^{k} \alpha_{r_1} \cdots \alpha_{r_n} D \left(A(r_1), \ldots, A(r_n) \right) \tag{4.16}$$

$$= \sum_{i_1,\ldots,i_k=0}^{n} \binom{n}{i_1,\ldots,i_k} \alpha_1^{i_1} \cdots \alpha_k^{i_k} D(A(1)[i_1], \ldots, A(k)[i_k]) \tag{4.17}$$

where the number in brackets indicates the number of repetitions of the corresponding entry.

Since D is symmetric in its arguments, the coefficients on the right-hand side are uniquely determined by the left-hand side. As an immediate consequence of (4.16) (or by a direct argument), we obtain for matrices $C_1, C_2 \in \mathbb{R}^{n,n}$ and $A(1), \ldots, A(n) \in \mathbb{R}^{n,n}$ that

$$D(C_1 A(1) C_2, \ldots, C_1 A(n) C_2) = \det(C_1) D(A(1), \ldots, A(n)) \det(C_2).$$

This fact can be used to show (by induction over n) that the mixed discriminant of positive semi-definite matrices is nonnegative.

Let $K_1, \ldots, K_{n-1} \in \mathcal{K}^n$ be convex bodies with support functions of class $\mathbf{C}^2$. In the following, we write

$$D(h_{K_1}, \ldots, h_{K_{n-1}})(u) := D \left(d^2 h_{K_1}(u)|u^\perp, \ldots, d^2 h_{K_{n-1}}(u)|u^\perp \right)$$

for the mixed discriminant of the Hessians of the support functions of $n - 1$ convex bodies, considered as $(n - 1) \times (n - 1)$ matrices, with respect to an orthonormal basis of $u^\perp$.

Theorem 4.11 *Let $K_1, \ldots, K_{n-1} \in \mathcal{K}^n$ be convex bodies with support functions of class $\mathbf{C}^2$. Then*

$$S(K_1, \ldots, K_{n-1}, \cdot) = \int_{\mathbb{S}^{n-1}} \mathbf{1}\{u \in \cdot\} D \left(h_{K_1}, \ldots, h_{K_{n-1}} \right)(u) \, \mathcal{H}^{n-1}(du).$$

Proof Let $\alpha_1, \ldots, \alpha_{n-1} \geq 0$ and $K = \alpha_1 K_1 + \cdots + \alpha_{n-1} K_{n-1}$. Since

$$d^2 h_K = \sum_{i=1}^{n-1} \alpha_i d^2 h_{K_i},$$

we obtain from (4.16)

$$\det\left(d^2 h_K(u)|u^{\perp}\right) = \det\left(\sum_{i=1}^{n-1}\alpha_i d^2 h_{K_i}(u)|u^{\perp}\right)$$

$$= \sum_{r_1=1}^{n-1}\cdots\sum_{r_{n-1}=1}^{n-1}\alpha_{r_1}\cdots\alpha_{r_{n-1}}D\left(h_{K_{r_1}},\ldots,h_{K_{r_{n-1}}}\right)(u).$$

Then the special case $j = n - 1$ of Theorem 4.9 and (4.11) yield

$$S_{n-1}\left(\sum_{i=1}^{n-1}\alpha_i K_i,\cdot\right) = \int_{\mathbb{S}^{n-1}} \mathbf{1}\{u \in \cdot\}\det\left(\sum_{i=1}^{n-1}\alpha_i d^2 h_{K_i}(u)|u^{\perp}\right)\mathcal{H}^{n-1}(du)$$

$$= \sum_{r_1=1}^{n-1}\cdots\sum_{r_{n-1}=1}^{n-1}\alpha_{r_1}\cdots\alpha_{r_{n-1}}$$

$$\times \int_{\mathbb{S}^{n-1}} \mathbf{1}\{u \in \cdot\}D\left(h_{K_{r_1}},\ldots,h_{K_{r_{n-1}}}\right)(u)\,\mathcal{H}^{n-1}(du).$$

A comparison of coefficients with the expansion (4.4) for $m = n - 1$ yields the assertion. □

Let $K \in \mathcal{K}^n$ be a convex body with support function of class C^2. Comparing Theorem 4.9 and the special case of Theorem 4.11 where $K_1 = \cdots = K_j = K$ and the remaining bodies are B^n, we obtain

$$R_j(K, u) = D(h_K[j], h_{B_n}[n - 1 - j])(u), \qquad u \in \mathbb{S}^{n-1}.$$

For $j = 1$, we have

$$R_1(K, u) = \frac{1}{n-1}\Delta h_K(u), \qquad u \in \mathbb{S}^{n-1}.$$

Let $\Delta_s f$ denote the spherical Laplace operator of a twice differentiable function $f : \mathbb{S}^{n-1} \to \mathbb{R}$. It can be obtained as the Euclidean Laplace operator of the extension of f to a function on $\mathbb{R}^n \setminus \{0\}$ which is homogeneous of degree zero. Thus we get

$$R_1(K, u) = h_K(u) + \frac{1}{n-1}\Delta_s h_K(u), \qquad u \in \mathbb{S}^{n-1}.$$

See Exercise 4.3.5 for further details.

Exercises and Supplements for Sect. 4.3

1. (a) Let $K \in \mathcal{K}_0^n$. Show that the restriction of σ_K to $\mathrm{reg}(K)$ is continuous.
   (b) Let K_i, $K \in \mathcal{K}^n$ with $o \in \mathrm{int}(K_i)$, $\mathrm{int}(K)$. Let $u \in \tilde{\pi}(\mathrm{reg}(K)) \cap \tilde{\pi}(\mathrm{reg}(K_i))$
       for $i \geq 1$. Show that if $K_i \to K$ as $i \to \infty$, then $\sigma_{K_i}(\rho_{K_i}(u)u) \to$
       $\sigma_K(\rho_K(u)u)$.
   (c) Show that the expressions in (4.10) are uniformly bounded.
2. Let $K \in \mathcal{K}_0^n$. Show that $\mathcal{H}^{n-1}(\partial K \setminus \mathrm{reg}(K)) = 0$.
       Hint: Exercises 2.2.7 and 2.2.13 can be used. More generally, one can show
   that the set of singular (that is, not regular) boundary points has σ-finite $(n-2)$-
   dimensional Hausdorff measure.
3. Show that if $a \in \mathbb{R}^n$, then

$$\det \left(I - a \cdot a^\top \right) = 1 - \|a\|^2.$$

4. Show that the mixed discriminant of positive semi-definite matrices is nonnegative.
5. For a map $f : \mathbb{S}^{n-1} \to \mathbb{R}$ we define the 0-homogeneous extension of f to $\mathbb{R}^n \setminus \{0\}$
   by $f_0(x) := f(\|x\|^{-1}x)$. If f is differentiable, then f_0 can be used to introduce
   the spherical gradient and the spherical Laplacian of f by

$$\nabla_s f(u) := \nabla f_0(u), \qquad \Delta_s f(u) := \Delta f_0(u), \qquad u \in \mathbb{S}^{n-1}.$$

   Here ∇ and Δ denote the Euclidean gradient and Laplace operator (with respect
   to an arbitrary orthonormal basis).
       Now let $f_1 : \mathbb{R}^n \setminus \{0\} \to \mathbb{R}$ be (positively) 1-homogeneous, that is, $f_1(tx) =$
   $t f_1(x)$ for $x \in \mathbb{R}^n \setminus \{0\}$ and $t > 0$. The restriction of f_1 to $\mathbb{S}^{n-1}$ is denoted by f
   and f_0 denotes the 0-homogenous extension of f, hence $f_0(x) = f(\|x\|^{-1}x) =$
   $f_1(\|x\|^{-1}x)$ for $x \in \mathbb{R}^n \setminus \{0\}$. Prove the following facts.

   (a) Suppose that f_1 is differentiable at $u \in \mathbb{S}^{n-1}$. Then f_0 is differentiable at u
       and

$$\nabla_s f(u) = \nabla f_1(u) - f(u)\, u.$$

   (b) Suppose that f_1 is twice differentiable at $u \in \mathbb{S}^{n-1}$. Then f_0 is twice
       differentiable at u and

$$\Delta_s f(u) = \Delta f_1(u) - (n-1)f(u).$$

6. Let $f, g : \mathbb{S}^{n-1}$ be of class $\mathbf{C}^2$. Prove the following integral formulas.

(a)

$$\int_{\mathbb{S}^{n-1}} \langle \nabla_s f(u), \nabla_s g(u) \rangle \, d\mathcal{H}^{n-1} = -\int_{\mathbb{S}^{n-1}} f \, \Delta_s g \, d\mathcal{H}^{n-1}.$$

(b)

$$\int_{\mathbb{S}^{n-1}} f \, \Delta_s g \, d\mathcal{H}^{n-1} = \int_{\mathbb{S}^{n-1}} g \, \Delta_s f \, d\mathcal{H}^{n-1}.$$

7. We continue to use the notation of Exercises 4.3.5 and 4.3.6. Let $K \in \mathcal{K}_0^n$ be a convex body with nonempty interior and support function $h = h_K$ on $\mathbb{R}^n \setminus \{0\}$. Since h is 1-homogeneous, we have $h_1 = h$ and also denote by h the restriction of h_K to $\mathbb{S}^{n-1}$. Clearly, h is differentiable at u as a function on $\mathbb{R}^n \setminus \{0\}$ if and only if the restriction of h to $\mathbb{S}^{n-1}$ is differentiable at u as a function on $\mathbb{S}^{n-1}$. Moreover, h is differentiable at $u \in \mathbb{S}^{n-1}$ if and only if h is differentiable at tu for some (and then also for all) $t > 0$.

If h is differentiable at $u \in \mathbb{S}^{n-1}$, then $\nabla_s h(u) = \nabla h(u) - h(u)u$ for $u \in \mathbb{S}^{n-1}$.

(a) Let $K_i \in \mathcal{K}_0^n$, $i \in \mathbb{N}$, be convex bodies with nonempty interiors and differentiable support functions $h_i = h_{K_i}$ such that $K_i \to K$ as $i \to \infty$. Show that if h is differentiable at $u \in \mathbb{S}^{n-1}$, then $\nabla_s h_i(u) \to \nabla_s h(u)$ as $i \to \infty$.
(b) For $\mathcal{H}^{n-1}$-almost all $u \in \mathbb{S}^{n-1}$, h is differentiable at $u \in \mathbb{S}^{n-1}$.
(c) Show that

$$V(K[2], B^n[n-2]) = \frac{1}{n} \int_{\mathbb{S}^{n-1}} h^2 - \frac{1}{n-1} \|\nabla_s h\|^2 \, d\mathcal{H}^{n-1}.$$

Hint: First, prove the assertion for a convex body K with a smooth support function. Then use (a), (b), the fact that an arbitrary convex body can be approximated by convex bodies with smooth support functions (see the proof of Theorem 4.16 and the reference given there) and the dominated convergence theorem.

4.4 Projection Bodies and Zonoids

For a convex body $K \in \mathcal{K}^n$ and a direction $u \in \mathbb{S}^{n-1}$, we define

$$v(K, u) := V_{n-1}(K \mid u^\perp),$$

the $(n-1)$-dimensional volume of the orthogonal projection $K|u^{\perp}$ of K onto the hyperplane through 0 and orthogonal to u. The function $v(K, \cdot)$ thus defined on $\mathbb{S}^{n-1}$ is called the *projection function* of K. We are interested in the information on the shape of K which can be deduced from the knowledge of its projection function $v(K, \cdot)$.

It is clear that for $K \in \mathcal{K}^n$, the translates $K + x$, $x \in \mathbb{R}^n$, of K have the same projection function. Furthermore, K and $-K$ have the same projection function. This shows that in general K is not determined by $v(K, \cdot)$ (not even up to translations). However, the question remains whether we get uniqueness up to translations and reflections. In order to give an answer, we first provide an analytic representation of $v(K, \cdot)$.

Theorem 4.12 *If $K \in \mathcal{K}^n$ and $u \in \mathbb{S}^{n-1}$, then*

$$v(K, u) = \frac{1}{2} \int_{\mathbb{S}^{n-1}} |\langle x, u \rangle|\, S_{n-1}(K, dx).$$

Proof Let $u \in \mathbb{S}^{n-1}$. An application of Fubini's theorem shows that

$$V(K + [-u, u]) = V(K) + 2v(K, u).$$

On the other hand, we have

$$V(K + [-u, u]) = \sum_{i=0}^{n} \binom{n}{i} V(K[i], [-u, u][n-i]).$$

From Exercise 3.3.1, we know that $V(K[i], [-u, u][n-i]) = 0$ holds for $i = 0, \ldots, n-2$, and hence

$$v(K, u) = \frac{n}{2} V(K, \ldots, K, [-u, u]). \tag{4.18}$$

The assertion now follows from Theorem 4.1, since the segment $[-u, u]$ has the support function $h_{[-u,u]} = |\langle \cdot, u \rangle|$. $\square$

Remark 4.5 Various properties of projection functions can be directly deduced from Theorem 4.12.

(1) We have $v(K, \cdot) = 0$ if and only if $\dim K \leq n - 2$.
(2) If $\dim K = n - 1$ and $K \subset x^{\perp}$, then

$$v(K, \cdot) = V_{n-1}(K) |\langle x, \cdot \rangle|.$$

(3) If $\dim K = n$ and K is not centrally symmetric, that is, if $S_{n-1}(K, \cdot) \neq S_{n-1}(-K, \cdot)$, then there is an infinite family of convex bodies with the same projection function. To see this, observe that Theorem 4.5 ensures that for $\alpha \in [0, 1]$ there is a body $K_\alpha \in \mathcal{K}^n$ with $\dim K_\alpha = n$ and

$$S_{n-1}(K_\alpha, \cdot) = \alpha S_{n-1}(K, \cdot) + (1 - \alpha) S_{n-1}(-K, \cdot).$$

For $\alpha \neq \beta$ the convex bodies K_α and K_β are not translates of each other. Moreover, $K_\alpha = -K_\beta$ if and only if $\alpha + \beta = 1$. On the other hand, for $\alpha \in [0, 1]$ we have

$$v(K_\alpha, \cdot) = \alpha v(K, \cdot) + (1 - \alpha) v(-K, \cdot) = v(K, \cdot).$$

This discussion also shows that there is always a centrally symmetric body, namely $K_{\frac{1}{2}}$, with the same projection function as K.

The convex body $K_{\frac{1}{2}}$ has maximal volume in the class $\mathcal{C} := \{K_\alpha : \alpha \in [0, 1]\}$ (by the Brunn–Minkowski theorem) and it is characterized by this fact. Hence $K_{\frac{1}{2}}$ is the unique convex body in $\mathcal{C}$ with maximal volume (see Exercise 4.2.2).

(4) Since $|\langle x, \cdot \rangle|$ is a support function, the function $v(K, \cdot)$ is a 'positive combination' of support functions, hence it is itself a support function of a convex body ΠK. More precisely,

$$h_{\Pi K} := v(K, \cdot) = \frac{1}{2} \int_{S^{n-1}} |\langle \cdot, x \rangle| \, S_{n-1}(K, dx) \tag{4.19}$$

is subadditive and positively homogeneous of degree 1 as a function on $\mathbb{R}^n$.

Definition 4.2 For $K \in \mathcal{K}^n$, the convex body ΠK defined by (4.19) is called the *projection body* of K.

Remark 4.6 The projection body ΠK of $K \in \mathcal{K}^n$ is always centrally symmetric to the origin and $\dim \Pi K = n$ if and only if $\dim K = n$.

Example 4.1 We determine the projection body of the unit cube $C^n := [-\frac{1}{2}, \frac{1}{2}]^n = \frac{1}{2} \sum_{i=1}^n [-e_i, e_i]$. Observe that C^n is a Minkowski sum of segments. Clearly, we have

$$S_{n-1}(C^n, \cdot) = \sum_{i=1}^n \delta_{e_i} + \sum_{i=1}^n \delta_{-e_i},$$

and therefore

$$h(\Pi C^n, v) = \frac{1}{2} \int_{\mathbb{S}^{n-1}} |\langle u, v \rangle| \, S_{n-1}(C^n, du) = \sum_{i=1}^{n} |\langle v, e_i \rangle|$$

$$= \sum_{i=1}^{n} h_{[-e_i, e_i]}(v) = h\left(\sum_{i=1}^{n} [-e_i, e_i], v \right)$$

$$= h(2C^n, v).$$

This shows that $\Pi C^n = 2C^n$.

Example 4.2 It is easy to see that $\Pi B^n = \kappa_{n-1} B^n$.

Remark 4.7 Examples 4.1 and 4.2 suggest to determine the (centrally symmetric) convex bodies for which the projection body is a multiple of the body itself.

Remark 4.8 Projection bodies of ellipsoids are ellipsoids and projection bodies of parallelotopes are parallelotopes. This is a consequence of Examples 4.1 and 4.2 and of the general relation

$$\Pi(\varphi K) = |\det \varphi| \varphi^{-\top} \Pi K, \tag{4.20}$$

which holds for any regular affine transformation φ of $\mathbb{R}^n$. Relation 4.20 can be seen from $h(\Pi K, u) = \frac{n}{2} V(K[n-1], [-u, u])$ for $u \in \mathbb{R}^n$ and the transformation properties of mixed volumes. Thus we obtain for an ellipsoid E with center 0 that

$$\Pi E = \frac{\kappa_{n-1}}{\kappa_n} V(E) \, E^\circ,$$

where $K^\circ = \{x \in \mathbb{R}^n : \langle x, z \rangle \le 1 \text{ for } z \in K\}$ is the polar body of $K \in \mathcal{K}^n$ with $0 \in \mathrm{int}\, K$ (see Exercises 1.1.14 and 2.3.2).

Example 4.3 In Examples 4.1 and 4.2, we determined the projection body of a symmetric body. Next we consider the projection body of a simplex $T \subset \mathbb{R}^n$ and show that

$$\Pi T = n V(T) \, (T - T)^\circ.$$

By the affine covariance expressed by (4.20), it is sufficient to verify this relation for a regular simplex T_n. Specifically, we can choose $a_0, \ldots, a_n \in \mathbb{R}^n$ and unit vectors $u_0, \ldots, u_n \in \mathbb{S}^{n-1}$ such that

$$T_n = \mathrm{conv}\{a_0, \ldots, a_n\} = \bigcap_{i=0}^{n} H^-(u_i, 1),$$

where $a_i = -nu_i$ for $i = 0, \ldots, n$, $\|a_i - a_j\|^2 = 2n(n + 1)$ for $i \neq j$, and $\langle a_i, a_j \rangle = n^2$ for $i = j$ and $\langle a_i, a_j \rangle = -n$ for $i \neq j$. In particular, $a_0 + \cdots + a_n = 0$ and $(n + 1)v = nV(T_n)$ if v is the $(n - 1)$-dimensional volume of the facets of T_n. Then

$$h(\Pi T_n, x) = \frac{1}{2} \int_{\mathbb{S}^{n-1}} |\langle x, u \rangle| \, S_{n-1}(T_n, du) = \sum_{k=0}^{n} \frac{1}{2} |\langle x, u_k \rangle| v$$

$$= \sum_{k=0}^{n} \frac{nV(T_n)}{2(n + 1)} |\langle x, u_k \rangle| = \sum_{k=0}^{n} \frac{nV(T_n)}{2(n + 1)} h_{[-u_k, u_k]}(x),$$

hence

$$\Pi T_n = \sum_{k=0}^{n} \frac{nV(T_n)}{n + 1} [-u_k/2, u_k/2] = \sum_{k=0}^{n} \frac{V(T_n)}{n + 1} [-a_k/2, a_k/2]$$

$$= \sum_{k=0}^{n} \frac{V(T_n)}{n + 1} [0, a_k],$$

where we used that $a_0 + \cdots + a_n = 0$. Using Exercises 1.19, 1.1.15 and 2.3.2, it follows that

$$(T_n - T_n)^\circ = \mathrm{conv}\{a_i - a_j : i \neq j\}^\circ = \bigcap_{i \neq j} H^-(a_i - a_j, 1).$$

Thus, it remains to be shown that

$$\frac{1}{n + 1} \sum_{k=0}^{n} [0, a_k] = \bigcap_{i \neq j} H^-(a_i - a_j, n).$$

If x is in the set on the left-hand side, then there are $\lambda_k \in [0, 1]$ such that

$$x = \sum_{k=0}^{n} \frac{\lambda_k}{n + 1} a_k.$$

Hence

$$\langle x, a_i - a_j \rangle = \sum_{k=0}^{n} \frac{\lambda_k}{n + 1} \langle a_k, a_i - a_j \rangle = \frac{\lambda_i - \lambda_j}{n + 1} n(n + 1) = n(\lambda_i - \lambda_j) \leq n$$

for $i \neq j$.

For the converse, suppose that x is in the set on the right-hand side. Any $x \in \mathbb{R}^n$ can be written in the form $x = \sum_{k=0}^{n} \lambda_k a_k$ with $\lambda_k \geq 0$ (this representation is not unique, of course). Then

$$n \geq \langle x, a_i - a_j \rangle = \sum_{k=0}^{n} \lambda_k \langle a_k, a_i - a_j \rangle = (\lambda_i - \lambda_j) n(n+1),$$

which shows that $\lambda_i - \lambda_j \leq 1/(n+1)$ for $i \neq j$. By symmetry, we may assume that $0 \leq \lambda_0 = \min\{\lambda_k : k = 0, \ldots, n\}$. Then

$$x = \sum_{l=1}^{n} (\lambda_l - \lambda_0) a_k \in \frac{1}{n+1} \sum_{k=0}^{n} [0, a_k],$$

which proves the reverse inclusion.

Example 4.3 modifies an argument by H. Martini and B. Weissbach [67]. It is natural to ask for further examples of convex polytopes whose projection bodies are proportional to the polar of their difference body. By a result of H. Martini [66], simplices are the only examples. Among all convex bodies, the example of a ball (or an ellipsoid) shows that there are further examples of convex bodies for which the projection body and the (difference body of the) given body are polars.

Before we continue to discuss projection functions, we describe projection bodies geometrically.

Definition 4.3 A finite sum of segments $Z := s_1 + \cdots + s_k$ is called a *zonotope*. A *zonoid* is a convex body which is the limit (in the Hausdorff metric) of a sequence of zonotopes.

Zonotopes are polytopes, since the Minkowski sum of convex hulls of arbitrary sets is equal to the convex hull of the Minkowski sum of these sets (see Exercise 1.1.8). Moreover, zonotopes are centrally symmetric. Namely, if $s_i = [-y_i, y_i] + x_i$ is the representation of the segment s_i with center x_i and endpoints $-y_i + x_i, y_i + x_i$, then

$$Z = \sum_{i=1}^{k} [-y_i, y_i] + \sum_{i=1}^{k} x_i.$$

Hence, $x := \sum_{i=1}^{k} x_i$ is the center of Z. Zonoids, as limits of zonotopes, are also centrally symmetric. In the following, we assume w.l.o.g. that the center of zonotopes and zonoids is the origin and denote the corresponding set of zonoids by $\mathcal{Z}^n$.

The following result is the key step in showing that zonoids and projection bodies are closely related. In particular, it provides an analytic description of the support function of a zonoid.

Theorem 4.13 *Let $K \in \mathcal{K}^n$. Then K is a zonoid if and only if there exists an even Borel measure $\mu(K, \cdot)$ on $\mathbb{S}^{n-1}$ such that*

$$h_K(u) = \int_{\mathbb{S}^{n-1}} |\langle x, u \rangle| \, \mu(K, dx), \quad u \in \mathbb{R}^n. \tag{4.21}$$

For a zonoid K, a measure $\mu(K, \cdot)$ such that (4.21) is satisfied is called a *generating measure* of K. It follows from Theorem 4.14 that $\mu(K, \cdot)$ is uniquely determined by K (if $\mu(K, \cdot)$ is even).

Proof (of Theorem 4.13) Suppose that

$$h_K(u) = \int_{\mathbb{S}^{n-1}} |\langle x, u \rangle| \, \mu(K, dx), \quad u \in \mathbb{R}^n,$$

where $\mu(K, \cdot)$ is an even Borel measure on $\mathbb{S}^{n-1}$. Symmetrizing the discrete measures obtained from Lemma 4.2, we find a sequence of even, discrete measures

$$\mu_j = \frac{1}{2} \sum_{i=1}^{k(j)} \alpha_{ij} (\delta_{u_{ij}} + \delta_{-u_{ij}}), \quad u_{ij} \in \mathbb{S}^{n-1}, \alpha_{ij} \geq 0,$$

on $\mathbb{S}^{n-1}$ such that $\mu_j \to \mu(K, \cdot)$. Then

$$Z_j := \sum_{i=1}^{k(j)} [-\alpha_{ij} u_{ij}, \alpha_{ij} u_{ij}]$$

is a zonotope and

$$h_{Z_j}(u) = \int_{\mathbb{S}^{n-1}} |\langle x, u \rangle| \, \mu_j(dx) \to \int_{\mathbb{S}^{n-1}} |\langle x, u \rangle| \, \mu(K, dx) = h_K(u),$$

uniformly in $u \in \mathbb{S}^{n-1}$ as $j \to \infty$. Therefore, $Z_j \to K$ as $j \to \infty$, that is, K is a zonoid.

Conversely, assume that $K = \lim_{j \to \infty} Z_j$, where Z_j is a zonotope for $j \in \mathbb{N}$. Then,

$$Z_j = \sum_{i=1}^{k(j)} [-y_{ij}, y_{ij}]$$

with suitable points $y_{ij} \in \mathbb{R}^n$. Consequently,

$$h_{Z_j}(u) = \sum_{i=1}^{k(j)} |\langle y_{ij}, u \rangle| = \int_{\mathbb{S}^{n-1}} |\langle x, u \rangle| \, \mu_j(dx),$$

where

$$\mu_j := \frac{1}{2} \sum_{i=1}^{k(j)} \|y_{ij}\| (\delta_{u_{ij}} + \delta_{-u_{ij}})$$

and

$$u_{ij} := \frac{y_{ij}}{\|y_{ij}\|} \quad \text{for } y_{ij} \neq 0;$$

if $y_{ij} = 0$, then we can choose an arbitrary unit vector for u_{ij}.

Next we show that the sequence $(\mu_j)_{j \in \mathbb{N}}$ has a weakly convergent subsequence. From (3.17) and the continuity of intrinsic volumes it follows that

$$\int_{\mathbb{S}^{n-1}} h_{Z_j}(u)\,\sigma(du) = \kappa_{n-1} V_1(Z_j) \to \kappa_{n-1} V_1(K)$$

(or using the fact that $h_{Z_j} \to h_K$ uniformly on $\mathbb{S}^{n-1}$), and hence we obtain that this sequence of integrals is bounded. On the other hand, by Fubini's theorem and Theorem 4.12 (for the unit ball), we get

$$\int_{\mathbb{S}^{n-1}} h_{Z_j}(u)\,\sigma(du) = \int_{\mathbb{S}^{n-1}} \int_{\mathbb{S}^{n-1}} |\langle x, u \rangle|\,\sigma(du)\,\mu_j(dx) = 2\kappa_{n-1}\mu_j(\mathbb{S}^{n-1}).$$

Hence, there is a constant C such that $\mu_j(\mathbb{S}^{n-1}) \leq C$ for $j \in \mathbb{N}$. Now we use the fact that the set $\mathcal{M}_C$ of all Borel measures ρ on $\mathbb{S}^{n-1}$ with $\rho(\mathbb{S}^{n-1}) \leq C$ is weakly sequentially compact (see [14, p. 37], [33, p. 344, Satz 8.4.13]). Therefore, $(\mu_j)_{j \in \mathbb{N}}$ has a convergent subsequence. We may assume that $(\mu_j)_{j \in \mathbb{N}}$ converges to a limit measure, which we denote by $\mu(K, \cdot)$. The weak convergence implies that

$$h_K(u) = \lim_{j \to \infty} h_{Z_j}(u) = \lim_{j \to \infty} \int_{\mathbb{S}^{n-1}} |\langle x, u \rangle|\,\mu_j(dx) = \int_{\mathbb{S}^{n-1}} |\langle x, u \rangle|\,\mu(K, dx)$$

for $u \in \mathbb{S}^{n-1}$, which yields the asserted result. $\square$

Remark 4.9 It follows from Eq. (4.21) that $\dim K = n$ if and only if $\dim \mu(K, \cdot) = n$. Moreover, since

$$\int_{\mathbb{S}^{n-1}} |\langle x, u \rangle|\,\mu(dx) = \int_{\mathbb{S}^{n-1}} |\langle x, u \rangle|\,\mu^*(dx), \quad u \in \mathbb{R}^n,$$

where $\mu^*(A) := \frac{1}{2}(\mu(A) + \mu(-A))$ for Borel sets $A \subset \mathbb{S}^{n-1}$, a convex body K is a zonoid if and only if there exists a (not necessarily even) Borel measure μ on $\mathbb{S}^{n-1}$

such that

$$h_K(u) = \int_{\mathbb{S}^{n-1}} |\langle x, u \rangle| \, \mu(dx), \quad u \in \mathbb{R}^n.$$

However, requiring the measure μ to be even will allow us to show that the generating measure is unique.

Finally, we connect zonoids to projection bodies.

Corollary 4.3 *The projection body ΠK of a convex body K is a zonoid. Conversely, if Z is a zonoid with $\dim Z = n$, then there is a convex body K with $\dim K = n$, centrally symmetric with respect to the origin and such that $Z = \Pi K$.*

Proof The first result follows from Theorems 4.12, 4.13 and Remark 4.9.

For the second, let Z be a zonoid. Then Theorem 4.13 shows that

$$h_Z(u) = \int_{\mathbb{S}^{n-1}} |\langle x, u \rangle| \, \mu(Z, dx)$$

with an even n-dimensional measure $\mu(Z, \cdot)$. By Theorem 4.5, there exists an n-dimensional convex body $K \in \mathcal{K}^n$ such that $2\mu(Z, \cdot) = S_{n-1}(K, \cdot)$. Hence

$$h_Z(u) = \frac{1}{2} \int_{\mathbb{S}^{n-1}} |\langle x, u \rangle| \, S_{n-1}(K, dx) = h(\Pi K, u), \quad u \in \mathbb{R}^n,$$

that is, $Z = \Pi K$. By Corollary 4.2, K is centrally symmetric. $\qquad\square$

Now we want to show that the generating measure of a zonoid is uniquely determined. We start with two auxiliary lemmas. If A is the $(n \times n)$-matrix of an injective linear mapping in $\mathbb{R}^n$, we define $AZ := \{Ax : x \in Z\}$ and denote by $A\mu$, for a Borel measure μ on $\mathbb{S}^{n-1}$, the image measure of

$$\int_{\mathbb{S}^{n-1}} \mathbf{1}\{x \in \cdot\} \|Ax\| \, \mu(dx)$$

under the mapping

$$x \mapsto \frac{Ax}{\|Ax\|}, \quad x \in \mathbb{S}^{n-1}.$$

In other words,

$$(A\mu)(\cdot) = \int_{\mathbb{S}^{n-1}} \mathbf{1}\left\{ \frac{Ax}{\|Ax\|} \in \cdot \right\} \|Ax\| \, \mu(dx).$$

Clearly, if μ is an even measure, so is $A\mu$.

Lemma 4.6 *If $Z \in \mathcal{K}^n$ is a zonoid and*

$$h_Z = \int_{\mathbb{S}^{n-1}} |\langle x, \cdot \rangle| \, \mu(Z, dx),$$

then AZ is a zonoid and

$$h_{AZ} = \int_{\mathbb{S}^{n-1}} |\langle x, \cdot \rangle| \, (A\mu(Z, \cdot))(dx).$$

Proof We have

$$h_{AZ}(u) = \sup_{x \in AZ} \langle u, x \rangle = \sup_{x \in Z} \langle u, Ax \rangle = \sup_{x \in Z} \langle A^\top u, x \rangle = h_Z(A^\top u)$$

$$= \int_{\mathbb{S}^{n-1}} |\langle x, A^\top u \rangle| \, \mu(Z, dx) = \int_{\mathbb{S}^{n-1}} |\langle Ax, u \rangle| \, \mu(Z, dx)$$

$$= \int_{\mathbb{S}^{n-1}} \left| \left\langle \frac{Ax}{\|Ax\|}, u \right\rangle \right| \|Ax\| \, \mu(Z, dx) = \int_{\mathbb{S}^{n-1}} |\langle y, u \rangle| \, (A\mu(Z, \cdot))(dy),$$

which proves all assertions.   □

Let $\mathcal{V}$ denote the vector space of functions

$$f = \int_{\mathbb{S}^{n-1}} |\langle x, \cdot \rangle| \, \nu_1(dx) - \int_{\mathbb{S}^{n-1}} |\langle x, \cdot \rangle| \, \nu_2(dx)$$

on the unit sphere $\mathbb{S}^{n-1}$, where ν_1, ν_2 vary among the finite even Borel measures on $\mathbb{S}^{n-1}$. Then $\mathcal{V}$ is a subspace of the Banach space $\mathbf{C}_e(\mathbb{S}^{n-1})$ of even continuous functions on $\mathbb{S}^{n-1}$ with the maximum norm.

Lemma 4.7 *The vector space $\mathcal{V}$ is dense in $\mathbf{C}_e(S^{n-1})$.*

Proof First, observe that $\mathrm{cl}\,\mathcal{V}$ is also a vector space.

Choosing for μ a nonnegative multiple of the spherical Lebesgue measure and $\rho = 0$ (or vice versa), we see that $\mathcal{V}$ contains the constant functions.

By Lemma 4.6, for a regular $(n \times n)$-matrix A the support function h_{AB^n} lies in $\mathcal{V}$. Since $h_{AB^n}(u) = \|A^\top u\|$, $u \in \mathbb{R}^n$, and since $\|A^\top u\| = \langle AA^\top u, u \rangle^{1/2}$, it follows that $\mathcal{V}$ contains all functions

$$f_B(u) =:= \sqrt{\langle Bu, u \rangle} = \left(\sum_{i,j=1}^{n} B_{ij} u_i u_j \right)^{\frac{1}{2}}, \quad u \in \mathbb{S}^{n-1},$$

where $B \in \mathbb{R}^{n,n}$ is a symmetric, positive definite matrix with entries B_{ij}, $i, j \in \{1, \ldots, n\}$, and where $u = (u_1, \ldots, u_n)^\top$.

Let $i_0, j_0 \in \{1, \ldots, n\}$ be fixed for the moment. Let $\Delta_{i_0 j_0} \in \mathbb{R}^{n,n}$ denote the symmetric matrix which has entry 1 in the positions (i_0, j_0) and (j_0, i_0) and zero in all other positions of the matrix. Then we define

$$B_{i_0 j_0}(\varepsilon) := B + \varepsilon \, \Delta_{i_0 j_0} \in \mathbb{R}^{n,n}, \quad \varepsilon \geq 0,$$

hence $B_{i_0 j_0}(\varepsilon)$ is symmetric, $B_{i_0 j_0}(0) = B$, and $B_{i_0 j_0}(\varepsilon)$ is positive definite if $\varepsilon \geq 0$ is sufficiently small. Next we consider

$$F_{i_0 j_0}(\varepsilon, u) := \sqrt{\langle B_{i_0 j_0}(\varepsilon) u, u \rangle}, \quad u \in \mathbb{S}^{n-1}, \, \varepsilon \geq 0 \text{ sufficiently small.}$$

By the mean value theorem, if $\varepsilon \geq 0$ is sufficiently small, then there are $\theta(\varepsilon) \in (0, 1)$ such that

$$\mathcal{V} \ni \frac{F_{i_0 j_0}(\varepsilon, u) - F_{i_0 j_0}(0, u)}{\varepsilon} = \frac{\partial F_{i_0 j_0}}{\partial \varepsilon}(\theta(\varepsilon)\varepsilon, u)$$

$$\rightarrow \frac{\partial F_{i_0 j_0}}{\partial \varepsilon}(0, u) = \begin{cases} \dfrac{u_{i_0} u_{j_0}}{f_B(u)}, & \text{for } i_0 \neq j_0, \\[3mm] \dfrac{u_{i_0}^2}{2 f_B(u)}, & \text{for } i_0 = j_0, \end{cases}$$

as $\varepsilon \downarrow 0$, uniformly in $u \in \mathbb{S}^{n-1}$, since $F_{i_0 j_0}(\varepsilon, u) > 0$ is uniformly bounded from below if $\varepsilon \geq 0$ is sufficiently small.

This shows that

$$f_B^{(1)}(u) := u_{i_0} u_{j_0} f_B(u)^{-1}, \quad u \in \mathbb{S}^{n-1},$$

defines a function $f_B^{(1)} \in \operatorname{cl} \mathcal{V}$. Repeating the previous argument with all possible pairs of indices, for $k \in \mathbb{N}$ we obtain

$$f_B^{(k)}(u) := u_1^{i_1} \cdots u_n^{i_n} f_B(u)^{-k}, \quad u \in \mathbb{S}^{n-1},$$

where $i_1, \ldots, i_n \in \mathbb{N}_0$ are such that $i_1 + \cdots + i_n = 2k$, which defines a function in $\operatorname{cl} \mathcal{V}$.

Now we choose B to be the unit matrix. Then $f_B \equiv 1$, hence the restriction to the unit sphere of every even polynomial is in $\operatorname{cl} \mathcal{V}$. The Stone–Weierstrass theorem (see Exercise 4.1.1) now implies that $\operatorname{cl} \mathcal{V} = \mathbf{C}_e(\mathbb{S}^{n-1})$ (see Exercise 4.4.1). $\square$

Lemma 4.7 can be expressed by saying that finite linear combinations of functions of the form $|\langle x_i, \cdot \rangle|$ on $\mathbb{S}^{n-1}$, where $x_i \in \mathbb{S}^{n-1}$, are dense in $\mathbf{C}_e(\mathbb{S}^{n-1})$. The statement of Lemma 4.7 is formally stronger, but the special case implies that the stronger assertion is true, since finite measures can be approximated by discrete measures.

Theorem 4.14 *For a zonoid* $Z \in \mathcal{K}^n$, *the (even) generating measure of* Z *is uniquely determined.*

Equivalently, if μ, ρ *are even finite Borel measures on* $\mathbb{S}^{n-1}$ *such that*

$$\int_{\mathbb{S}^{n-1}} |\langle x, \cdot \rangle| \, \mu(dx) = \int_{\mathbb{S}^{n-1}} |\langle x, \cdot \rangle| \, \rho(dx),$$

then $\mu = \rho$.

Proof Let us suppose that we have two even measures μ and ρ on $\mathbb{S}^{n-1}$ with

$$\int_{\mathbb{S}^{n-1}} |\langle x, \cdot \rangle| \, \mu(dx) = \int_{\mathbb{S}^{n-1}} |\langle x, \cdot \rangle| \, \rho(dx).$$

Then

$$\int_{\mathbb{S}^{n-1}} \int_{\mathbb{S}^{n-1}} |\langle x, u \rangle| \, \mu(dx) \, \nu(du) = \int_{\mathbb{S}^{n-1}} \int_{\mathbb{S}^{n-1}} |\langle x, u \rangle| \, \rho(dx) \, \nu(du),$$

for all measures ν on $\mathbb{S}^{n-1}$. Applying Fubini's theorem and taking differences of the corresponding equations obtained for two measures ν_1, ν_2, we obtain

$$\int_{\mathbb{S}^{n-1}} f(x) \, \mu(dx) = \int_{\mathbb{S}^{n-1}} f(x) \, \rho(dx),$$

for all functions $f \in \mathcal{V}$. Lemma 4.7 shows that this implies that $\mu = \rho$.   □

Remark 4.10 Let $\mathcal{M}_e(\mathbb{S}^{n-1})$ denote the set of all finite, even Borel measures on the unit sphere. The map $C : \mathcal{M}_e(\mathbb{S}^{n-1}) \to \mathbf{C}_e(\mathbb{S}^{n-1})$, $\mu \mapsto C(\mu)$, given by

$$C(\mu)(u) := \int_{\mathbb{S}^{n-1}} |\langle u, x \rangle| \, \mu(dx), \quad u \in \mathbb{S}^{n-1},$$

is called the *cosine transform* on the space $\mathcal{M}_e(\mathbb{S}^{n-1})$, and $C(\mu)$ is the cosine transform of the measure μ. Theorem 4.14 expresses the fact that the cosine transform C is injective on $\mathcal{M}_e(\mathbb{S}^{n-1})$.

Corollary 4.4 *A centrally symmetric convex body* $K \in \mathcal{K}^n$ *with* $\dim K = n$ *is uniquely determined (up to translations) by its projection function* $v(K, \cdot)$.

Proof Let K, L be n-dimensional, centrally symmetric convex bodies satisfying $v(K, \cdot) = v(L, \cdot)$. Then, by Theorem 4.12 it follows that $S_{n-1}(K, \cdot)$ and $S_{n-1}(L, \cdot)$ are even measures with equal cosine transforms. Now Theorem 4.14 shows that $S_{n-1}(K, \cdot) = S_{n-1}(L, \cdot)$. By Theorem 4.3, this in turn implies that K and L are translates of each other, since both sets are n-dimensional. This completes the argument.   □

The cosine transform C is an injective map from $\mathcal{M}_e(\mathbb{S}^{n-1})$ to $\mathbf{C}_e(\mathbb{S}^{n-1})$. However, it is not surjective, since the homogeneous extension of $C(\mu)$, for some $\mu \in \mathcal{M}_e(\mathbb{S}^{n-1})$, is a support function and hence all directional derivatives exist. In other words, the cosine transform has some regularizing effect on a measure. In order to obtain a bijective correspondence, it is therefore natural to either extend (say, to distributions) or restrict (say, to $\mathbf{C}_e^\infty(\mathbb{S}^{n-1})$ functions) the domain and the co-domain suitably. Using expansions of functions on the sphere into spherical harmonics, the following result can be shown.

Theorem 4.15 *The cosine transform $C : \mathbf{C}_e^\infty(\mathbb{S}^{n-1}) \to \mathbf{C}_e^\infty(\mathbb{S}^{n-1})$, defined by*

$$(Cf)(u) := \int_{\mathbb{S}^{n-1}} |\langle u, x \rangle| f(x) \, \mathcal{H}^{n-1}(dx), \quad u \in \mathbb{S}^{n-1},$$

for $f \in \mathbf{C}_e^\infty(\mathbb{S}^{n-1})$, is a linear bijection and a self-adjoint linear operator in $L^2(\mathbb{S}^{n-1}, \mathcal{H}^{n-1})$.

For a proof, we refer to [78, Corollary 5.6], where the main work is accomplished in Appendix A, which offers a self-contained introduction to spherical harmonics and basic harmonic analysis on $\mathbb{S}^{n-1}$; see also the Appendix on spherical harmonics in [81] and [81, Theorem 3.5.4], which provides an alternative proof of the injectivity of the cosine transform on $\mathcal{M}_e(\mathbb{S}^{n-1})$ and further states that if $G \in \mathbf{C}_e^k(\mathbb{S}^{n-1})$ for an even integer $k \geq n + 2$, then there is some $g \in C_e(\mathbb{S}^{n-1})$ such that $G = C(g)$. This indicates that as the degree of smoothness of G increases, g also gets increasingly smooth.

Let us consider some geometric consequences. The class $\mathcal{Z}^n$ of zonoids in $\mathbb{R}^n$ has been defined as the closure with respect to the Hausdorff metric of the class of all finite Minkowski sums of segments. Hence, the zonoids form a closed subset in the space of centrally symmetric convex bodies. For $n = 2$, Exercise 4.4.2 shows that $\mathcal{Z}^2$ is just the set of all centrally symmetric convex bodies in the plane. It follows from Exercise 4.4.7 that $\mathcal{Z}^n$, for $n \geq 3$, is nowhere dense in the space of centrally symmetric convex bodies. In fact, any centrally symmetric convex body can be approximated by centrally symmetric convex polytopes which have a support set that is not centrally symmetric.

In order to arrive at a dense class of bodies, it is natural to consider "differences of zonoids".

Definition 4.4 A convex body $K \in \mathcal{K}^n$ is called a centred *generalized zonoid* if there are centred zonoids $Z_1, Z_2 \in \mathcal{Z}^n$ such that $Z_2 = K + Z_1$. Generalized zonoids are translates of centred generalized zonoids.

It is clear that zonoids are also generalized zonoids. Moreover, their support functions have an integral representation which extends the one for zonoids in a natural way.

Lemma 4.8 *A convex body $K \in \mathcal{K}^n$ is a centred generalized zonoid if and only if there is an even signed Borel measure ϱ on $\mathbb{S}^{n-1}$ such that*

$$h_K(u) = \int_{\mathbb{S}^{n-1}} |\langle u, x \rangle| \varrho(dx), \qquad u \in \mathbb{S}^{n-1}. \tag{4.22}$$

Proof Let $K \in \mathcal{K}^n$ be a centred generalized zonoid and $Z_2 = K + Z_1$ with centred $Z_1, Z_2 \in \mathcal{Z}^n$. Then there are measures $\varrho_1, \varrho_2 \in \mathcal{M}_e(\mathbb{S}^{n-1})$ such that

$$h_{Z_i}(u) = \int_{\mathbb{S}^{n-1}} |\langle u, x \rangle| \varrho_i(dx), \qquad u \in \mathbb{S}^{n-1}, \tag{4.23}$$

for $i = 1, 2$, and thus $h_K = h_{Z_2} - h_{Z_1}$ and (4.23) yield (4.22) with the finite even signed Borel measure $\varrho = \varrho_2 - \varrho_1$.

For the converse, we use the Hahn–Jordan decomposition of ϱ (see Theorem 4.1.4, Corollary 4.1.5 and the subsequent remark on p. 125 in [26]) which provides finite Borel measures ϱ_1, ϱ_2 on $\mathbb{S}^{n-1}$ with $\varrho = \varrho_2 - \varrho_1$. Since

$$\varrho_2(A) = \sup\{\varrho(B) : B \in \mathcal{B}(\mathbb{S}^{n-1}), B \subset A\},$$

it follows that ϱ_2 is even, and hence the same is true for ϱ_1. Defining Z_1, Z_2 by (4.23), we get $h_K = h_{Z_2} - h_{Z_1}$, and hence $Z_2 = K + Z_1$.   □

It is evident from the definition that generalized zonoids are centrally symmetric. Moreover it follows from Exercise 4.4.7 that all support sets of generalized zonoids are centrally symmetric. Hence a cross-polytope $K_2 := \{x \in \mathbb{R}^n : |x_1| + \cdots + |x_n| \leq 1\}$ (see Exercise 2.3.9) is an example of a centrally symmetric convex body which is not a generalized zonoid. In particular, for $n \geq 3$ the generalized zonoids constitute a genuine subset of the centrally symmetric convex bodies. If Z is a generalized zonoid and a polytope, then all faces of Z are centrally symmetric and hence Z is a zonotope by Exercise 4.4.3. However, we have the following important result.

Theorem 4.16 *The generalized zonoids form a dense subset in the space of centrally symmetric convex bodies.*

Proof The proof is based on two ingredients. The first is an approximation argument, which is of independent interest. Let $h = h_K$ be the support function of a convex body $K \in \mathcal{K}^n$. For $\varepsilon > 0$ let $\varphi : [0, \infty) \to [0, \infty)$ be of class $\mathbf{C}^\infty$ with support in $[\varepsilon/2, \varepsilon]$ and $\int_{\mathbb{R}^n} \varphi(\|z\|)\, dz = 1$. Define a convolution type transformation Th of h by

$$(Th)(x) := \int_{\mathbb{R}^n} h(x + \|x\|z)\varphi(\|z\|)\, dz, \qquad x \in \mathbb{R}^n.$$

Then Th is a support function and $Th \in \mathbf{C}^\infty(\mathbb{R}^n \setminus \{0\})$. Moreover, if TK is defined by $h_{TK} = Th_K$, then $d(K, TK) \leq R\varepsilon$ if $K \subset B(0, R)$, $R > 0$. For a detailed proof of these facts, we refer to [81, p. 183–5].

Clearly, TK is centrally symmetric with respect to 0 if this is the case for K. This shows that centrally symmetric convex bodies with support functions of class $\mathbf{C}^\infty$ are dense in the space of centrally symmetric convex bodies.

The second ingredient is Theorem 4.15. If combined with Lemma 4.8, it yields that centrally symmetric convex bodies with support functions of class $\mathbf{C}^\infty$ are generalized zonoids with a generating measure of the form $g\, d\mathcal{H}^{n-1}$ with a density function $g \in \mathbf{C}_e^\infty(\mathbb{S}^{n-1})$. $\qquad\square$

Exercises and Supplements for Sect. 4.4

1. Show that even polynomials restricted to $\mathbb{S}^{n-1}$ form a dense subspace of $\mathbf{C}_e(\mathbb{S}^{n-1})$.
2. Show that a planar centrally symmetric convex body $K \in \mathcal{K}^2$ is a zonoid.
3.* Let $n \geq 3$ and let $P \in \mathcal{K}^n$ be a polytope. Show that P is a zonotope if and only if all 2-faces of P are centrally symmetric. In fact, all faces of a zonotope are again zonotopes and hence centrally symmetric.

   One way to prove this is to establish first the following assertion. If $P \in \mathcal{P}^n$ is an n-polytope, $n \geq 3$, with centrally symmetric facets, then P is centrally symmetric.

   This can be deduced, for instance, by starting with the proof of the following fact.

   Let $P \in \mathcal{P}^n$ be an n-dimensional polytope, let $P_1, \ldots, P_m \subset \mathcal{P}^n$ be n-dimensional, centrally symmetric polytopes such that $P = P_1 \cup \ldots \cup P_m$, $P_i \cap P_j$ is the empty set or a face of both, P_i and P_j, for $i, j = 1, \ldots, m$. Further, assume that each facet of P is a facet of precisely one of the polytopes $P_1, \ldots, P_m$. Then P is centrally symmetric.
4. Let $Z \in \mathcal{K}^n$ be a zonoid and $u_1, \ldots, u_k \in \mathbb{S}^{n-1}$. Show that there exists a zonotope P which is the sum of at most k segments such that

   $$h_Z(u_i) = h_P(u_i), \quad i = 1, \ldots, k.$$

5. Let $P, Q \in \mathcal{P}^n$ be zonotopes, and let $K \in \mathcal{K}^n$ be a convex body such that $P = K + Q$. Show that K is also a zonotope.
6.* (a) Let $a, b, c \in \mathbb{R}$. Prove Hlawka's inequality

   $$|a+b| + |a+c| + |b+c| \leq |a| + |b| + |c| + |a+b+c|.$$

   (b) A positively homogeneous function $f : \mathbb{R}^n \to [0, \infty)$ is said to satisfy Hlawka's inequality if

   $$f(x+y) + f(x+z) + f(y+z) \leq f(x) + f(y) + f(z) + f(x+y+z)$$

for $x, y, z \in \mathbb{R}^n$. Show that the function $h : \mathbb{R}^n \to [0, \infty)$ given by

$$h(x) = \sum_{i=1}^{k} \alpha_i |\langle x, x_i \rangle|$$

with $k \in \mathbb{N}$, $\alpha_i \geq 0$ and $x_i \in \mathbb{R}^n$ for $i = 1, \ldots, k$ satisfies Hlawka's inequality.

(c) Show that if $P \in \mathcal{P}^n$ is a zonotope, then h_P satisfies Hlawka's inequality

$$(*) \; h_P(x) + h_P(y) + h_P(z) + h_P(x + y + z)$$

$$\geq h_P(x + y) + h_P(x + z) + h_P(y + z),$$

for $x, y, z \in \mathbb{R}^n$. (In fact, the support function of a zonoid also satisfies Hlawka's inequality.)

(d) Show the converse of (c) (Witsenhausen's Theorem): If $P \in \mathcal{P}^n$ and h_P satisfies Hlawka's inequality, then P is a zonotope.

7. Let $Z \in \mathcal{K}^n$ be a (generalized) zonoid. For $e \in \mathbb{S}^{n-1}$, we define the open hemisphere $\Omega_e^{n-1} := \{u \in \mathbb{S}^{n-1} : \langle u, e \rangle > 0\}$ and the equator $\mathbb{S}_e^{n-1} := \{u \in \mathbb{S}^{n-1} : \langle u, e \rangle = 0\}$.

(a) For $e \in \mathbb{S}^{n-1}$, show that the support set $Z(e)$ of Z is again a (generalized) zonoid and that

$$h_{Z(e)}(v) = \int_{\mathbb{S}_e^{n-1}} |\langle v, u \rangle| \, \mu(Z, du) + \langle x_e, v \rangle, \qquad v \in \mathbb{S}^{n-1},$$

where

$$x_e := 2 \int_{\Omega_e^{n-1}} u \, \mu(Z, du).$$

In particular, this shows that Z has centrally symmetric faces.

(b) Use (a) to show that a (generalized) zonoid which is a polytope must be a zonotope.

8.* Let $f : \mathbb{S}^{n-1} \to \mathbb{R}$ be an even function which is integrable with respect to spherical Lebesgue measure $\sigma = \mathcal{H}^{n-1}$ on $\mathbb{S}^{n-1}$. Define the function $h : \mathbb{R}^n \to \mathbb{R}$ by

$$h(x) = \int_{\mathbb{S}^{n-1}} |\langle x, u \rangle| f(u) \, \sigma(du), \qquad x \in \mathbb{R}^n.$$

(a) Show that h is continuously differentiable on $\mathbb{R}^n \setminus \{0\}$ with

$$\nabla h(x) = \int_{\mathbb{S}^{n-1}} \mathrm{sgn}(\langle x, u \rangle) u f(u) \sigma(du), \qquad x \in \mathbb{R}^n \setminus \{0\},$$

where $\mathrm{sgn}(0) = 0$, $\mathrm{sgn}(a) = 1$ if $a > 0$, and $\mathrm{sgn}(a) = -1$ if $a < 0$.

(b) If f is continuous, then $h \in \mathbf{C}^2(\mathbb{R}^n \setminus \{0\})$ and

$$d^2 h(e; x, y) = \frac{2}{\|e\|} \int_{\mathbb{S}_e^{n-1}} \langle u, x \rangle \langle u, y \rangle f(u) \, \mathcal{H}^{n-2}(du),$$

for $e \in \mathbb{R}^n \setminus \{0\}$ and $x, y \in \mathbb{R}^n$.

(c) Part (b) yields a result due to Lindquist [61] stating that for a continuous function f the function $h = h_f$ is a support function if and only if

$$\int_{\mathbb{S}_e^{n-1}} \langle u, x \rangle^2 f(u) \, \mathcal{H}^{n-2}(du) \geq 0$$

for all $e \in \mathbb{S}^{n-1}$ and all $x \in \mathbb{S}_e^{n-1}$.

9. (a) Show that the map $K \mapsto \Pi K$ is continuous on $\mathcal{K}^n$.

(b) Let $K \in \mathcal{K}^2$ with $K = -K$. Show that

$$\Pi K = \vartheta(2K),$$

where ϑ is the rotation by the angle $\frac{\pi}{2}$.

(c) Reconsider Exercise 4.4.2.

10. Let $K, L \in \mathcal{K}^n$. Show that

$$V(K, \ldots, K, \Pi L) = V(L, \ldots, L, \Pi K).$$

11. Let $K, M \in \mathcal{K}^n$ and $L := \Pi M$. If

$$V_{n-1}(K | u^\perp) \leq V_{n-1}(L | u^\perp) \quad \text{for } u \in \mathbb{S}^{n-1},$$

then

$$V_n(K) \leq V_n(L).$$

12. Let $K, L \in \mathcal{K}^n$. Show that the following statements are equivalent.

(a) $V_{n-1}(K | u^\perp) \leq V_{n-1}(L | u^\perp)$ for $u \in \mathbb{S}^{n-1}$.

(b) $F(\varphi K) \leq F(\varphi L)$ for all regular affine transformations φ of $\mathbb{R}^n$.

4.5   Valuations

Let S denote a family of subsets of $\mathbb{R}^n$ which is intersectional, that is, for any two sets their intersection belongs to S. In the following, we assume that $\emptyset \in S$. Examples of intersectional families are $\mathcal{K}^n$, $\mathcal{P}^n$, or boxes with parallel axes. We do not change our notation for $\mathcal{K}^n$, $\mathcal{P}^n$, although the empty set should be included, in this section.

Definition 4.5 Let S be an intersectional class in $\mathbb{R}^n$. A functional φ on S with values in some abelian group is called a *valuation* (or is said to be *additive*) if

$$\varphi(K \cup L) + \varphi(K \cap L) = \varphi(K) + \varphi(L)$$

whenever $K, L, K \cup L \in S$. Moreover, we require $\varphi(\emptyset) = 0$.

Examples of valuations are measures (volume, volume with density), intrinsic volumes (obtained via the Steiner formula for the volume of parallel sets). Moreover, general mixed volumes give rise to a variety of valuations, as stated in Exercise 4.5.6. There are many more examples such as the support function, mixed area measures, the identity map on convex bodies, the polar map on convex bodies containing the origin in their interiors.

In the definition of a valuation it is required that the union is still in the domain so that the functional can be applied. On the other hand, it would be desirable to extend the domain of a valuation so that together with two convex bodies (say) it also contains the union and so that additivity is preserved on the larger domain. This leads to the set $U(\mathcal{K}^n)$ of finite unions of convex bodies (*polyconvex sets*). We call $U(\mathcal{K}^n)$ the *convex ring* of $\mathbb{R}^n$ and write $\mathcal{R}^n$ for this set. More generally the question arises whether a valuation on an intersectional family S in $\mathbb{R}^n$ can be extended as an additive functional to the (possibly) larger domain $U(S)$ of finite unions of elements of S. It would be natural to define, for instance,

$$\varphi(K \cup L) = \varphi(K) + \varphi(L) - \varphi(K \cap L), \quad K, L \in S.$$

However, it is not clear whether the right-hand side is independent of the particular representation of $K \cup L$ in terms of K and L.

The general extension problem on $\mathcal{K}^n$ (without any additional hypothesis on the functional) is open. However, the problem can be completely resolved on $\mathcal{P}^n$, and this can be used to establish an extendability result for functionals on $\mathcal{K}^n$ which have some weak continuity property. In considering extensions to finite unions of elements from an intersectional class S, the following notion is natural. We denote by $S(m)$, $m \in \mathbb{N}$, the set of all nonempty subsets of $\{1, \ldots, m\}$, for $v \in S(m)$ we write $|v|$ for the cardinality of v, and $K_v := \cap_{i \in v} K_i$ for given $K_1, \ldots, K_m \in S$.

Definition 4.6 Let S be an intersectional class in $\mathbb{R}^n$. A functional φ on S with values in some abelian group is called *fully additive* if

$$\varphi(K_1 \cup \cdots \cup K_m) = \sum_{v \in S(m)} (-1)^{|v|-1} \varphi(K_v), \qquad (4.24)$$

whenever $m \in \mathbb{N}$, $K_1, \ldots, K_m \in S$, and $K_1 \cup \cdots \cup K_m \in S$.

It is clear that a functional φ which is additive on $U(S)$ has to be fully additive, that is, it satisfies the inclusion-exclusion principle as stated in the definition (and more generally also for sets in $U(S)$). In the following, we study the reverse conclusion and show that full additivity, as stated in the definition, yields extendability.

The following theorem is stated for a general intersectional class, since we will apply it for both, polytopes and general convex bodies.

Theorem 4.17 *Let φ be a functional on an intersectional class S with values in some abelian group. If φ is fully additive, then φ can be additively extended to $U(S)$.*

Proof Let φ be a fully additive function on S. Let $K \in U(S)$ have representations

$$K = K_1 \cup \cdots \cup K_k = L_1 \cup \cdots \cup L_l$$

with $K_i, L_j \in S$, and set $\tau = (K_1, \ldots, K_k)$ and $\sigma = (L_1, \ldots, L_l)$. Let $v \in S(k)$. Since

$$K_v = \bigcup_{j=1}^{l} (K_v \cap L_j) \in S \quad \text{with } K_v \cap L_j \in S$$

and φ is fully additive, we get

$$\varphi(K_v) = \sum_{w \in S(l)} (-1)^{|w|-1} \varphi(K_v \cap L_w).$$

Hence,

$$\varphi(K, \tau) := \sum_{v \in S(k)} (-1)^{|v|-1} \varphi(K_v)$$

$$= \sum_{v \in S(k)} \sum_{w \in S(l)} (-1)^{|v|+|w|} \varphi(K_v \cap L_w)$$

$$= \varphi(K, \sigma),$$

by symmetry. Therefore, we can unambiguously define

$$\varphi(K) := \varphi(K, \tau),$$

which yields a consistent extension of φ from $\mathcal{S}$ to $U(\mathcal{S})$. It remains to prove that φ is additive on $U(\mathcal{S})$. Hence, let $K, L \in U(\mathcal{S})$ be given with representations

$$K = K_1 \cup \cdots \cup K_k, \quad L = L_1 \cup \cdots \cup L_l$$

and $K_i, L_j \in \mathcal{S}$. Then

$$K \cup L = \bigcup_{i=1}^{k} K_i \cup \bigcup_{j=1}^{l} L_j$$

and

$$K \cap L = \bigcup_{i=1}^{k} \bigcup_{j=1}^{l} (K_i \cap L_j).$$

Then we have

$$\varphi(K) = \sum_{v \in S(k)} (-1)^{|v|-1} \varphi(K_v),$$

$$\varphi(L) = \sum_{w \in S(l)} (-1)^{|w|-1} \varphi(L_w),$$

$$\varphi(K \cup L) = \sum_{v \in S(k)} (-1)^{|v|-1} \varphi(K_v) + \sum_{w \in S(l)} (-1)^{|w|-1} \varphi(L_w)$$

$$+ \sum_{a \in S(k), b \in S(l)} (-1)^{|a|+|b|-1} \varphi(K_a \cap L_b).$$

We shortly write $S(k, l)$ for $S(\{1, \ldots, k\} \times \{1, \ldots, l\})$. For a set $c \in S(k, l)$, we define $\pi_1(c)$ as the set of all $i \in \{1, \ldots, k\}$ for which there is some $j \in \{1, \ldots, l\}$ such that $(i, j) \in c$, and $\pi_2(c)$ is similarly defined. Then we have

$$\varphi(K \cap L) = \sum_{c \in S(k,l)} (-1)^{|c|-1} \varphi\left(K_{\pi_1(c)} \cap L_{\pi_2(c)}\right)$$

$$= \sum_{a \in S(k), b \in S(l)} \sum_{\substack{c \in S(k,l), \\ \pi_1(c)=a, \pi_2(c)=b}} (-1)^{|c|-1} \varphi(K_a \cap L_b).$$

To conclude the proof, we need the following purely combinatorial result.

Claim For $a \in S(k), b \in S(l)$,

$$\sum_{\substack{c \in S(k,l), \\ \pi_1(c)=a, \pi_2(c)=b}} (-1)^{|c|-1} = (-1)^{|a|+|b|}.$$

Proof (of Claim) We proceed by induction over $|b| \geq 1$. If $|b| = 1$ and $b = \{b_0\}$, then $c = \{(i, b_0) : i \in a\}$ is uniquely determined and the assertion is clear. Now we assume that a, b are given with $|b| \geq 2$ and the assertion is true for a', b' with $|b'| < |b|$. By symmetry we can assume that $a = \{1, \ldots, a_0\}$, $b = \{1, \ldots, b_0, b_0 + 1\}$ with $a_0, b_0 \geq 1$. Let $c \in S(k, l)$ with $\pi_1(c) = a$ and $\pi_2(c) = b$. Then there is some $k \in \{1, \ldots, a_0\}$ and there are indices $1 \leq i_1 < \ldots < i_k \leq a_0$ such that $(i, b_0 + 1) \in c$ if and only if $i \in \{i_1, \ldots, i_k\}$. Let c' denote the set obtained from c by removing these k pairs $(i, b_0 + 1)$. There are $\binom{a_0}{k}$ possible choices for these indices. Then $c' \in S(k, l)$ and $\pi_1(c') = a$, $\pi_2(c') = b' := \{1, \ldots, b_0\}$. Conversely, any choice of such a set c' and of k indices, as just described, leads to a set c. Thus we obtain

$$\sum_{\substack{c \in S(k,l), \\ \pi_1(c)=a, \pi_2(c)=b}} (-1)^{|c|-1} = \sum_{\substack{c' \in S(k,l), \\ \pi_1(c')=a, \pi_2(c')=b'}} \sum_{k=1}^{a_0} \binom{a_0}{k} (-1)^{|c'|+k-1}$$

$$= \sum_{\substack{c' \in S(k,l), \\ \pi_1(c')=a, \pi_2(c')=b'}} (-1)^{|c'|-1} \sum_{k=1}^{a_0} \binom{a_0}{k} (-1)^k$$

$$= \sum_{\substack{c' \in S(k,l), \\ \pi_1(c')=a, \pi_2(c')=b'}} (-1)^{|c'|-1} \cdot (-1)$$

$$= (-1)^{|a|+|b'|} \cdot (-1) = (-1)^{|a|+|b|},$$

where the induction hypothesis was used for the second to last equation. □

This completes the proof of the theorem. □

Next we consider additive functionals on $S = \mathcal{P}^n$ and study the extendability problem. It turns out that for this particular intersectional class a condition weaker than additivity always implies that full additivity is satisfied. Finally, we prove a (slightly stronger) version of Groemer's extendability result for $S = \mathcal{K}^n$, which uses σ-continuity of the functional as an additional hypothesis.

Definition 4.7 Let φ be a functional on $\mathcal{P}^n$ with values in some abelian group. Then φ is said to be *weakly additive* if for $P \in \mathcal{P}^n$ and all hyperplanes H with corresponding closed halfspaces H^+, H^-, bounded by H, the relation

$$\varphi(P) = \varphi(P \cap H^+) + \varphi(P \cap H^-) - \varphi(P \cap H)$$

is satisfied.

Clearly, additivity yields weak additivity on $\mathcal{P}^n$. The following stronger converse is surprising.

Theorem 4.18 *Every weakly additive function on $\mathcal{P}^n$ with values in some abelian group is fully additive on $\mathcal{P}^n$.*

Proof Let φ be a weakly additive function on $\mathcal{P}^n$. For convex polytopes $P, P_1, \ldots, P_m \in \mathcal{P}^n$ with $P = P_1 \cup \cdots \cup P_m$ we use the notations $(P_1, \ldots, P_m) =: \tau$ and

$$\varphi(P, \tau) := \sum_{v \in S(m)} (-1)^{|v|-1} \varphi(P_v). \tag{4.25}$$

Then we have to show that

$$\varphi(P, \tau) = \varphi(P). \tag{4.26}$$

For the proof, we use induction over n. For $n = 0$, the assertion is clear. Hence, we assume that $n \geq 1$ and that the assertion is true in spaces of smaller dimension. Without loss of generality, we may assume that $\dim P = n$. As a first special case, we assume that one of the polytopes $P_1, \ldots, P_m$, say P_1, is equal to P. Then any summand

$$(-1)^{|v|-1} \varphi(P_v)$$

with $v = \{i_1, \ldots, i_r\}$, $1 \leq i_1 < \cdots < i_r \leq m$, and $i_1 > 1$ in the right-hand side of (4.25) is cancelled by the summand

$$(-1)^{|v|} \varphi(P_1 \cap P_v) = (-1)^{|v|} \varphi(P_v).$$

Thus, only the summand $\varphi(P_1) = \varphi(P)$ remains. This shows that (4.26) holds and henceforth this first case is already confirmed.

For given n, we prove (4.26) by induction over m. The case $m = 1$ is clear. Hence we can assume that $m \geq 2$ and the assertion is true for all representations of P as unions of fewer than m polytopes. As a second special case, we assume that one of the polytopes $P_1, \ldots, P_m$, say P_m, is of dimension less than n. Since P is the closure of its interior, we must have $P = P_1 \cup \cdots \cup P_{m-1}$ and $P_m \subset P$. Hence $P_m = (P_1 \cap P_m) \cup \cdots \cup (P_{m-1} \cap P_m)$. Since these are unions of fewer than m polytopes, the induction hypothesis gives

$$\varphi(P) = \sum_{v \in S(m-1)} (-1)^{|v|-1} \varphi(P_v)$$

and

$$0 = \varphi(P_m) - \sum_{v \in S(m-1)} (-1)^{|v|-1} \varphi(P_v \cap P_m).$$

Adding these two inequalities, we get (4.26). Also this special case is confirmed.

The n-polytope P_1 is properly contained in P (having excluded the two special cases). Hence it has a facet whose affine hull H meets int P. The two closed halfspaces, bounded by H, are denoted by H^+, H^-, where $P_1 \subset H^+$. Since φ is weakly additive, we have

$$\varphi(P) = \varphi(P \cap H^+) + \varphi\left(\bigcup_{i=1}^{m}(P_i \cap H^-)\right) - \varphi\left(\bigcup_{i=1}^{m}(P_i \cap H)\right).$$

Relation (4.26) can be applied to the second term on the right-hand side, since $\dim(P_1 \cap H^-) < n$ (second special case). The relation also holds for the third term, since it holds in smaller dimensions. Furthermore, for any $v \in S(m)$ we have

$$\varphi(P_v \cap H^-) - \varphi(P_v \cap H) = \varphi(P_v) - \varphi(P_v \cap H^+)$$

by the assumed weak additivity. Rearranging terms, we get

$$\varphi(P) - \varphi(P, \tau) = \varphi(P \cap H^+) - \varphi(P \cap H^+, \tau^+),$$

where $\tau^+ = (P_1 \cap H^+, \dots, P_m \cap H^+)$. We have $P_1 \subset P \cap H^+$. If P_1 is properly contained in $P \cap H^+$, then we can repeat the procedure with $P, P_1, \dots, P_m$ replaced by their intersection with H^+. After finitely many such repetitions, the remaining polytope is P_1. But then we are finished by the first special case. This completes the double induction and the proof.   □

Combining Theorems 4.17 and 4.18, we obtain the following result.

Corollary 4.5 *Every weakly additive functional on $\mathcal{P}^n$ with values in some abelian group has an additive extension to $U(\mathcal{P}^n)$.*

For the following result, we define a weak form of continuity. A function φ from $\mathcal{K}^n$ into some topological (Hausdorff) vector space is called *σ-continuous* if for every decreasing sequence $(K_i)_{i \in \mathbb{N}}$ in $\mathcal{K}^n$ the condition

$$\lim_{i \to \infty} \varphi(K_i) = \varphi\left(\bigcap_{i \in \mathbb{N}} K_i\right)$$

is satisfied. Clearly, continuity yields σ-continuity, since a decreasing sequence of convex bodies converges in the Hausdorff metric to the intersection (exercise, see the argument for the completeness of the space of convex bodies). Hence, the following theorem in particular shows that a continuous valuation on $\mathcal{K}^n$ has an additive extension to $U(\mathcal{K}^n)$.

The following result is a mild extension of a result due to Groemer.

Theorem 4.19 *Let φ be a function on $\mathcal{K}^n$ with values in a topological vector space. If φ is weakly additive on $\mathcal{P}^n$ and is σ-continuous on $\mathcal{K}^n$, then φ has an additive extension to $U(\mathcal{K}^n)$.*

Proof Let φ satisfy the assumptions of Theorem 4.19. Let $K_1, \ldots, K_m \in \mathcal{K}^n$ be convex bodies such that $K_1 \cup \cdots \cup K_m \in \mathcal{K}^n$. We apply Exercise 3.1.15 on the simultaneous approximation of convex bodies and their convex union by convex polytopes. Specifically, we use this exercise with K_i replaced by $K_i + 2^{-k}B^n$, $k \in \mathbb{N}$, and with $\varepsilon = 2^{-k}$. Note that $\bigcup_{i=1}^{m}(K_i + 2^{-k}B^n) = (\bigcup_{i=1}^{m} K_i) + 2^{-k}B^n$ is convex. Then we get polytopes $P_1^{(k)}, \ldots, P_m^{(k)}$ with convex union and such that $K_i + 2^{-k}B^n \subset P_i^{(k)} \subset K_i + 2^{1-k}B^n$. Each sequence $(P_i^{(k)})_{k\in\mathbb{N}}$ is decreasing. By Theorem 4.18, the function φ is fully additive on $\mathcal{P}^n$, hence

$$\varphi(P_1^{(k)} \cup \cdots \cup P_m^{(k)}) = \sum_{v\in S(m)} (-1)^{|v|-1}\varphi(P_v^{(k)}).$$

Since

$$\bigcap_{k\in\mathbb{N}} \left(P_1^{(k)} \cup \cdots \cup P_m^{(k)}\right) = K_1 \cup \cdots \cup K_m$$

and

$$\bigcap_{k\in\mathbb{N}} P_v^{(k)} = K_v \quad \text{if} \quad K_v \neq \emptyset,$$

the σ-continuity of φ yields

$$\varphi(K_1 \cup \cdots \cup K_m) = \sum_{v\in S(m)} (-1)^{|v|-1}\varphi(K_v).$$

Thus, φ is fully additive on $\mathcal{K}^n$. By Theorem 4.17, φ has an additive extension to $U(\mathcal{K}^n)$. This proves the result.   □

The following characterization theorem is a key result in the theory of valuations with many applications (as we shall see in Chap. 5). At some point of the argument, Theorem 4.19 will be used.

Theorem 4.20 (Hadwiger's Characterization Theorem) *Let $\varphi : \mathcal{K}^n \to \mathbb{R}$ be an additive, continuous functional which is invariant under proper rigid motions. Then there are constants $c_0, \ldots, c_n \in \mathbb{R}$ such that*

$$\varphi = \sum_{i=0}^{n} c_i V_i.$$

The main point of the proof consists in establishing the following special case. A subsequent induction argument over the dimension then yields the general result. The current approach is due to Dan Klain [51, 53]. For the classical approach due to Hugo Hadwiger, see [43].

A functional $\varphi : \mathcal{K}^n \to \mathbb{R}$ is said to be *simple* if $\varphi(K) = 0$ whenever $\dim(K) < n$.

Proposition 4.1 *Let $\varphi : \mathcal{K}^n \to \mathbb{R}$ be a simple, additive, continuous functional which is invariant under proper rigid motions. Suppose that $\varphi(C^n) = 0$ for the unit cube C^n. Then $\varphi = 0$.*

Proof We proceed by induction over the dimension n of the space. For $n = 0$ there is nothing to show. Let $n = 1$. Then φ is zero on one-pointed sets and on (closed line) segments of unit length. Define $f(a) := \varphi(I_a)$, where I_a is a segment of length a; this is independent of the position, since φ is translation invariant. By additivity and concatenation of segments, it follows that f satisfies Cauchy's functional equation $f(a + b) = f(a) + f(b)$ for $a, b \geq 0$. Moreover, f is continuous, since φ is continuous. Hence f is linear. Since $f(1) = 0$, we get $f = 0$. This shows that φ vanishes in segments.

Let $n \geq 2$ and suppose the assertion is established in smaller dimensions. Let $H \subset \mathbb{R}^n$ be a hyperplane and I a segment of length 1 which is orthogonal to H. For convex bodies $K \subset H$, we define $\psi(K) := \varphi(K + I)$. Then ψ is additive (Exercise 4.5.3) and continuous. If $H = \mathbb{R}^{n-1} = \mathbb{R}^{n-1} \times \{0\}$, then ψ is also rigid motion invariant and vanishes if $\dim K < n - 1$ or K is the unit cube. Hence $\psi = 0$ by the induction hypothesis, first for the special choice of H, but then for all H by the rigid motion invariance. Hence $\varphi(K + I) = 0$ whenever I is a unit segment orthogonal to H. Arguing as for $n = 1$ we obtain that $\varphi(K + S) = 0$ for all closed segments S orthogonal to H. This shows that φ vanishes on orthogonal cylinders with convex base.

Let $K \subset H$ be a convex body again, and let $S = \operatorname{conv}\{0, s\}$ be a segment not parallel to H. If $m \in N$ is sufficiently large, then the cylinder $Z := K + mS$ can be cut by a hyperplane H_0 orthogonal to S so that the two closed halfspaces H_0^+, H_0^- bounded by H_0 satisfy $K \subset H_0^-$ and $K + ms \subset H_0^+$. Then

$$Z_0 := \left((Z \cap H_0^-) + ms\right) \cup (Z \cap H_0^+)$$

is an orthogonal cylinder, and we deduce that

$$m\varphi(K + S) = \varphi(Z) = \varphi(Z_0) = 0.$$

Hence φ vanishes on arbitrary cylinders with convex base.

Let $P \in \mathcal{P}^n$ and let S be a segment. Then the polytope $P + S$ has a decomposition

$$P + S = \bigcup_{i=1}^{k} P_i,$$

where $P_1 = P$ and P_i, for $i > 1$, is a convex cylinder such that $\dim(P_i \cap P_j) < n$ for $i \neq j$. Thus, using Corollary 4.5 we get $\varphi(K + S) = \varphi(K)$, first for a polytope and then for a general convex body K. But then $\varphi(K + Z) = \varphi(K)$, first for a zonotope, but then also for a zonoid Z. In particular, this also yields $\varphi(Z) = 0$ for a zonoid Z. Now let K be a generalized zonoid. Hence, $Z_2 = K + Z_1$ with zonoids Z_1, Z_2. But then $\varphi(K) = \varphi(K + Z_1) = \varphi(Z_2) = 0$. Since generalized zonoids are dense in the centrally symmetric convex bodies, we finally get $\varphi(K) = 0$ whenever K is centrally symmetric.

Now let $\triangle \subset \mathbb{R}^n$ be a simplex, say $\triangle = \operatorname{conv}\{0, v_1, \ldots, v_n\}$. Furthermore, let $v := v_1 + \cdots + v_n$ and $\triangle = \operatorname{conv}\{v, v - v_1, \ldots, v - v_n\}$, so that $\triangle_0 = -\triangle + v$. The vectors $v_1, \ldots, v_n$ span a parallelotope P. It is the union of the simplices $\triangle_0, \triangle$ and the part of P, denoted by Q, that lies between the hyperplanes spanned by $v_1, \ldots, v_n$ and $v - v_1, \ldots, v - v_n$, respectively. The polytope Q is centrally symmetric, and $\triangle \cap Q, \triangle_0 \cap Q$ are of dimension $n - 1$. Hence we get

$$0 = \varphi(P) = \varphi(\triangle) + \varphi(Q) + \varphi(\triangle_0),$$

thus we conclude $\varphi(-\triangle) = -\varphi(\triangle)$.

If the dimension n is even, then $\triangle_0$ is obtained from $\triangle$ by a proper rigid motion (a reflection in the origin), and the invariance of φ under proper rigid motions thus yields $\varphi(\triangle) = 0$.

If the dimension $n > 1$ is odd, we decompose φ as follows (see [80, p. 17]). Let z be the center of the inscribed ball of $\triangle$, and let p_i be the point where this ball touches the facet F_i of $\triangle$ for $i = 1, \ldots, n + 1$. For $i \neq j$, let Q_{ij} be the convex hull of the face $F_i \cap F_j$ and the points z, p_i, p_j. The polytope Q_{ij} is invariant under reflection in the hyperplane spanned by $F_i \cap F_j$ and z. If $Q_1, \ldots, Q_m$ are the polytopes Q_{ij} for $1 \leq 1 < j \leq n + 1$ in any order, then $\triangle = Q_1 \cup \cdots \cup Q_m$ and any two of these polytopes have a lower-dimensional intersection. Since $-Q_r$ is the image of Q_r under a proper rigid motion, namely a reflection in a hyperplane followed by a reflection in a point, we have

$$\varphi(-\triangle) = \sum_{r=1}^{m} \varphi(-Q_r) = \sum_{r=1}^{m} \varphi(Q_r) = \varphi(\triangle).$$

This finally yields $\varphi(\triangle) = 0$.

Since any polytope can be decomposed into simplices (Exercise 4.5.2) and polytopes are dense in the convex bodies, this completes the induction argument, and hence the proof is finished. $\square$

Proof (of Theorem 4.20) We again use induction with respect to the dimension n. For $n = 0$, the assertion is clear. Suppose that $n \geq 1$ and the assertion has been proved in dimensions less than n. Let $H \subset \mathbb{R}^n$ be a hyperplane. If $H = \mathbb{R}^{n-1} \times \{0\}$, then the restriction of φ to the convex bodies lying in H is additive, continuous and invariant under proper rigid motions of H into itself. By the induction hypothesis, there are constants $c_0, \ldots, c_{n-1}$ so that $\varphi(K) = \sum_{i=0}^{n-1} c_i V_i(K)$ for convex bodies

$K \subset H$. The intrinsic volume $V_i(K)$ is independent of the subspace in which K lies. By the rigid motion invariance of φ and V_i, the relation $\varphi(K) = \sum_{i=0}^{n-1} c_i V_i(K)$ holds for all hyperplanes, and hence for all $K \in \mathcal{K}^n$ with dim $K \leq n - 1$. But then the function ψ, defined by

$$\psi(K) := \varphi(K) - \sum_{i=0}^{n} c_i V_i(K)$$

for $K \in \mathcal{K}^n$, where c_n is chosen so that ψ vanishes at a fixed unit cube, satisfies the assumptions of Proposition 4.1. This yields the result. $\square$

A valuation $\varphi : \mathcal{K}^n \to \mathbb{R}$ is called *even* if $\varphi(-K) = \varphi(K)$ for $K \in \mathcal{K}^n$, it is called *odd* if $\varphi(-K) = -\varphi(K)$ for $K \in \mathcal{K}^n$. Each valuation $\varphi : \mathcal{K}^n \to \mathbb{R}$ can be decomposed into an even and an odd part by

$$\varphi(K) = \frac{1}{2}(\varphi(K) + \varphi(-K)) + \frac{1}{2}(\varphi(K) - \varphi(-K)), \qquad K \in \mathcal{K}^n.$$

Apparently, the intrinsic volumes are all even valuations. Hence, a valuation which is rigid motion invariant and continuous is necessarily even. This is a straightforward consequence of Hadwiger's characterization theorem. In fact, the proof of Hadwiger's characterization theorem provides further insights into the structure of translation invariant, continuous valuations. This is discussed in the exercises and supplements below.

Hadwiger has also proved a characterization result where the assumption of continuity is replaced by monotonicity.

Theorem 4.21 (Hadwiger's Second Characterization Theorem) *Let* $\varphi : \mathcal{K}^n \to \mathbb{R}$ *be an additive, increasing functional which is invariant under proper rigid motions. Then there are nonnegative constants* $c_0, \ldots, c_n \in \mathbb{R}$ *such that*

$$\varphi = \sum_{i=0}^{n} c_i V_i.$$

It was shown by McMullen [70] that a translation invariant, increasing valuation on $\mathcal{K}^n$ is continuous. Hence, Hadwiger's second characterization theorem is a straightforward consequence of Theorem 4.20. By McMullen's observation, monotonicity is the stronger assumption as compared to continuity in this context. This raises the question whether an even more elementary proof of Hadwiger's second theorem is possible. In fact, it is not hard to show that if $\varphi : \mathcal{K}^n \to \mathbb{R}$ is a translation invariant, increasing, simple additive functional, then φ is a positive multiple of the volume (see Exercise 4.5.4). This is exactly the counterpart to Proposition 4.1. However, it is not clear how this could be used in the induction step as in the proof of Theorem 4.20, since the difference of increasing functionals need not be increasing in general.

Exercises and Supplements for Sect. 4.5

1. Show that a decreasing sequence of convex bodies converges in the Hausdorff metric to the intersection.
2. Show that any polytope can be decomposed into simplices.
3. Let $\varphi : \mathcal{K}^n \to \mathbb{R}$ be additive, continuous, and rigid motion invariant. Let $H = \mathbb{R}^{n-1}$ be a hyperplane and I a segment of length 1 which is orthogonal to H. For convex bodies $K \subset H$, define $\psi(K) := \varphi(K + I)$. Show that ψ is additive, continuous and rigid motion invariant with respect to H.
4.* Let $f : \mathcal{P}^n \to \mathbb{R}$ be translation invariant, simple, additive and increasing. Show that there is a constant $c \geq 0$ such that $f = c\, V_n$.
5. Let $K, K' \in \mathcal{K}^n$ and $K \cup K' \in \mathcal{K}^n$. Show that

   (a) $(K \cap K') + (K \cup K') = K + K'$,
   (b) $(K \cap K') + M = (K + M) \cap (K' + M)$, for all $M \in \mathcal{K}^n$,
   (c) $(K \cup K') + M = (K + M) \cup (K' + M)$, for all $M \in \mathcal{K}^n$.

6. Let $\varphi(K) := V(K[j], M_{j+1}, \ldots, M_n)$, where $K \in \mathcal{K}^n$, and where the convex bodies $M_{j+1}, \ldots, M_n \in \mathcal{K}^n$ are fixed. Show that φ is additive.
7. Show that the mappings $K \mapsto S_j(K, A)$ (and similarly for more general mixed area measures) are additive on $\mathcal{K}^n$, for all $j \in \{0, \ldots, n\}$ and all Borel sets $A \subset \mathbb{S}^{n-1}$.
8. Show that the convex ring $U(\mathcal{K}^n)$ is dense in $\mathcal{C}^n$ in the Hausdorff metric.
9. Let T_2 denote an equilateral triangle in $\mathbb{R}^2$. For $K \in \mathcal{K}^2$, define the functional $\varphi(K) := V(K + T_2) - V(K + (-T_2))$. Show that $\varphi \neq 0$ and φ is a translation invariant, continuous, simple, odd, and additive functional.
10. Let $\varphi : \mathcal{K}^n \to \mathbb{R}$ be a translation invariant, continuous and simple valuation. Then there is a constant $c \in \mathbb{R}$ such that $\varphi(K) + \varphi(-K) = c V_n(K)$ for $K \in \mathcal{K}^n$. Hence, if φ is also even, then φ is proportional to the volume functional.
11. Let $\varphi : \mathcal{K}^n \to \mathbb{R}$ be a translation invariant, continuous and simple valuation which is odd. Then there is an odd continuous function $f \in \mathbf{C}(\mathbb{S}^{n-1})$ such that

$$\varphi(K) = \int_{\mathbb{S}^{n-1}} f(u)\, S_{n-1}(K, du), \qquad K \in \mathcal{K}^n.$$

The function f is uniquely determined up to a linear function.

12. By Exercises 4.5.10 and 4.5.11, a characterization of all translation invariant, continuous and simple valuations on $\mathcal{K}^n$ is obtained. More is known about translation invariant, weakly continuous (simple) valuations on polytopes.

Chapter 5
Integral-Geometric Formulas

In this chapter, we discuss various integral formulas for intrinsic volumes, which are based on projections, sections or sums of convex bodies. We shall also discuss some applications of a stereological nature.

As a motivation, we start with the formula for the projection function $v(K, \cdot)$ from Theorem 4.12. Integrating $v(K, u)$ over all $u \in \mathbb{S}^{n-1}$ with respect to the spherical Lebesgue measure $\sigma = \mathcal{H}^{n-1}$ on $\mathbb{S}^{n-1}$ and using Fubini's theorem, we obtain

$$
\int_{\mathbb{S}^{n-1}} v(K, u)\, \sigma(du) = \int_{\mathbb{S}^{n-1}} \frac{1}{2} \int_{\mathbb{S}^{n-1}} |\langle u, x \rangle|\, S_{n-1}(K, dx)\, \sigma(du)
$$

$$
= \int_{\mathbb{S}^{n-1}} \frac{1}{2} \int_{\mathbb{S}^{n-1}} |\langle u, x \rangle|\, \sigma(du)\, S_{n-1}(K, dx)
$$

$$
= \int_{\mathbb{S}^{n-1}} v(B^n, x)\, S_{n-1}(K, dx)
$$

$$
= 2\kappa_{n-1} V_{n-1}(K).
$$

Since $v(K, u) = V_{n-1}(K|u^\perp)$, we may replace the integration over $\mathbb{S}^{n-1}$ by an integration over the space $G(n, n-1)$ of $(n-1)$-dimensional linear subspaces of $\mathbb{R}^n$, by considering the normalized image measure ν_{n-1} of spherical Lebesgue measure σ under the mapping $u \mapsto u^\perp$. Then we get

$$
\int_{G(n,n-1)} V_{n-1}(K|L)\, \nu_{n-1}(dL) = \frac{2\kappa_{n-1}}{n\kappa_n} V_{n-1}(K),
$$

where $K|L$ denotes the orthogonal projection of K to L. This formula is known as Cauchy's surface area formula for convex bodies. Our first goal is to generalize this projection formula to other intrinsic volumes V_j and to projections to subspaces of lower dimensions. This requires a natural measure ν_q on the space of q-dimensional

© Springer Nature Switzerland AG 2020
D. Hug, W. Weil, *Lectures on Convex Geometry*, Graduate Texts in Mathematics 286, https://doi.org/10.1007/978-3-030-50180-8_5

linear subspaces. Subsequently, we shall consider integrals over sections of K with affine flats, where the integration is carried out with respect to a natural measure μ_q on affine q-flats. To prepare these integrations, we first explain how the measures ν_q and μ_q can be introduced in an elementary way. Moreover, we take the opportunity to introduce some basic results and concepts related to invariant measures.

5.1   Invariant Measures

In the following, we shall introduce a natural measure on the space $G(n, k)$ of k-dimensional linear subspaces of $\mathbb{R}^n$, endowed with a suitable topology, via the operation of the topological group $SO(n)$ on $G(n, k)$. For this reason, we first briefly consider topological groups and their invariant measures.

Let $(G, \circ)$ be a *topological group*. This means that G is a topological space, $\circ$ is a binary operation on G such that $(G, \circ)$ is a group and composition $G \times G \to G$, $(g, h) \mapsto g \circ h$, and inversion $G \to G$, $g \mapsto g^{-1}$, are continuous operations. In the following, we always assume that the topological group $(G, \circ)$ is a locally compact topological space, which should subsume the Hausdorff separation property. Then, for $g \in G$, the map $l_g : G \to G$, $x \mapsto g \circ x$, the inversion map, and the map $r_g : G \to G$, $x \mapsto x \circ g$, are homeomorphisms and therefore Borel measurable. For a set $A \subset G$, we simply write $gA := l_g(A)$, $Ag := r_g(A)$, and $A^{-1} := \{a^{-1} : a \in A\}$. Thus, if A is a Borel set, so are these transforms. If the group operation is clear from the context, it will be omitted.

A Borel measure μ on G (more precisely, on the Borel sets of G) which is nonzero and finite on compact sets, is called a *left Haar measure* on G if $\mu(gA) = \mu(A)$ for $g \in G$ and $A \in \mathcal{B}(G)$. It is called a *right Haar measure* on G if $\mu(Ag) = \mu(A)$ for $g \in G$ and $A \in \mathcal{B}(G)$, and it is called *inversion invariant* if $\mu(A^{-1}) = \mu(A)$ for $A \in \mathcal{B}(G)$. We say that μ is a Haar measure on G if it is left, right, and inversion invariant. In particular, in this case we have $l_g(\mu) = r_g(\mu) = \mu$ for $g \in G$. It is not difficult to prove that if μ is a left Haar measure on G, then $\nu(A) := \mu(A^{-1})$ for $A \in \mathcal{B}(G)$ is a right Haar measure on G. For the sake of completeness, we state the following general result on Haar measures (see [26, Chapter 9]).

Theorem 5.1 *Let $(G, \circ)$ be a locally compact topological group. Then there exists a left Haar measure on $(G, \circ)$. It is unique up to a scalar multiple. The same is true for right Haar measures.*

For the case of the (proper) orthogonal group $SO(n)$, we can provide a direct construction which proves the existence of a Haar measure on $SO(n)$. In the following, we consider elements of $SO(n)$ as orientation preserving orthogonal linear maps of $\mathbb{R}^n$ or as quadratic matrices, as convenient. Thus the topology on $SO(n)$ is the subspace topology of the space of regular $n \times n$ matrices in $\mathbb{R}^{n,n}$.

Lemma 5.1 *There is a Haar probability measure ν on $SO(n)$.*

Proof We consider the set $LU_n \subset (\mathbb{S}^{n-1})^n$ of linearly independent n-tuples. Then LU_n is an open subset of $(\mathbb{S}^{n-1})^n$ and the complement has measure zero with respect to the n-fold product measure σ^n of spherical Lebesgue measure σ. On LU_n we define the mapping T onto $SO(n)$ by

$$T(x_1, \ldots, x_n) := \left(\frac{y_1}{\|y_1\|}, \ldots, \frac{y_n}{\|y_n\|} \right), \tag{5.1}$$

where $(y_1, \ldots, y_n) \in \mathbb{R}^{n,n}$ is the n-tuple of vectors (considered as a square matrix) obtained from $(x_1, \ldots, x_n) \in \mathbb{R}^{n,n}$ by the Gram–Schmidt orthogonalization procedure (and where, in addition, the sign of y_n is chosen such that the matrix on the right-hand side of (5.1) has determinant 1). Up to the sign of y_n, we thus have

$$y_k := x_k - \sum_{i=1}^{k-1} \langle x_k, y_i \rangle \frac{y_i}{\|y_i\|^2}, \qquad k = 2, \ldots, n,$$

and $y_1 := x_1$. Clearly, T is almost everywhere defined with respect to σ^n and continuous. Let $\overline{\nu}$ be the image measure of σ^n under T, that is,

$$\overline{\nu}(A) = \int_{LU_n} \mathbf{1}\{T(x_1, \ldots, x_n) \in A\}\, \sigma^n(d(x_1, \ldots, x_n)),$$

for a Borel set $A \subset SO(n)$. Since $T(\varrho x_1, \ldots, \varrho x_n) = \varrho T(x_1, \ldots, x_n)$ for $\varrho \in SO(n)$ and $(x_1, \ldots, x_n) \in LU_n$ and $\varrho(x_1, \ldots, x_n) = (\varrho x_1, \ldots, \varrho x_n)$, the left invariance of $\overline{\nu}$ follows from the rotation invariance of σ and Fubini's theorem. Since $\overline{\nu}(LU_n) = (n\kappa_n)^n$, normalization yields a left invariant probability measure.

While the construction of ν is specific for $SO(n)$, the remaining properties are implied by the following general lemma. □

Lemma 5.2 *A left Haar measure on a compact topological group is also a right Haar measure and inversion invariant.*

Proof Let ν be a left Haar measure on the compact topological group G. Since ν is positive and finite on G, we can assume that $\nu(G) = 1$. Let $f \in \mathbf{C}(G)$ and $y \in G$.

We first establish the inversion invariance. Using the left invariance of ν and writing $\overline{f}(x) := f(x^{-1})$ for $x \in G$, we get

$$\int_G f(x^{-1}y)\, \nu(dx) = \int_G f\left(\left(y^{-1}x\right)^{-1} \right) \nu(dx)$$

$$= \int_G \overline{f}(y^{-1}x)\, \nu(x) = \int_G \overline{f}(x)\, \nu(dx)$$

$$= \int_G f(x^{-1})\, \nu(dx). \tag{5.2}$$

Hence, integrating (5.2) with respect to $\nu(dy)$ and by Fubini's theorem, we obtain

$$\int_G f(x^{-1}) \, \nu(dx) = \int_G \int_G f(x^{-1}y) \, \nu(dx) \, \nu(dy)$$

$$= \int_G \int_G f(x^{-1}y) \, \nu(dy) \, \nu(dx)$$

$$= \int_G f(y) \, \nu(dy),$$

again by the left invariance and since $\nu(G) = 1$.

Finally, the right invariance follows from relation (5.2) and the inversion invariance. □

The translation group $\mathbb{R}^n$ has a canonical topological (Euclidean) structure and the Lebesgue measure is up to a constant factor the unique translation invariant Haar measure on $\mathbb{R}^n$. Let $G(n)$ denote the group of proper rigid motions of $\mathbb{R}^n$ with the composition as the group operation. The map $\gamma : \mathbb{R}^n \times SO(n) \to G(n)$, $(t, \sigma) \mapsto \gamma(t, \sigma)$ with $\gamma(t, \sigma)(x) := \sigma x + t$, for $x \in \mathbb{R}^n$, is bijective and used to transfer the product topology from $\mathbb{R}^n \times SO(n)$ to $G(n)$. Thus $G(n)$ becomes a locally compact topological group with countable base.

A Borel measure μ on $\mathcal{B}(G(n))$ is introduced by

$$\mu := \int_{\mathbb{R}^n} \int_{SO(n)} \mathbf{1}\{\gamma(t, \rho) \in \cdot\} \, \nu(d\rho) \, \lambda_n(dt).$$

Lemma 5.3 *The Borel measure μ on $G(n)$ is a Haar measure.*

Proof Let $A \in \mathcal{B}(G(n))$, and let $g = \gamma(s, \sigma) \in G(n)$ with $s \in \mathbb{R}^n$ and $\sigma \in SO(n)$. Then $g \circ \gamma(t, \rho) = \gamma(\sigma t + s, \sigma \circ \rho)$ and therefore

$$\mu(g^{-1}A) = \int_{\mathbb{R}^n} \int_{SO(n)} \mathbf{1}\{g \circ \gamma(t, \rho) \in A\} \, \nu(d\rho) \, \lambda_n(dt)$$

$$= \int_{\mathbb{R}^n} \int_{SO(n)} \mathbf{1}\{\gamma(\sigma t + s, \sigma \circ \rho) \in A\} \, \nu(d\rho) \, \lambda_n(dt)$$

$$= \int_{\mathbb{R}^n} \int_{SO(n)} \mathbf{1}\{\gamma(\sigma t + s, \rho) \in A\} \, \nu(d\rho) \, \lambda_n(dt)$$

$$= \int_{SO(n)} \int_{\mathbb{R}^n} \mathbf{1}\{\gamma(\sigma t + s, \rho) \in A\} \, \lambda_n(dt) \, \nu(d\rho)$$

$$= \int_{SO(n)} \int_{\mathbb{R}^n} \mathbf{1}\{\gamma(t, \rho) \in A\} \, \lambda_n(dt) \, \nu(d\rho) = \mu(A),$$

where we used the left invariance of ν, the $SO(n)$ invariance and translation invariance of λ_n and Fubini's theorem (repeatedly). This shows that μ is left invariant.

Using that $\gamma(t, \rho) \circ g = \gamma(\rho s + t, \rho \circ \sigma)$, the right invariance of ν, the $SO(n)$ invariance and translation invariance of λ_n and Fubini's theorem, we also see that μ is right invariant. (This is also implied by the inversion invariance, which we show next.)

To establish the inversion invariance, we first observe that $\gamma(t, \sigma)^{-1} = \gamma(-\rho^{-1}t, \rho^{-1})$. Then

$$\mu(A^{-1}) = \int_{\mathbb{R}^n} \int_{SO(n)} \mathbf{1}\{\gamma(t, \rho)^{-1} \in A\} \, \nu(d\rho) \, \lambda_n(dt)$$

$$= \int_{\mathbb{R}^n} \int_{SO(n)} \mathbf{1}\{\gamma(-\rho^{-1}t, \rho^{-1}) \in A\} \, \nu(d\rho) \, \lambda_n(dt)$$

$$= \int_{\mathbb{R}^n} \int_{SO(n)} \mathbf{1}\{\gamma(-\rho t, \rho) \in A\} \, \nu(d\rho) \, \lambda_n(dt)$$

$$= \int_{\mathbb{R}^n} \int_{SO(n)} \mathbf{1}\{\gamma(t, \rho) \in A\} \, \nu(d\rho) \, \lambda_n(dt) = \mu(A),$$

where we used the inversion invariance of ν and the isometry invariance of the Lebesgue measure λ_n.   □

The following general lemma implies that up to a constant factor, the measure μ on $\mathcal{B}(G(n))$ is the unique left Haar measure (and then also the unique Haar measure) on $G(n)$.

Lemma 5.4 *Let $(G, \circ)$ be a locally compact group. Let ν be a Haar measure and let μ be a left Haar measure on G. Then there is a constant $c > 0$ such that $\nu = c\mu$.*

Proof Let $f, g : G \to [0, \infty)$ be measurable. Repeatedly using Fubini's theorem and (in this order) that ν is right invariant, μ is left invariant, ν is inversion invariant, and ν is right invariant, we obtain

$$\int_G f \, d\nu \cdot \int_G g \, d\mu = \int_G \int_G f(x)g(y) \, \nu(dx) \, \mu(dy)$$

$$= \int_G \int_G f(x \circ y)g(y) \, \nu(dx) \, \mu(dy)$$

$$= \int_G \int_G f(x \circ y)g(y) \, \mu(dy) \, \nu(dx)$$

$$= \int_G \int_G f(y)g(x^{-1} \circ y) \, \mu(dy) \, \nu(dx)$$

$$= \int_G f(y) \int_G g(x^{-1} \circ y) \, \nu(dx) \, \mu(dy)$$

$$= \int_G f(y) \int_G g(x \circ y)\, \nu(dx)\, \mu(dy)$$

$$= \int_G f(y) \int_G g(x)\, \nu(dx)\, \mu(dy)$$

$$= \int_G f\, d\mu \cdot \int_G g\, d\nu.$$

Since $\nu \neq 0$, there is a compact set $A_0 \subset G$ with $\nu(A_0) > 0$. Let $A \in \mathcal{B}(G)$ be arbitrary. The choice $f = \mathbf{1}_{A_0}$ and $g = \mathbf{1}_A$ yields $\mu(A) = (\mu(A_0)/\nu(A_0)) \cdot \nu(A)$, which proves the assertion. $\square$

In the remainder of this section, we study invariant measures on spaces on which groups operate. In particular, we are interested in the (linear) *Grassmann space* $G(n, k)$ with the operation of the proper rotation group $SO(n)$ and the (affine) Grassmann space $A(n, k)$ with the operation of the proper rigid motion group $G(n)$.

Let $G(n, k)$ be the set of k-dimensional linear subspaces of $\mathbb{R}^n$ with $k \in \{0, \ldots, n\}$. For a fixed $U_0 \in G(n, k)$ the map $\beta_k : SO(n) \to G(n, k)$, $\varrho \mapsto \varrho U_0$, is surjective. We endow $G(n, k)$ with the finest topology such that β_k is continuous. Hence, a map $f : G(n, k) \to T$, where T is an arbitrary topological space, is continuous if and only if $f \circ \beta_k$ is continuous. Thus $G(n, k)$ becomes a topological space.

Lemma 5.5

(a) *The topological space $G(n, k)$ is compact with countable base.*
(b) *The topology is independent of the particular choice of $U_0 \in G(n, k)$.*
(c) *The map β_k is open.*
(d) *The operation $SO(n) \times G(n, k) \to G(n, k)$, $(\varrho, U) \mapsto \varrho U$, is continuous and transitive.*

Proof See Exercise 5.1.6. $\square$

Thus, more generally, we are in the following situation. We have a (locally compact or even compact) topological group G and a (locally compact or even compact) topological space X on which G acts, that is, there is a continuous map $\varphi : G \times X \to X$ such that $\varphi(e, x) = x$ for $x \in X$, if e is the neutral element of G, and $\varphi(g, \varphi(h, x)) = \varphi(g \circ h, x)$ for $g, h \in G$ and $x \in X$. This operation (map) is called *transitive* if for any $x, y \in X$ there is some $g \in G$ such that $\varphi(g, x) = y$. Then the map $\varphi(g, \cdot) : X \to X$ is a homeomorphism.

In the above situation, we additionally know that the map $\varphi(\cdot, x_0) : G \to X$ is open for some $x_0 \in X$. It can be shown, also in the general framework, that then this map is open for any $x_0 \in X$. Let $S_x := \{g \in G : \varphi(g, x) = x\}$ be the stabilizer subgroup of x. Introducing on the factor space G/S_x the topology induced by the projection map $G \to G/S_x$, it follows that G/S_x and X are homeomorphic. In this case, (X, φ) is called a homogeneous G-space.

In the following, we simply write gx instead of $\varphi(g, x)$ if the operation φ is clear from the context. Moreover, for $g \in G$ and $A \subset X$, we write gA for $\{\varphi(g, x) : x \in A\} = \{gx : x \in A\}$. Then a Borel measure μ on $\mathcal{B}(X)$ is called G-invariant if $\mu(gA) = \mu(A)$ for $A \in \mathcal{B}(X)$ and $g \in G$. A G-invariant, regular Borel measure $\mu \neq 0$ on $\mathcal{B}(X)$ is called a *Haar measure*.

In the following, X will always be a locally compact space with a countable base. Then a Borel measure μ on $\mathcal{B}(X)$ that is finite on compact sets is regular (see [26, Proposition 7.2.3]) and σ-finite (see [26, Proposition 7.2.5]). As usual, a Borel measure μ on X is called regular if

1. $\mu(K) < \infty$ for compact sets $K \subset X$;
2. $\mu(A) = \inf\{\mu(U) : A \subset U, U \text{ open}\}$ for $A \in \mathcal{B}(X)$;
3. $\mu(U) = \inf\{\mu(K) : K \subset U, K \text{ compact}\}$ for $U \in X$ open.

In other words, the first condition implies the other two, that is, μ is outer and inner regular in the sense of these conditions whenever it is finite on compact sets.

Theorem 5.2 *Let G be a locally compact topological group, let X be a locally compact topological space, and let $\varphi : G \times X \to X$ be a continuous, transitive operation such that $\varphi(\cdot, x) : G \to X$ is open for some $x \in X$ (i.e., (X, φ) is a homogeneous G-space). Then any two (G-invariant) Haar measures on X are proportional to each other.*

In the specific cases considered in the following, the uniqueness assertion can be proved directly. However, it is useful to know the general result behind these special cases. Similarly, in all cases needed below, the existence of a Haar measure on a homogeneous G-space will be established in a direct way. There are also general existence results. The existence of a Haar measure on a homogeneous G-space (X, φ) can be ensured, for instance, if G is compact, or, more generally, if G and S_x are unimodular (for some $x \in X$). In fact, the existence of a Haar measure is equivalent to the condition that $\triangle_{S_x}(h) = \triangle_G(h)$ for $h \in S_x$, where $\triangle_{S_x}$ and $\triangle_G$ are the modular functions of the topological groups S_x and G. We refer to the exercises and [26, p. 313] for modular functions of topological groups (and their basic properties) and to [73] for proofs of facts not established here.

For a compact topological group G, and a topological Hausdorff space X on which G acts continuously and transitively, we can provide a simple construction of a G-invariant Haar measure on X, based on a Haar measure on G.

Theorem 5.3 *Let G be a compact topological space with a Haar probability measure ν. Let X be a topological Hausdorff space, and let $\varphi : G \times X \to X$ be a continuous and transitive operation. Then X is compact and, for any fixed $x_0 \in X$,*

$$\mu(A) = \int_G \mathbf{1}\{\varphi(g, x_0) \in A\} \, \nu(dg), \qquad A \in \mathcal{B}(X),$$

defines a G-invariant Haar probability measure on X. It is the unique G-invariant Haar probability measure on X.

Proof The map $\varphi(\cdot, x_0) : G \to X$ is continuous and surjective, hence $X = \varphi(G, x_0)$ is compact. Let $h \in G$ and $A \in \mathcal{B}(X)$. Since $\varphi(g, x_0) \in hA$ if and only if $\varphi(h^{-1} \circ g, x_0) \in A$, we get

$$\mu(hA) = \int_G \mathbf{1}\{\varphi(g, x_0) \in hA\}\, \nu(dg)$$

$$= \int_G \mathbf{1}\{\varphi(h^{-1} \circ g, x_0) \in A\}\, \nu(dg)$$

$$= \int_G \mathbf{1}\{\varphi(g, x_0) \in A\}\, \nu(dg)$$

$$= \mu(A),$$

since ν is left invariant. Moreover, $\mu(X) = \nu(G) = 1$, hence μ is regular.

Suppose that $\bar{\mu}$ is another Haar probability measure on X. Let $f : X \to [0, \infty)$ be measurable. Then, for $g \in G$, we have

$$\int_X f(x)\, \bar{\mu}(dx) = \int_X f(\varphi(g, x))\, \bar{\mu}(dx)$$

$$= \int_G \int_X f(\varphi(g, x))\, \bar{\mu}(dx)\, \nu(dg)$$

$$= \int_X \int_G f(\varphi(g, x))\, \nu(dg)\, \bar{\mu}(dx).$$

For given $x, x_0 \in X$, there is some $h \in G$ such that $\varphi(h, x_0) = x$. Since ν is also right invariant, we get

$$\int_G f(\varphi(g, x))\, \nu(dg) = \int_G f(\varphi(g, \varphi(h, x_0)))\, \nu(dg)$$

$$= \int_G f(\varphi(g \circ h, x_0))\, \nu(dg)$$

$$= \int_G f(\varphi(g, x_0))\, \nu(dg),$$

and hence

$$\int_X f(x)\, \bar{\mu}(dx) = \int_X \int_G f(\varphi(g, x_0))\, \nu(dg)\, \bar{\mu}(dx)$$

$$= \int_X f(x)\, \mu(dx),$$

which shows that $\mu = \bar{\mu}$. $\square$

Corollary 5.1 *Let $k \in \{0, \ldots, n\}$ and $U_0 \in G(n, k)$. Then*

$$\nu_k := \int_{SO(n)} \mathbf{1}\{\varrho U_0 \in \cdot\} \, \nu(d\varrho)$$

is the uniquely determined $SO(n)$-invariant Haar probability measure on $G(n, k)$. In particular, the definition is independent of the choice of the linear subspace $U_0 \in G(n, k)$.

Now we address the affine Grassmannian $A(n, k)$ on which the motion group $G(n)$ operates. Since these spaces are not compact, we have to adjust the approach described for the linear Grassmannian. First, we describe how the usual topology can be introduced. Let $U_0 \in G(n, k)$ be fixed for the moment. Consider the map

$$\gamma_k : U_0^\perp \times SO(n) \to A(n, k), \qquad (x, \varrho) \mapsto \varrho(U_0 + x).$$

We endow $A(n, k)$ with the finest topology such that γ_k is continuous.

Lemma 5.6

(a) *The topological space $A(n, k)$ is locally compact with countable base.*
(b) *The topology is independent of the particular choice of $U_0 \in G(n, k)$.*
(c) *The map γ_k is open.*
(d) *The operation $G(n) \times A(n, k) \to A(n, k)$, $(g, E) \mapsto gE := \{gx : x \in E\}$, is continuous and transitive.*

Proof See Exercise 5.1.7. □

On $A(n, k)$ we introduce the Borel measure

$$\mu_k := \gamma_k \left((\lambda_{n-k} | U_0^\perp) \otimes \nu \right)$$

$$= \int_{SO(n)} \int_{U_0^\perp} \mathbf{1}\{\varrho(U_0 + x) \in \cdot\} \lambda_{n-k}(dx) \, \nu(d\varrho)$$

$$= \int_{G(n,k)} \int_{U^\perp} \mathbf{1}\{U + z \in \cdot\} \lambda_{n-k}(dz) \, \nu(dU), \qquad (5.3)$$

which is independent of the choice of $U_0 \in G(n, k)$. For a set $C \subset \mathbb{R}^n$, we write $A_C := \{E \in A(n, k) : E \cap C \neq \emptyset\}$, hence $A_{B^n} = \gamma_k((B^n \cap U_0^\perp) \times SO(n))$. The analogue of Corollary 5.1 for the affine Grassmannian is stated in the following theorem.

Theorem 5.4 *The Borel measure μ_k on $\mathcal{B}(A(n, k))$ is the unique $G(n)$-invariant Haar measure on $A(n, k)$ with $\mu_k(A_{B^n}) = \kappa_{n-k}$.*

Proof Let $A \subset A(n, k)$ be a compact set. The system of sets

$$\gamma_k(\{x \in U_0^\perp : \|x\| < m\} \times SO(n)), \qquad m \in \mathbb{N},$$

is an open cover of A, hence A is contained in one of these sets. It follows that $\mu_q(A) < \infty$, which shows that $\mu_q \neq 0$ is a regular Borel measure on $A(n, k)$.

Next we prove the invariance property. Let $g = \gamma(t, \sigma) \in G(n)$ with $t \in \mathbb{R}^n$ and $\sigma \in SO(n)$. Let $f : A(n, k) \to [0, \infty)$ be measurable. Then, using the translation invariance of λ_{n-k} on $U_0^\perp$ and the rotation invariance of ν on $SO(n)$, we obtain

$$\int_{A(n,k)} f(gE)\, \mu_k(dE)$$

$$= \int_{SO(n)} \int_{U_0^\perp} f(\gamma(t, \sigma)(\varrho(U_0 + x)))\, \lambda_{n-k}(dx)\, \nu(d\varrho)$$

$$= \int_{SO(n)} \int_{U_0^\perp} f(\sigma \circ \varrho(U_0 + x + [(\sigma \circ \varrho)^{-1}t]|U_0^\perp)\, \lambda_{n-k}(dx)\, \nu(d\varrho)$$

$$= \int_{SO(n)} \int_{U_0^\perp} f(\sigma \circ \varrho(U_0 + x)\, \lambda_{n-k}(dx)\, \nu(d\varrho)$$

$$= \int_{SO(n)} \int_{U_0^\perp} f(\varrho(U_0 + x)\, \lambda_{n-k}(dx)\, \nu(d\varrho)$$

$$= \int_{A(n,k)} f(E)\, \mu_k(dE).$$

Let μ be a translation invariant regular Borel measure on $A(n, k)$. By the following general theorem, there is a finite Borel measure τ on $G(n, k)$ such that

$$\mu = \int_{G(n,k)} \int_{U^\perp} \mathbf{1}\{U + z \in \cdot\}\, \lambda_{n-k}(dz)\, \tau(dU).$$

Suppose that μ is also $SO(n)$-invariant. For $C \in \mathcal{B}(G(n, k))$, we define $B_C := \{U + z \in A(n, k) : z \in U^\perp \cap B^n, U \in C\}$ and observe that $\varrho B_C = B_{\varrho C}$ for $\varrho \in SO(n)$. Hence

$$\kappa_{n-k}\tau(C) = \mu(B_C) = \mu(\varrho B_C) = \mu(B_{\varrho C}) = \kappa_{n-k}\tau(\varrho C),$$

which shows that τ is $SO(n)$-invariant. Therefore, $\tau = c\nu_k$ for some constant $c \geq 0$. Requiring that $\mu(A_{B^n}) = \kappa_{n-k}$, we get $c = 1$, and thus $\mu = \mu_k$, which proves the uniqueness assertion.   □

For the proof of the following theorem, we need to describe the distortion of Lebesgue measure under a linear orthogonal projection map $\pi_{U^\perp} : L \to U^\perp$ from a linear subspace $L \in G(n, n-k)$ on another linear subspace $U^\perp \in G(n, n-k)$, where $U \in G(n, k)$ and U, L are complementary, that is, $U \oplus L = \mathbb{R}^n$. In order to introduce a quantity which describes the relative position of complementary subspaces (called the *subspace determinant*), we choose orthonormal bases

$$a_1, \ldots, a_k \text{ of } U, \quad c_1, \ldots, c_{n-k} \text{ of } U^\perp,$$

$$b_1, \ldots, b_{n-k} \text{ of } L, \quad d_1, \ldots, d_k \text{ of } L^\perp.$$

Then, independently of the chosen specific orthonormal bases, we can define

$$[U, L] := |\det(a_1, \ldots, a_k, b_1, \ldots, b_{n-k})|$$

$$= |\det(\langle a_i, d_j \rangle_{i,j=1}^k)|$$

$$= |\det(c_1, \ldots, c_{n-k}, d_1, \ldots, d_k)|$$

$$= [U^\perp, L^\perp].$$

Let $\pi_{U^\perp} : L \to U^\perp$ denote the orthogonal projection of L onto $U^\perp$, where we still assume that $U \oplus L = \mathbb{R}^n$. Then $\pi_{U^\perp}$ is a homeomorphism and

$$(\pi_{U^\perp})\lambda_L = [U, L]^{-1}\lambda_{U^\perp}, \qquad (\pi_{U^\perp})^{-1}\lambda_{U^\perp} = [U, L]\lambda_L. \tag{5.4}$$

To verify this, we first observe that $\ker(\pi_{U^\perp}) = \{0\}$. Since the kernel of $\pi_{U^\perp}$ is trivial, the linear map $\pi_{U^\perp}$ is injective. Since $\dim U^\perp = \dim L$, the linear map is bijective and hence a homeomorphism. Define $\nu(\cdot) := \lambda_{U^\perp}(\pi_{U^\perp}(\cdot))$ on $\mathcal{B}(L)$. Since ν is invariant under translations in L and locally finite, we obtain $\nu = c_{U,L}\lambda_L$. To determine the constant, we consider $W := [0, b_1] + \cdots + [0, b_{n-k}]$, which is a unit cube in L with $\lambda_L(W) = 1$. Then we get

$$\pi_{U^\perp}(W) = [0, b_1|_{U^\perp}] + \cdots + [0, b_{n-k}|_{U^\perp}],$$

and therefore

$$\lambda_{U^\perp}(\pi_{U^\perp}(W)) = |\det(b_1|_{U^\perp}, \ldots, b_{n-k}|_{U^\perp})|$$

$$= |\det(\langle b_i, c_j \rangle_{i,j=1}^{n-k})|$$

$$= |\det(b_1, \ldots, b_{n-k}, a_1, \ldots, a_k)|$$

$$= [U, L].$$

This proves the assertion.

We shall use (5.4) below in (5.5).

Theorem 5.5 *Let μ be a regular, translation invariant Borel measure on $A(n, k)$. Then there is a unique finite Borel measure τ on $G(n, k)$ such that*

$$\mu = \int_{G(n,k)} \int_{U^\perp} \mathbf{1}\{U + x \in \cdot\} \lambda_{n-k}(dx)\, \tau(dU).$$

Proof In the first part of the argument, we fix $L \in G(n, n - k)$ arbitrarily. Then we define

$$G_L := \{U \in G(n, k) : U \cap L = \{0\}\},$$

$$A_L := \{U + x \in A(n, k) : U \in G_L, x \in L\}.$$

Since $L \oplus U = \mathbb{R}^n$ for $U \in G_L$, we have $A_L + t = A_L$ for $t \in \mathbb{R}^n$, that is, A_L is the translation invariant class of k-dimensional affine subspaces whose direction space is complementary to L. The map

$$\varphi_L : L \times G_L \to A_L, \quad (x, U) \mapsto U + x,$$

is a homeomorphism.

Let $C \subset G_L$ be a Borel set. A locally finite (that is, finite on compact sets), translation invariant Borel measure on L is defined by

$$\eta_C(B) := \mu(\varphi_L(C \times B)), \qquad B \in \mathcal{B}(L).$$

Hence, we get $\eta_C(B) = \rho_L(C)\lambda_L(B)$, or $\mu(\varphi_L(B \times C)) = \lambda_L(B)\rho_L(C)$. But then ρ_L is a finite Borel measure on G_L, which implies that

$$\mu(\varphi_L(B \times C)) = (\lambda_L \otimes \rho_L)(B \times C),$$

and therefore

$$(\varphi_L)^{-1}\mu = \lambda_L \otimes \rho_L, \qquad \mu|_{A_L} = \varphi_L(\lambda_L \otimes \rho_L).$$

For a measurable function $f : A(n, k) \to [0, \infty)$, we thus get

$$\int_{A_L} f \, d\mu = \int_{G_L} \int_L f(U + x) \, \lambda_L(dx) \, \rho_L(dU),$$

and then it follows from (5.4) that

$$\int_{A_L} f \, d\mu = \int_{G_L} \int_L f(U + \pi_{U^\perp}(x)) \, \lambda_L(dx) \rho_L(dU)$$

$$= \int_{G_L} \int_{U^\perp} f(U + z) \, \lambda_{U^\perp}(dz) \, [L, U]^{-1} \rho_L(dU) \qquad (5.5)$$

$$= \int_{G(d,k)} \int_{U^\perp} f(U + z) \, \lambda_{U^\perp}(dz) \, \tau_L(dU),$$

where $\tau_L := [L, U]^{-1} \rho_L|_{G_L}$.

Since G_L is open, for $L \in G(n, n - k)$, and $G(n, k)$ is compact, there are $L_1, \ldots, L_r \in G(n, n - k)$ such that

$$G(n, k) = \bigcup_{i=1}^{r} G_{L_i} \quad \text{and hence} \quad A(n, k) = \bigcup_{i=1}^{r} A_{L_i}.$$

Furthermore, A_{L_i} is invariant with respect to translations, for $i = 1, \ldots, r$. Since these unions are not disjoint, we recursively define the sets

$$A_j := A_{L_j} \setminus (A_1 \cup \cdots \cup A_{j-1}), \quad A_0 := \emptyset, \tag{5.6}$$

for $j = 1, \ldots, r$, which are translation invariant and satisfy

$$A(n, k) = A_1 \cup \cdots \cup A_r.$$

The restriction of μ to A_i, which we write as $\mu \llcorner A_i$, is translation invariant. As above, we obtain a measure $\tau_i := (\mu \llcorner A_i)_{L_i}$ on $G(n, k)$, which is concentrated on G_{L_i}, so that

$$\int_{A(n,k)} f \, d\mu = \sum_{i=1}^{r} \int_{A_i} f \, d\mu$$

$$= \sum_{i=1}^{r} \int_{A_{L_i}} f \, d(\mu \llcorner A_i)$$

$$= \sum_{i=1}^{r} \int_{G(n,k)} \int_{U^\perp} f(U + z) \lambda_{U^\perp}(dz) \tau_i(dU)$$

$$= \int_{G(n,k)} \int_{U^\perp} f(U + z) \lambda_{U^\perp}(dz) \, \tau(dU)$$

with $\tau := \tau_1 + \cdots + \tau_r$.

Finally, we address the uniqueness assertion. For $C \in \mathcal{B}(G(n, k))$, let

$$B_C := \{U + z \in A(n, k) : z \in U^\perp \cap B^n, U \in C\}.$$

Then, we get $\kappa_{n-k} \tau(C) = \mu(B_C)$, which shows that the finite measure τ is uniquely determined by μ. $\qquad\square$

Exercises and Supplements for Sect. 5.1

1. Show that $G(n)$ is a locally compact topological group with countable base.
2. Let μ be a left Haar measure on a locally compact topological group. Show that μ is inversion invariant if and only if μ is right invariant.
3. Let μ be a left Haar measure on a locally compact topological group G. Show that G is compact if and only if μ is finite.
4. Discuss properties of the modular function.
5. Show that up to a constant factor the spherical Lebesgue measure is the uniquely determined $SO(n)$-invariant finite Borel measure on $\mathbb{S}^{n-1}$.
6.* Prove Lemma 5.5.
7. Prove Lemma 5.6.
8. Let $k \in \{0, \dots, n\}$. Show that

$$\nu_{n-k} = \int_{G(n,k)} \mathbf{1}\{U^\perp \in \cdot\} \, \nu_k(dU).$$

9. Let $0 \le q \le k \le n$. For $L \in G(n,k)$, define

$$G(L, q) := \{U \in G(n, q) : U \subset L\}.$$

Let ν_q^L denote the $SO(L)$-invariant Haar measure on $G(L, q)$ with respect to L as the ambient space, where

$$SO(L) := \{\rho \in SO(n) : \rho(x) = x \text{ for } x \in L^\perp\}.$$

Let $f : G(n, q) \to [0, \infty)$ be measurable. Then

$$\int_{G(n,q)} f(L) \, \nu_q(dL) = \int_{G(n,k)} \int_{G(L,q)} f(U) \, \nu_q^L(dU) \, \nu_k(dL).$$

10. Let $0 \le q \le k \le n$. For $F \in A(n, k)$, define

$$A(F, q) := \{E \in A(n, q) : E \subset F\}.$$

Further, let $F_0^\perp$ be the linear subspace orthogonal to $F_0 \in G(n, k)$, the linear subspace parallel to F, and

$$G(F) := \{g \in G(n) : g(x) = x \text{ for } x \in F_0^\perp\},$$

that is, $G(F) = \gamma_k(F_0 \times SO(F_0))$. For $F = F_0 + x \in A(n, k)$ with $x \in F_0^\perp$, let μ_q^F denote the $G(F)$-invariant Haar measure on $A(F, q)$ with respect to F as the ambient space which is given by

$$\mu_q^F := \int_{G(F_0,q)} \int_{F_0 \cap L^\perp} \mathbf{1}\{L + y + x \in \cdot\} \, \lambda_{k-q}(dy) \, \nu_q^{F_0}(dL).$$

Let $f : A(n, q) \to [0, \infty)$ be measurable. Then

$$\int_{A(n,q)} f(E)\, \mu_q(dE) = \int_{A(n,k)} \int_{A(F,q)} f(E)\, \mu_q^F(dE)\, \mu_k(dF).$$

5.2 Projection Formulas

In the introduction to this chapter, we have already seen a particular projection formula. We now first derive a version of such an integral-geometric formula for more general intrinsic volumes and projection subspaces of arbitrary dimension. From this result, we then deduce a version for certain mixed volumes and finally for certain mixed area measures. As a particular case, we thus obtain a projection formula for area measures of convex bodies.

The projection formula can be deduced from the special case considered in the introduction, by using expansion of parallel volumes and Exercise 5.1.9 for a recursive argument. This will be discussed in the exercises. Another approach, as we shall see now, is based on Hadwiger's characterization theorem.

In the following, we use the flag coefficients

$$\begin{bmatrix} m \\ j \end{bmatrix} := \binom{m}{j} \frac{\kappa_m}{\kappa_j \kappa_{m-j}}, \qquad j \in \{0, \ldots, m\},\ m \in \mathbb{N}_0,$$

which allow us to write constants in a systematic way. As before, for $K \in \mathcal{K}^n$ and a linear subspace $U \subset \mathbb{R}^n$, we denote by $K|U$ the orthogonal projection of K to U.

Theorem 5.6 (Cauchy–Kubota Formula) *Let $K \in \mathcal{K}^n$ be a convex body, let $k \in \{0, \ldots, n\}$, and let $i \in \{0, \ldots, k\}$. Then*

$$\int_{G(n,k)} V_i(K|U)\, \nu_k(dU) = \beta_{nik}\, V_i(K), \tag{5.7}$$

where

$$\beta_{nik} := \frac{\binom{k}{i}\kappa_k \kappa_{n-i}}{\binom{n}{i}\kappa_n \kappa_{k-i}} = \begin{bmatrix} n \\ i \end{bmatrix}^{-1} \begin{bmatrix} k \\ i \end{bmatrix}.$$

Proof For fixed parameters k, i, we consider the functional

$$\mathcal{P}(K) := \int_{G(n,k)} V_i(K|U)\, \nu_k(dU), \qquad K \in \mathcal{K}^n.$$

Since $U \mapsto V_i(K|U)$, $U \in G(n, k)$, is continuous, the integral exists and is finite.

For $t \in \mathbb{R}^n$ and $\sigma \in \mathrm{SO}(n)$, we have

$$V_i((\sigma K + t)|U) = V_i((\sigma K)|U + t|U) = V_i(K|(\sigma^{-1}U)),$$

where we used the translation and rotation invariance of V_i. Since v_k is rotation invariant, it follows that $\mathcal{P}$ is invariant under (proper) rigid motions.

For fixed $U \in \mathrm{G}(n,k)$, the map $K \mapsto V_i(K|U)$ is continuous. It is also uniformly bounded, if K ranges in a ball of some fixed radius. Hence, by the bounded convergence theorem, $\mathcal{P}$ is continuous.

Next we show that $\mathcal{P}$ is additive. To see this, it is sufficient to observe that

$$(K \cup L)|U = (K|U) \cup (L|U) \qquad \text{and} \qquad (K \cap L)|U = (K|U) \cap (L|U)$$

for $K, L \in \mathcal{K}^n$ with $K \cup L \in \mathcal{K}^n$ and $U \in \mathrm{G}(n,k)$. Then the additivity of $\mathcal{P}$ follows from the additivity of V_i.

Finally, we observe that $\mathcal{P}$ is positively homogeneous of degree i.

By Hadwiger's characterization theorem, there is a constant $c_i \in \mathbb{R}$ such that

$$\int_{\mathrm{G}(n,k)} V_i(K|U)\, v_k(dU) = c_i\, V_i(K), \qquad K \in \mathcal{K}^n.$$

To determine the constant c_i, we choose $K = B^n$. Then, by the invariance properties of intrinsic volumes, the symmetry of Euclidean balls, and by the independence of the intrinsic volumes of the ambient space, we obtain

$$V_i(B^k) = \int_{\mathrm{G}(n,k)} V_i(B^n|U)\, v_k(dU) = c_i\, V_i(B^n),$$

and therefore

$$c_i = \frac{V_i(B^k)}{V_i(B^n)} = \frac{\binom{k}{i}}{\kappa_{k-i}} \kappa_k \cdot \left(\frac{\binom{n}{i}}{\kappa_{n-i}} \kappa_n \right)^{-1},$$

which yields the constant as asserted.   □

Remark 5.1 For $k = i$, the Cauchy–Kubota formulas yield

$$V_i(K) = \frac{1}{\beta_{nii}} \int_{\mathrm{G}(n,i)} V_i(K|U)\, v_i(dU),$$

hence $V_i(K)$ is proportional to the mean content of the projections of K onto i-dimensional subspaces. Since $V_i(K|U)$ is also the content of the base of the cylinder circumscribed to K (with direction space U), $V_i(K|U)$ was called the 'quermass' of K in direction $U^\perp$. This explains the name 'quermassintegral' for the functionals W_{n-i}, which are proportional to V_i.

Remark 5.2 For $k = i = 1$, we obtain

$$V_1(K) = \frac{1}{\beta_{n11}} \int_{G(n,1)} V_1(K|U)\, \nu_1(dU).$$

This shows again that $V_1(K)$ is proportional to the mean width of K.

Remark 5.3 The right-hand side of (5.7) has an additive extension to $U(\mathcal{K}^n)$. The same is true for the integrand $K \mapsto V_i(K|U)$, $K \in \mathcal{K}^n$, for each $U \in G(n,k)$, and therefore also for the integral mean. It should be noted, however, that even if $K_1 \cup \cdots \cup K_m \in \mathcal{K}^n$ for convex bodies $K_1, \ldots, K_m$, in general the projection of an intersection $[\cap_{i \in I} K_i]|U$ is not equal to the intersection of the projections $\cap_{i \in I}[K_i|U]$, for arbitrary subsets $I \subset \{1, \ldots, m\} =: [m]$. Still we have

$$\int_{G(n,k)} V_i((K_1 \cup \cdots \cup K_m)|U)\, \nu_k(dU)$$

$$= \int_{G(n,k)} \sum_{\emptyset \neq I \subset [m]} (-1)^{|I|-1} V_i\left(\left[\bigcap_{i \in I} K_i\right]\bigg|U\right) \nu_k(dU)$$

$$= \sum_{\emptyset \neq I \subset [m]} (-1)^{|I|-1} V_i\left(\bigcap_{i \in I} K_i\right)$$

$$= V_i(K_1 \cup \cdots \cup K_m).$$

Hence, with the correct evaluation of the intrinsic volume under the integral, the projection formula extends to polyconvex sets.

The situation is simpler when we consider Crofton formulas and intersectional kinematic formulas in the following Sects. 5.3 and 5.4, where intersections of a fixed convex body with a "random" k-flat or with another "random" convex body are studied (but again care is required in the context of rotation sum formulas).

The projection formula for intrinsic volumes can be applied to Minkowski combinations of convex bodies. This first yields projection formulas for mixed volumes, and then also a local version of mixed area measures.

Theorem 5.7 *Let $k \in \{0, \ldots, n\}$. If $K_1, \ldots, K_k \in \mathcal{K}^n$, then*

$$\int_{G(n,k)} V^{(k)}(K_1|U, \ldots, K_k|U)\, \nu_k(dU) = \frac{\kappa_k}{\kappa_n} V(K_1, \ldots, K_k, B^n[n-k]).$$

Proof Let $k \in \{0, \ldots, n\}$, $K_1, \ldots, K_k \in \mathcal{K}^n$, and $\rho_1, \ldots, \rho_k \geq 0$. Then we have

$$\int_{G(n,k)} V_k \left(\left(\sum_{j=1}^{k} \rho_j K_j \right) | U \right) \nu_k(dU)$$

$$= \int_{G(n,k)} V_k \left(\sum_{j=1}^{k} \rho_j (K_j | U) \right) \nu_k(dU)$$

$$= \int_{G(n,k)} \sum_{r_1,\ldots,r_k=0}^{k} \binom{k}{r_1, \ldots, r_k} \rho_1^{r_1} \cdots \rho_k^{r_k}$$

$$\times V^{(k)}((K_1|U)[r_1], \ldots, (K_k|U)[r_k]) \, \nu_k(dU)$$

$$= \sum_{r_1,\ldots,r_k=0}^{k} \binom{k}{r_1, \ldots, r_k} \rho_1^{r_1} \cdots \rho_k^{r_k}$$

$$\times \int_{G(n,k)} V^{(k)}((K_1|U)[r_1], \ldots, (K_k|U)[r_k]) \, \nu_k(dU).$$

On the other hand, we also have

$$\int_{G(n,k)} V_k \left(\left(\sum_{j=1}^{k} \rho_j K_j \right) | U \right) \nu_k(dU)$$

$$= \frac{\kappa_k \, \kappa_{n-k}}{\binom{n}{k} \kappa_n} V_k \left(\sum_{j=1}^{k} \rho_j K_j \right)$$

$$= \frac{\kappa_k \, \kappa_{n-k}}{\binom{n}{k} \kappa_n} \frac{\binom{n}{k}}{\kappa_{n-k}} V \left(\sum_{j=1}^{k} \rho_j K_j[k], B^n[n-k] \right)$$

$$= \frac{\kappa_k}{\kappa_n} \sum_{r_1,\ldots,r_k=0}^{k} \binom{k}{r_1 \ldots r_k} \rho_1^{r_1} \cdots \rho_k^{r_k} V(K_1[r_1], \ldots, K_k[r_k], B^n[n-k]).$$

Comparing coefficients for $r_1 = \cdots = r_k = 1$, we obtain the assertion. $\square$

For $k \in \{1, \ldots, n\}$ and $U \in G(n, k)$, we write $S^{(U)}(M_1, \ldots, M_{k-1}, \cdot)$ for the mixed area measure of convex bodies $M_1, \ldots, M_{k-1} \subset U$ with respect to U as the ambient space. Here it should be observed that in contrast to the intrinsic volumes the area measures are not independent of the ambient space.

Combining the representation of mixed volumes from Theorem 4.1 with the integral-geometric projection formula of Theorem 5.7, we get

$$\int_{G(n,k)} \frac{1}{k} \int_{\mathbb{S}^{n-1} \cap U} h(K_k|U, u)\, S^{(U)}(K_1|U, \ldots, K_{k-1}|U, du)\, \nu_k(dU)$$

$$= \frac{\kappa_k}{\kappa_n} \frac{1}{n} \int_{\mathbb{S}^{n-1}} h(K_k, u)\, S(K_1, \ldots, K_{k-1}, B^n[n-k], du),$$

and hence,

$$\int_{G(n,k)} \int_{\mathbb{S}^{(U)}_{k-1}} h(K_k, u)\, S^{(U)}(K_1|U, \ldots, K_{k-1}|U, du)\, \nu_k(dU)$$

$$= \frac{k\kappa_k}{n\kappa_n} \int_{\mathbb{S}^{n-1}} h(K_k, u)\, S(K_1, \ldots, K_{k-1}, B^n[n-k], du).$$

Here we used that $h(K_k|U, u) = h(K_k, u)$ for $u \in U$.

Since differences of support functions are dense in $C(\mathbb{S}^{n-1})$, we obtain the following result.

Theorem 5.8 *Let $k \in \{1, \ldots, n\}$. If $K_1, \ldots, K_{k-1} \in \mathcal{K}^n$, then*

$$\int_{G(n,k)} S^{(U)} K_1|U, \ldots, K_{k-1}|U, \cdot \cap U)\, \nu_k(dU)$$

$$= \frac{k\kappa_k}{n\kappa_n} S(K_1, \ldots, K_{k-1}, B^n[n-k], \cdot).$$

The following special case is of particular relevance.

Theorem 5.9 *Let $k \in \{1, \ldots, n\}$ and $i \in \{1, \ldots, k\}$. If $K \in \mathcal{K}^n$, then*

$$\int_{G(n,k)} S_i^{(U)}(K|U, \cdot \cap U)\, \nu_k(dU) = \frac{k\kappa_k}{n\kappa_n} S_i(K, \cdot).$$

Exercises and Supplements for Sect. 5.2

1. Provide an alternative proof of Theorem 5.6, following the suggestion at the beginning of this section.
2. Show that

$$(K \cup L)|U = (K|U) \cup (L|U) \qquad \text{and} \qquad (K \cap L)|U = (K|U) \cap (L|U)$$

for $K, L \in \mathcal{K}^n$ with $K \cup L \in \mathcal{K}^n$ and $U \in G(n, k)$.

5.3   Section Formulas

Instead of projecting a given convex body to a "random uniform subspace", we now intersect it with a "random uniform flat". We start with the special case where the functional V_0 is applied to the intersection. In fact, the corresponding result can be derived from a projection formula. Let $k \in \{0, \dots, n\}$ and $K \in \mathcal{K}^n$. Then

$$\int_{A(n,k)} V_0(K \cap E) \, \mu_k(dE)$$

$$= \int_{G(n,k)} \int_{U^\perp} V_0(K \cap (U + x)) \, \lambda_{n-k}(dx) \, \nu_k(dU)$$

$$= \int_{G(n,k)} V_{n-k}(K|U^\perp) \, \nu_k(dU)$$

$$= \int_{G(n,n-k)} V_{n-k}(K|W) \, \nu_{n-k}(dW)$$

$$= \begin{bmatrix} n \\ k \end{bmatrix}^{-1} V_{n-k}(K).$$

For the second equality, we used that $K \cap (U + x) \neq \emptyset$ if and only if $x \in K|U^\perp$, provided that $x \in U^\perp$. Moreover, for the third equality we used Exercise 5.1.8.

The following result states a general section formula of this type for intrinsic volumes. It turns out that such a formula is not only connected to the projection formula, as we have just seen, but also to the kinematic formula which will be considered in the following section. We shall prove the result by using Hadwiger's characterization theorem. Another approach will be discussed in the exercises.

Theorem 5.10 (Crofton Formula) *Let $k \in \{0, \dots, n\}$ and $i \in \{0, \dots, k\}$. If $K \in \mathcal{K}^n$, then*

$$\int_{A(n,k)} V_i(K \cap E) \, \mu_k(dE) = \alpha_{nik} \, V_{n+i-k}(K)$$

with

$$\alpha_{nik} := \frac{\binom{k}{i} \kappa_k \kappa_{n+i-k}}{\binom{n}{k-i} \kappa_n \kappa_i} = \begin{bmatrix} n \\ k-i \end{bmatrix}^{-1} \begin{bmatrix} k \\ i \end{bmatrix}.$$

Proof For fixed parameters k, i, we define the functional

$$
\mathcal{C}(K) := \int_{\mathrm{A}(n,k)} V_i(K \cap E)\, \mu_k(dE)
$$

$$
= \int_{\mathrm{G}(n,k)} \int_{U^\perp} V_i(K \cap (U + x))\, \lambda_{n-k}(dx)\, \nu_k(dU) \qquad (5.8)
$$

for $K \in \mathcal{K}^n$. If $g \in \mathrm{G}(n)$, then $V_i((gK) \cap E) = V_i(K \cap (g^{-1}E))$ for $E \in \mathrm{A}(n,k)$ and $K \in \mathcal{K}^n$, since V_i is rigid motion invariant. Since μ_k is invariant under proper rigid motions, we conclude that $\mathcal{C}$ is also invariant under proper rigid motions.

Let $K, L \in \mathcal{K}^n$ with $K \cup L \in \mathcal{K}^n$, and let $E \in \mathrm{A}(n,k)$. Then obviously we have

$$
(K \cup L) \cap E = (K \cap E) \cup (L \cap E) \quad \text{and} \quad (K \cap L) \cap E = (K \cap E) \cap (L \cap E).
$$

Using the additivity of V_i and the linearity of the integral, we see that $\mathcal{C}$ is additive.

Next we show that $\mathcal{C}$ is continuous. For this, let $K_j \to K$ as $j \to \infty$. If $E \cap \mathrm{int}\, K \neq \emptyset$ or $E \cap K = \emptyset$, then $K_j \cap E \to K \cap E$ as $j \to \infty$ by Exercise 3.1.8 (b). Moreover, the set $\mathrm{A}_K(n,k)$ of all $E \in \mathrm{A}(n,k)$ such that $E \cap \mathrm{int}\, K = \emptyset$ and $E \cap K \neq \emptyset$ satisfies $\mu_k(\mathrm{A}_K(n,k)) = 0$ (see Exercise 5.3.3). Now the required continuity follows from the dominated convergence theorem.

The functional $\mathcal{C}$ is positively homogeneous of degree $n - k + i$. This can be seen from (5.8) by applying the transformation $x \mapsto \lambda x$, $x \in U^\perp$, for some fixed $\lambda > 0$, in the inner integral. Since this transformation has the Jacobian λ^{n-k}, the assertion follows from the homogeneity of degree i of V_i.

Thus Hadwiger's characterization theorem implies that

$$
\int_{\mathrm{G}(n,k)} \int_{U^\perp} V_i(K \cap (U + x))\, \lambda_{n-k}(dx)\, \nu_k(dU) = c_{n-k+i}\, V_{n-k+i}(K),
$$

for $K \in \mathcal{K}^n$, for some constant $c_{n-k+i} \in \mathbb{R}$. To determine the constant, we choose $K = B^n$ and determine both sides of the resulting relation. Using the symmetry of B^n and basic properties of V_i, we obtain for the left-hand side

$$
\int_{\mathrm{G}(n,k)} \int_{U^\perp} V_i(B^n \cap (U + x))\, \lambda_{n-k}(dx)\, \nu_k(dU)
$$

$$
= \int_{U^\perp} \sqrt{1 - \|x\|^2}^{\,i}\, V_i(B^k)\, \lambda_{n-k}(dx)
$$

$$
= V_i(B^k)(n - k)\kappa_{n-k} \int_0^1 \sqrt{1 - t^2}^{\,i}\, t^{n-k+i}\, dt
$$

$$= \frac{\binom{k}{i}}{\kappa_{k-i}} \kappa_k \, \kappa_{n-k} \frac{\Gamma\left(\frac{i}{2}+1\right)\Gamma\left(\frac{n-k}{2}+1\right)}{\Gamma\left(\frac{n-k+i}{2}+1\right)}$$

$$= \binom{k}{i} \frac{\kappa_{n-k+i}\kappa_k}{\kappa_i \kappa_{k-i}}.$$

Since $V_{n-k+i}(B^n) = \binom{n}{k-i}\frac{1}{\kappa_{k-i}}\kappa_n$, it follows that $c_{n-k+i} = \alpha_{nik}$.   □

Remark 5.4 Theorem 5.10 remains true for polyconvex sets $K \in U(\mathcal{K}^n)$. This follows from the additivity of the intrinsic volumes and the linearity of the integral. Here we also have that $(K \cap M) \cap E = (K \cap E) \cap (M \cap E)$, where $K, M \in \mathcal{K}^n$ and E is an affine subspace.

Remark 5.5 Replacing the pair (i, k) by $(0, n - i)$, we obtain

$$V_i(K) = \frac{1}{\alpha_{n0(n-i)}} \int_{A(n,n-i)} V_0(K \cap E) \, \mu_{n-i}(dE)$$

$$= \frac{1}{\alpha_{n0(n-i)}} \mu_{n-j}(\{E \in A(n, n-i) : K \cap E \neq \emptyset\}).$$

Hence, up to a constant $V_i(K)$ is the measure of all $(n - i)$-flats which intersect the convex body K.

Remark 5.6 We can give another interpretation of the intrinsic volume $V_i(K)$, for $i \in \{0, \dots, n - 1\}$, in terms of flats touching K. Namely, consider the set

$$A_i(K, \varepsilon) := \{E \in A(n, n-i-1) : K \cap E = \emptyset, (K + B^n(\varepsilon)) \cap E \neq \emptyset\}.$$

These are the $(n - i - 1)$-flats E which hit the parallel body $K + B^n(\varepsilon)$ but not K. If the limit

$$\lim_{\varepsilon \to 0+} \frac{1}{\varepsilon} \mu_{n-i-1}(A_i(K, \varepsilon))$$

exists, we can interpret it as the measure of all $(n - i - 1)$-flats touching K. We write $\omega_j := j\kappa_j$ for the surface area of a $(j - 1)$-dimensional unit sphere. Now Remark 5.5 and the Steiner formula for the intrinsic volumes (see Exercise 3.3.7) show that

$$\frac{1}{\varepsilon} \mu_{n-i-1}(A_i(K, \varepsilon)) = \frac{\alpha_{n0(n-i-1)}}{\varepsilon} [V_{i+1}(K + B^n(\varepsilon)) - V_{i+1}(K)]$$

$$= \frac{\alpha_{n0(n-i-1)}}{\varepsilon} \sum_{j=0}^{i} \varepsilon^{i+1-j} \binom{n-j}{n-i-1} \frac{\kappa_{n-j}}{\kappa_{n-i-1}} V_j(K)$$

$$\rightarrow \alpha_{n0(n-i-1)}(n-i)\frac{\kappa_{n-i}}{\kappa_{n-i-1}}V_i(K)$$

$$= \frac{\kappa_{i+1}\kappa_{n-i}}{\kappa_n}\frac{n-i}{\binom{n}{i+1}}V_i(K) = \frac{\omega_{i+1}\kappa_{n-i}}{\kappa_n}\binom{n}{i}^{-1}V_i(K),$$

as $\varepsilon \rightarrow 0+$. In the special case $i = n - 1$, we obtain $\varepsilon^{-1}\mu_0(A_0(K,\varepsilon)) \rightarrow 2V_{n-1}(K) = F(K)$ as $\varepsilon \rightarrow 0+$, which is just again the content of Remark 3.17. On the other hand, for $i = 0$ we obtain $\varepsilon^{-1}\mu_{n-1}(A_{n-1}(K,\varepsilon)) \rightarrow 2V_0(K) = 2$ as $\varepsilon \rightarrow 0+$.

Remark 5.7 We can use Remark 5.5 to solve some problems of *Geometrical Probability*. Namely, if $K, K_0 \in \mathcal{K}^n$ are such that $K \subset K_0$ and $V(K_0) > 0$, then we can restrict the rigid motion invariant measure μ_k to $\{E \in A(n,k) : K_0 \cap E \neq \emptyset\}$ and normalize it to get a probability measure. A random k-flat X_k with this distribution is called a *random k-flat in K_0*. We then get

$$\mathrm{Prob}(X_k \cap K \neq \emptyset) = \frac{V_{n-k}(K)}{V_{n-k}(K_0)}.$$

As an example, we mention the Buffon needle problem. Originally the problem was formulated in the following way: Given an array of parallel lines in the plane $\mathbb{R}^2$ with distance 1, what is the probability that a randomly thrown needle of length $L < 1$ intersects one of the lines? If we consider the disc of radius $\frac{1}{2}$ around the center of the needle, there will be almost surely exactly one line of the array intersecting this disc. Hence, the problem can be formulated in an equivalent way. Assume the needle N is fixed with center at 0. What is the probability that a random line X_1 in $B^2(\frac{1}{2})$ intersects the needle N? The answer is

$$\mathrm{Prob}(X_1 \cap N \neq \emptyset) = \frac{V_1(N)}{V_1(B^2(\frac{1}{2}))} = \frac{L}{\pi/2} = \frac{2L}{\pi}.$$

Remark 5.8 In continuation of Remark 5.7, we can consider, for $K, K_0 \in \mathcal{K}^n$ with $K \subset K_0$ and $V(K_0) > 0$ and for a random k-flat X_k in K_0, the expected ith intrinsic volume of $K \cap X_k$, $i \in \{0, \ldots, k\}$. We get

$$\mathbb{E}V_i(K \cap X_k) = \frac{\int V_i(K \cap E)\,\mu_k(dE)}{\int V_0(K_0 \cap E)\,\mu_k(dE)}$$

$$= \frac{\alpha_{nik}V_{n+i-k}(K)}{\alpha_{n0k}V_{n-k}(K_0)}.$$

This shows that if K_0 is assumed to be known (and K is unknown) and if $V_i(K \cap X_k)$ is observable, then

$$\frac{\alpha_{n0k} V_{n-k}(K_0)}{\alpha_{nik}} V_i(K \cap X_k)$$

is an unbiased estimator of $V_{n+i-k}(K)$. Varying k, we get in this way three estimators for the volume $V(K)$, two for the surface area $F(K)$ and one for the mean width $\overline{B}(K)$ of K.

This estimation procedure is indeed of practical interest, since we do not have to assume that the set K under consideration is convex (recall that the Crofton formula remains true for polyconvex sets, which form a dense subclass in the compact sets of $\mathbb{R}^n$). For instance, it can be used in practical situations to estimate the surface area of a complicated tissue A in, say, a cubical specimen K_0 by measuring the boundary length $L(A \cap X_2)$ of a planar section $A \cap X_2$. Since the latter quantity is still complicated to obtain, one uses the Crofton formulas again and estimates $L(A \cap X_2)$ by counting intersections with random lines X_1 in $K_0 \cap X_2$. Such *stereological formulas* are used and have been developed further in many applied sciences, including medicine, biology, geology, metallurgy and materials science.

In the remaining part of this section, we show how the Crofton formulas can be combined with an idea of Hadwiger to see in a rather direct way that the intrinsic volumes have an additive extension to $U(\mathcal{K}^n)$. Of course, we already know the assertion itself, since V_j is additive and continuous on $\mathcal{K}^n$.

Theorem 5.11 *For $j = 0, \ldots, n$, there is a unique additive extension of V_j to the convex ring $\mathcal{R}^n$.*

Proof It remains to show the existence.

We begin with the Euler characteristic V_0 and prove the existence of an additive extension by induction on the dimension n, $n \geq 0$.

It is convenient to start with the case of dimension $n = 0$ since $U(\mathcal{K}^0) = \{\emptyset, \{0\}\} = \mathcal{K}^0$. Since $V_0(\emptyset) = 0$ and $V_0(\{0\}) = 1$, V_0 is additive on $U(\mathcal{K}^0)$.

For the step from dimension $n - 1$ to dimension n, $n \geq 1$, we choose a fixed direction $u_0 \in \mathbb{S}^{n-1}$ and consider the family of hyperplanes $H_\alpha := H(u_0, \alpha)$ for $\alpha \in \mathbb{R}$. Let $A \in U(\mathcal{K}^n)$ have a representation $A = \bigcup_{i=1}^k K_i$ with $K_i \in \mathcal{K}^n$. Then we have

$$A \cap H_\alpha = \bigcup_{i=1}^k (K_i \cap H_\alpha),$$

and by induction hypothesis the additive extension $V_0(A \cap H_\alpha)$ exists. From the inclusion-exclusion formula (see (4.24)) we obtain that the function $f_A : \alpha \mapsto V_0(A \cap H_\alpha)$ is integer-valued and bounded from below and above. Therefore, f_A is piecewise constant and (4.24) shows that the value of $f_A(\alpha)$ can only change if the hyperplane H_α supports one of the convex bodies K_v, $v \in S(k)$. We define the 'jump function'

$$g_A(\alpha) := f_A(\alpha) - \lim_{\beta \searrow \alpha} f_A(\beta), \qquad \alpha \in \mathbb{R},$$

and put

$$V_0(A) := \sum_{\alpha \in \mathbb{R}} g_A(\alpha).$$

This definition makes sense since $g_A(\alpha) \neq 0$ only for finitely many values of α. Moreover, for $k = 1$, that is $A = K \in \mathcal{K}^n$, $K \neq \emptyset$, we have $V_0(K) = 0 + 1 = 1$, hence V_0 is an extension of the Euler characteristic. By induction hypothesis, $A \mapsto f_A(\alpha)$ is additive on $U(\mathcal{K}^n)$ for each α. Therefore, as a limit, $A \mapsto g_A(\alpha)$ is additive and so V_0 is additive. The uniqueness, which is clear from (4.24), shows that this construction does not depend on the choice of the direction u_0.

Now we consider the case $j > 0$. Let $A \in U(\mathcal{K}^n)$ with $A = \bigcup_{i=1}^k K_i$ and $K_i \in \mathcal{K}^n$. Then, for $\alpha > 0$ and $x \in \mathbb{R}^n$, we have

$$A \cap (B^n(\alpha) + x) = \bigcup_{i=1}^k (K_i \cap (B^n(\alpha) + x)).$$

Therefore, (4.24) implies that

$$V_0(A \cap (B^n(\alpha) + x)) = \sum_{v \in S(k)} (-1)^{|v|-1} V_0(K_v \cap (B^n(\alpha) + x)).$$

Since $V_0(K_v \cap (B^n(\alpha) + x)) = 1$ if and only if $x \in K_v + B^n(\alpha)$, we then get from the Steiner formula

$$\int_{\mathbb{R}^n} V_0(A \cap (B^n(\alpha) + x))\, dx = \sum_{v \in S(k)} (-1)^{|v|-1} \int_{\mathbb{R}^n} V_0(K_v \cap (B^n(\alpha) + x))\, dx$$

$$= \sum_{v \in S(k)} (-1)^{|v|-1} V_n(K_v + B^n(\alpha))$$

$$= \sum_{v \in S(k)} (-1)^{|v|-1} \left(\sum_{j=0}^n \alpha^{n-j} \kappa_{n-j} V_j(K_v) \right)$$

$$= \sum_{j=0}^n \alpha^{n-j} \kappa_{n-j} \left(\sum_{v \in S(k)} (-1)^{|v|-1} V_j(K_v) \right).$$

If we define

$$V_j(A) := \sum_{v \in S(k)} (-1)^{|v|-1} V_j(K_v),$$

then

$$\int_{\mathbb{R}^n} V_0(A \cap (B^n(\alpha) + x)) \, dx = \sum_{j=0}^{n} \alpha^{n-j} \kappa_{n-j} V_j(A).$$

Since this equation holds for all $\alpha > 0$, the values $V_j(A)$ for $j = 0, \ldots, n$ depend only on A and not on the special representation, and moreover V_j is additive. $\square$

Remark 5.9 The formula

$$\int_{\mathbb{R}^n} V_0(A \cap (B^n(\alpha) + x)) \, dx = \sum_{j=0}^{n} \alpha^{n-j} \kappa_{n-j} V_j(A),$$

which we derived and used in the above proof, is a *generalized Steiner formula*; it reduces to the classical Steiner formula if $A \in \mathcal{K}^n$.

Remark 5.10 The extended Euler characteristic V_0 (also called the Euler–Poincaré characteristic) plays an important role in topology. In $\mathbb{R}^2$ and for $A \in U(\mathcal{K}^2)$, the integer $V_0(A)$ can be interpreted as the number of connected components minus the number of 'holes' in A.

Remark 5.11 On $U(\mathcal{K}^n)$, the functional V_n is still the volume (Lebesgue measure) and $F = 2V_{n-1}$ can still be interpreted as the surface area (at least for sets which are the closure of their interior). The other (extended) intrinsic volumes V_j do not have a direct geometric interpretation.

Exercises and Supplements for Sect. 5.3

1. Provide an alternative proof of Theorem 5.10 which is based on the special case $i = 0$ (already proved in the introduction of this section) and an application of the recursion given in Exercise 5.1.10.
2. Calculate the probability that a random secant of $B^2(1)$ is longer than $\sqrt{3}$. (According to the interpretation of a 'random secant', one might get here the values $\frac{1}{2}$, $\frac{1}{3}$ or $\frac{1}{4}$. Explain why $\frac{1}{2}$ is the right, 'rigid motion invariant' answer.)
3. Let $A_K(n, k)$ be the set of all $E \in A(n, k)$ such that $E \cap \text{int } K = \emptyset$ and $E \cap K \neq \emptyset$. Show that $\mu_k(A_K(n, k)) = 0$.

4. Hadwiger's idea of proving the additivity of the Euler characteristic V_0 by induction can be adjusted to other situations. For instance, it was observed by P. Mani that

$$h(K, u) := \sum_{\alpha \in \mathbb{R}} \alpha \left[V_0(K \cap H(u, \alpha)) \quad \lim_{\beta \downarrow \alpha} V_0(K \cap H(u, \beta)) \right]$$

for $K \in U(\mathcal{K}^n)$ and $u \in \mathbb{R}^n \setminus \{0\}$ is properly defined and provides the additive extension of the support function.

5.4 Kinematic Formulas

Intersecting a convex body $K \in \mathcal{K}^n$ with the image gL of a convex body $L \in \mathcal{K}^n$ under a proper rigid motion $g \in G(n)$, we obtain $K \cap gL$, which is again a convex body in $\mathbb{R}^n$ or the empty set. We may then apply a functional $\varphi : \mathcal{K}^n \to \mathbb{R}$ to this intersection, for instance, we can choose the ith intrinsic volume $\varphi = V_i$, and average over all $g \in G(n)$ with respect to the Haar measure μ on $G(n)$, which has been introduced in Sect. 5.1. The following theorem shows that the result can be expressed as a sum of products of intrinsic volumes of K and L (for $\varphi = V_i$). Thus we obtain a closed and complete system of kinematic formulas, which can also be iterated.

We prepare the theorem with two auxiliary results.

Lemma 5.7 *Let* $K, L \in \mathcal{K}^n$. *If* $\varphi : \mathcal{K}^n \to [0, \infty)$ *is continuous or increasing (on nonempty convex bodies), then* $g \mapsto \varphi(K \cap gL)$, $g \in G(n)$, *is measurable and bounded. If* $G^*(n) := \{g \in G(n) : K \cap gL \neq \emptyset, \mathrm{int}(K) \cap gL = \emptyset\}$, *then* $\mu(G^*(n)) = 0$.

Proof We have

$$\{K \cap gL : g \in G(n), K \cap gL \neq \emptyset\} \subset \{M \in \mathcal{K}^n : M \subset K, M \neq \emptyset\}.$$

Since the set on the right-hand side is compact, this yields that $g \mapsto \varphi(K \cap gL)$, $g \in G(n)$, is bounded if φ is continuous. The assertion is obvious, if φ is increasing. The map $g \mapsto \varphi(K \cap gL)$, $g \in G^*(n)$, is continuous by Exercise 3.1.8, since $G(n)$ operates continuously on $\mathcal{K}^n$. Hence, it remains to show that $\mu(G^*(n)) = 0$. For this, observe that for $g = \gamma(t, \varrho) \in G(n)$, we have $g \in G^*(n)$ if and only if $t \in \mathrm{bd}(K + \varrho L^*)$, where $L^* = -L$, as follows from a separation argument. Hence, we obtain

$$\mu(G^*(n)) = \int_{\mathrm{SO}(n)} \lambda_n(\mathrm{bd}(K + \varrho L^*)) \, \nu(d\varrho) = 0,$$

which completes the proof. $\qquad\square$

Lemma 5.8 *Let* $0 \leq k \leq l \leq n$. *If* $K, L \in \mathcal{K}^n$, *then*

$$\int_{SO(n)} V(K[k], \varrho L[l - k], B^n[n - l]) \, \nu(d\varrho)$$

$$= \begin{bmatrix} n \\ k \end{bmatrix}^{-1} \begin{bmatrix} n \\ l - k \end{bmatrix}^{-1} \frac{\kappa_n}{\kappa_k \kappa_{l-k}} V_k(K) V_{l-k}(L).$$

Proof We define the functional

$$\mathcal{R}_L(K) := \int_{SO(n)} V(K[k], \varrho L[l - k], B^n[n - l]) \, \nu(d\varrho)$$

for $K \in \mathcal{K}^n$ and fixed $L \in \mathcal{K}^n$. Then $\mathcal{R}_L$ is translation invariant, rotation invariant, continuous, and additive (for the latter, see Exercise 4.5.6). Since $\mathcal{R}_L$ is homogeneous of degree k, Hadwiger's characterization theorem yields that $\mathcal{R}_L(K) = c_k(L) V_k(K)$ for $K, L \in \mathcal{K}^n$. Thus, choosing $K = B^n$ we obtain

$$c_k(L) = V_k(B^n)^{-1} \mathcal{R}_L(B^n)$$

$$= \frac{\kappa_{n-k}}{\kappa_n} \binom{n}{k}^{-1} \int_{SO(n)} V(\varrho L[l - k], B^n[n - l + k]) \, \nu(d\varrho)$$

$$= \frac{\kappa_{n-k}}{\kappa_n} \binom{n}{k}^{-1} \kappa_{n-l+k} \binom{n}{l - k}^{-1} V_{l-k}(L)$$

$$= \begin{bmatrix} n \\ k \end{bmatrix}^{-1} \begin{bmatrix} n \\ l - k \end{bmatrix}^{-1} \frac{\kappa_n}{\kappa_k \kappa_{l-k}} V_{l-k}(L),$$

which completes the argument.   □

Theorem 5.12 (Principal Kinematic Formula) *Let* $i \in \{0, \ldots, n\}$ *and* $K, L \in \mathcal{K}^n$. *Then*

$$\int_{G(n)} V_i(K \cap gL) \, \mu(dg) = \sum_{k+l=n+i} \alpha_{nik} V_k(K) V_l(L),$$

where

$$\alpha_{nik} = \begin{bmatrix} n \\ k - i \end{bmatrix}^{-1} \begin{bmatrix} k \\ i \end{bmatrix} = \begin{bmatrix} n \\ l \end{bmatrix}^{-1} \begin{bmatrix} k \\ i \end{bmatrix}$$

is the same constant as in the Crofton formula.

Proof We first consider the case $i = 0$. We put again $L^* = -L$ and write $\mathcal{K}_i(K, L)$ for the kinematic integral. Then

$$
\mathcal{K}_0(K, L) = \int_{SO(n)} \int_{\mathbb{R}^n} V_0(K \cap (\varrho L + t))\, \lambda_n(dt)\, \nu(d\varrho)
$$

$$
= \int_{SO(n)} V_n(K + \varrho L^*)\, \nu(d\varrho)
$$

$$
= \sum_{j=0}^{n} \binom{n}{j} \int_{SO(n)} V(K[j], \varrho L^*[n - j])\, \nu(d\varrho),
$$

since $K \cap (\varrho L + t) \neq \emptyset$ if and only if $t \in K + \varrho L^*$. Using Lemma 5.8 with $l = n$ and $k = j$, we get

$$
\mathcal{K}_0(K, L) = \sum_{j=0}^{n} \binom{n}{j} \begin{bmatrix} n \\ j \end{bmatrix}^{-1} \begin{bmatrix} n \\ n-j \end{bmatrix}^{-1} \frac{\kappa_n}{\kappa_j \kappa_{n-j}}\, V_j(K) V_{n-j}(L)
$$

$$
= \sum_{j=0}^{n} \begin{bmatrix} n \\ j \end{bmatrix}^{-1} V_j(K) V_{n-j}(L),
$$

which completes the argument in the case $i = 0$.

For the derivation of the case $i > 0$, we start by using the Crofton formula, then we apply Fubini's theorem, we use the already established special case $i = 0$, and finally we again apply Crofton's formula. Proceeding in this order, we get

$$
\mathcal{K}_i(K, L) = \int_{G(n)} \begin{bmatrix} n \\ i \end{bmatrix} \int_{A(n,n-i)} V_0((K \cap gL) \cap E)\, \mu_{n-i}(dE)\, \mu(dg)
$$

$$
= \begin{bmatrix} n \\ i \end{bmatrix} \int_{A(n,n-i)} \int_{G(n)} V_0((K \cap E) \cap gL)\, \mu(dg)\, \mu_{n-i}(dE)
$$

$$
= \begin{bmatrix} n \\ i \end{bmatrix} \int_{A(n,n-i)} \sum_{k=0}^{n} \begin{bmatrix} n \\ k \end{bmatrix}^{-1} V_k(K \cap E) V_{n-k}(L)\, \mu_{n-i}(dE)
$$

$$
= \sum_{k=0}^{n-i} \begin{bmatrix} n \\ i \end{bmatrix} \begin{bmatrix} n \\ k \end{bmatrix}^{-1} V_{n-k}(L) \int_{A(n,n-i)} V_k(K \cap E)\, \mu_{n-i}(dE)
$$

$$
= \sum_{k=0}^{n-i} \begin{bmatrix} n \\ i \end{bmatrix} \begin{bmatrix} n \\ k \end{bmatrix}^{-1} \begin{bmatrix} n-i \\ k \end{bmatrix} \begin{bmatrix} n \\ n-i-k \end{bmatrix}^{-1} V_{n-k}(L) V_{k+i}(K)
$$

$$
= \sum_{k=0}^{n-i} \begin{bmatrix} k+i \\ i \end{bmatrix} \begin{bmatrix} n \\ k \end{bmatrix}^{-1} V_{k+i}(K) V_{n-k}(L),
$$

which is the asserted formula. $\square$

Remark 5.12 The principal kinematic formula is additive in K and L and extends to polyconvex sets. Here we also have that $(K \cap M) \cap gL = (K \cap gL) \cap (M \cap gL)$, where $K, M, L \in \mathcal{K}^n$ and $g \in G(n)$.

If we replace the intersection in the principal kinematic formula by Minkowski addition, the resulting integral will be unbounded over $G(n)$. For this reason, we drop the translative part of the operation. We shall see that then there is still an integral-geometric formula for the mean values $V_i(K + \varrho L)$ if $\varrho \in SO(n)$ and the integration is with respect to the Haar measure ν on the compact rotation group $SO(n)$.

Theorem 5.13 *Let $i \in \{0, \dots, n\}$. If $K, L \in \mathcal{K}^n$, then*

$$\int_{SO(n)} V_i(K + \varrho L)\, \nu(d\varrho) = \sum_{k=0}^{i} \begin{bmatrix} n \\ i-k \end{bmatrix}^{-1} \begin{bmatrix} n-k \\ i-k \end{bmatrix} V_k(K) V_{i-k}(L).$$

Proof We write $\mathcal{R}_i(K, L)$ for the rotation sum integral. Expanding the integrand, we obtain

$$\mathcal{R}_i(K, L) = \kappa_{n-i}^{-1} \binom{n}{i} \int_{SO(n)} V((K + \varrho L)[i], B^n[n-i])\, \nu(d\varrho)$$

$$= \kappa_{n-i}^{-1} \binom{n}{i} \sum_{k=0}^{i} \binom{i}{k} \int_{SO(n)} V(K[k], \varrho L[i-k], B^n[n-i])\, \nu(d\varrho)$$

$$= \kappa_{n-i}^{-1} \binom{n}{i} \sum_{k=0}^{i} \binom{i}{k} \begin{bmatrix} n \\ k \end{bmatrix}^{-1} \begin{bmatrix} n \\ i-k \end{bmatrix}^{-1} \frac{\kappa_n}{\kappa_k \kappa_{i-k}} V_k(K) V_{i-k}(L),$$

where we used Lemma 5.8 in the last step. The assertion of the theorem is obtained by simplifying the constant. $\qquad\qquad\square$

A more general version of Theorem 5.13 can be derived if in that theorem K is replaced by $K + \lambda M$, for $\lambda \geq 0$ and $M \in \mathcal{K}^n$. In the resulting integral formula, we can expand in λ. Then the following theorem is obtained by comparison of coefficients. To simplify the constants, we express the result in terms of $\overline{S}_j(L) := S_j(L, \mathbb{S}^{n-1})$, $L \in \mathcal{K}^n$, instead of $V_j(L)$.

Theorem 5.14 *Let $0 \leq l \leq i \leq n$. If $K, L, M \in \mathcal{K}^n$, then*

$$\int_{SO(n)} V((K + \varrho L)[i-l], M[l], B^n[n-i])\, \nu(d\varrho)$$

$$= \frac{1}{n\kappa_n} \sum_{k=0}^{i} \binom{i-l}{i-k} V(K[k-l], M[l], B^n[n-k])\, \overline{S}_{i-k}(L).$$

From the special case $l = 1$ of Theorem 5.14 we obtain the following local version of a rotation sum formula for area measures.

Corollary 5.2 *Let* $i \in \{0, \ldots, n-1\}$. *If* $K, L \in \mathcal{K}^n$, *then*

$$\int_{SO(n)} S_i(K + \varrho L, \cdot) \, \nu(d\varrho) = \frac{1}{n\kappa_n} \sum_{j=0}^{i} \binom{i}{j} S_j(K, \cdot) \overline{S}_{i-j}(L).$$

This corollary is a special case of a rotation sum formula which involves the area measures of both bodies (see [81, Theorem 4.4.6]).

Exercises and Supplements for Sect. 5.4

1. For a variant of the proof of the principal kinematic formula, one can consider the kinematic integral as a functional $\mathcal{K}_i(K, L)$ of convex bodies $K, L \in \mathcal{K}^n$. First, fixing L and considering the dependence on K, one applies Hadwiger's characterization theorem to $K \mapsto \mathcal{K}_i(K, L)$ and thus finds that

$$\mathcal{K}_i(K, L) = \sum_{j=0}^{n} c_j(L) V_j(K)$$

for all $K, L \in \mathcal{K}^n$. Since $V_0, \ldots, V_n$ is a basis of the vector space of motion invariant, continuous valuations and by the properties of $L \mapsto \mathcal{K}_i(K, L)$ for fixed $K \in \mathcal{K}^n$, it follows that Hadwiger's characterization theorem can also be applied to the coefficient functionals c_j. Thus, we obtain

$$\mathcal{K}_i(K, L) = \sum_{j,k=0}^{n} c_{jk} V_j(K) V_k(L)$$

for all $K, L \in \mathcal{K}^n$. By a homogeneity argument, it follows that $c_{jk} = 0$ unless $j + k = n + i$. For $i = 0$, the coefficients can be determined by choosing $K = B^n$ and $L = rB^n$ with $r > 0$. A direct determination of the coefficients in the case $i > 0$ does not seem to be so easy.

2. Hadwiger's general integral-geometric formula. Let $\varphi : \mathcal{K}^n \to \mathbb{R}$ be additive and continuous. Then

$$\int_{G(n)} \varphi(K \cap gM) \, \mu(dg) = \sum_{k=0}^{n} \varphi_{n-k}(K) V_k(M), \quad K, M \in \mathcal{K}^n,$$

where

$$\varphi_{n-k}(K) := \int_{A(n,k)} \varphi(K \cap F)\, \mu_k(dF), \quad K \in \mathcal{K}^n,\, k = 0, \dots, n.$$

3. An iterated version of the principal kinematic formula. For $K_0, \dots, K_k \in \mathcal{K}^n$ and $j \in \{0, \dots, n\}$, prove that

$$\int_{(G(n))^k} V_j(K_0 \cap g_1 K_1 \cap \dots \cap g_k K_k)\, \mu^k(d(g_1, \dots, g_k))$$

$$= \sum_{\substack{m_0,\dots,m_k=j \\ m_0+\dots+m_k=kn+j}}^{n} c_j^n \prod_{i=0}^{k} c_n^{m_i} V_{m_i}(K_i),$$

where $c_j^k := \frac{k!\kappa_k}{j!\kappa_j}$. Note that the case $k = 1$ is the principal kinematic formula.

There is also an iterated version of the principal kinematic formula for a general functional as in Exercise 5.4.2.

4. Check the details of the derivation of Theorem 5.14 and of Corollary 5.2.
5. The integral-geometric formulas obtained in this chapter have been extended in various directions. We have already seen that some of the integral-geometric results for intrinsic volumes can be localized and stated for area measures. The area measures are Borel measures on the unit sphere. For the purposes of integral geometry, another sequence of measures, the curvature measures, have proved to be useful. These curvature measures are Borel measures on $\mathbb{R}^n$ and can also be considered as local generalizations of the intrinsic volumes. For instance, a complete system of kinematic and Crofton formulas is known for these curvature measures. Another direction of research concerns integral-geometric formulas for the translation group. In this case, the integral average can often be expressed in terms of mixed functionals or measures. We refer to [81, Section 4.4], [82], and to [84, Part II] for a systematic introduction to and many other aspects of integral geometry and its applications. For an introduction to a deeper study of the connections between integral geometry and the theory of valuations, we also recommend the Lecture Notes [3] of an advanced course on integral geometry and valuations, the survey [12], as well as the Lecture Notes [48].
6. Show that Theorems 5.13, 5.14 and Corollary 5.2 have additive extensions to the convex ring.
7. Hints to the literature: Applications of integral geometry to stochastic geometry are systematically described and developed in [83, 84].

Chapter 6
Solutions of Selected Exercises

6.1 Solutions of Exercises for Chap. 1

Exercise 1.1.3

We proceed similarly as for **Exercise 1.1.1**, (e) $\Rightarrow$ (a). Let $\lambda_1, \ldots, \lambda_m \in \mathbb{R}$ with $\lambda_1 + \cdots + \lambda_m = 0$ and $\lambda_1 x_1 + \cdots + \lambda_m x_m = 0$. Choose arbitrary positive constants $\mu_1, \ldots, \mu_m > 0$ such that $\mu_i + \lambda_i > 0$ for $i = 1, \ldots, m$. We define the constant $\Lambda := \sum_{i=1}^m (\mu_i + \lambda_i) > 0$. Then

$$1 - \sum_{i=1}^m \frac{\mu_i + \lambda_i}{\Lambda} - \sum_{i=1}^m \frac{\mu_i}{\Lambda} + \sum_{i=1}^m \frac{\lambda_i}{\Lambda} = \sum_{i=1}^m \frac{\mu_i}{\Lambda}$$

and

$$\sum_{i=1}^m \frac{\mu_i + \lambda_i}{\Lambda} x_i = \sum_{i=1}^m \frac{\mu_i}{\Lambda} x_i + \sum_{i=1}^m \frac{\lambda_i}{\Lambda} x_i = \sum_{i=1}^m \frac{\mu_i}{\Lambda} x_i.$$

By the assumed uniqueness of convex combinations, we deduce that $(\lambda_i + \mu_i)/\Lambda = \mu_i/\Lambda$ for $i = 1, \ldots, m$, and hence $\lambda_1 = \cdots = \lambda_m = 0$.

Exercise 1.1.7

"$\subset$": Since $\operatorname{conv} A \subset \operatorname{conv}(A \cup B)$ and $\operatorname{conv} B \subset \operatorname{conv}(A \cup B)$ we have $\operatorname{conv} A \cup \operatorname{conv} B \subset \operatorname{conv}(A \cup B)$. Therefore

$$\operatorname{conv}(\operatorname{conv} A \cup \operatorname{conv} B) \subset \operatorname{conv}(\operatorname{conv}(A \cup B)) = \operatorname{conv}(A \cup B).$$

© Springer Nature Switzerland AG 2020

D. Hug, W. Weil, *Lectures on Convex Geometry*, Graduate Texts
in Mathematics 286, https://doi.org/10.1007/978-3-030-50180-8_6

"⊃":   Since $A \subset \text{conv}\, A$ and $B \subset \text{conv}\, B$ we have $A \cup B \subset \text{conv}\, A \cup \text{conv}\, B$.
        Therefore $\text{conv}(A \cup B) \subset \text{conv}(\text{conv}\, A \cup \text{conv}\, B)$.

Exercise 1.1.8

"⊃":   Since $A \subset \text{conv}\, A$ and $B \subset \text{conv}\, B$ we have $A + B \subset \text{conv}\, A + \text{conv}\, B$.
        A sum of convex sets is again convex. Therefore $\text{conv}\, A + \text{conv}\, B$ is convex
        and we have $\text{conv}(A + B) \subset \text{conv}(\text{conv}\, A + \text{conv}\, B) = \text{conv}\, A + \text{conv}\, B$.

"⊂":   Let $x \in \text{conv}\, A + \text{conv}\, B$, i.e. $x = \sum_{i=1}^{k} \alpha_i a_i + \sum_{j=1}^{l} \beta_j b_j$, where $k, l \in$
        $\mathbb{N}$, $a_1, \ldots, a_k \in A$, $b_1, \ldots, b_l \in B$, $\alpha_1, \ldots, \alpha_k, \beta_1, \ldots, \beta_l \in [0, 1]$ with
        $\sum_{i=1}^{k} \alpha_i = 1 = \sum_{j=1}^{l} \beta_j$. Then

$$x = \sum_{i=1}^{k} \alpha_i a_i + \sum_{j=1}^{l} \beta_j b_j$$

$$= \sum_{j=1}^{l} \beta_j \sum_{i=1}^{k} \alpha_i a_i + \sum_{i=1}^{k} \alpha_i \sum_{j=1}^{l} \beta_j b_j$$

$$= \sum_{j=1}^{l} \sum_{i=1}^{k} \beta_j \alpha_i \underbrace{(a_i + b_j)}_{\in A + B}$$

and $\sum_{j=1}^{l} \sum_{i=1}^{k} \beta_j \alpha_i = 1$. Therefore $x \in \text{conv}(A + B)$.

Exercise 1.1.13

First, suppose that $A \subset \mathbb{R}^n$ is convex, closed and unbounded.

Claim If $0 \in A$, then there is a $u \in \mathbb{S}^{n-1}$ such that $[0, \infty)u \subset A$.

Let $0 \in A$. Since A is unbounded, there is a sequence $(x_k)_{k \in \mathbb{N}} \subset A$ such that
$\|x_k\| \to \infty$ as $k \to \infty$. Define $u_k := \frac{x_k}{\|x_k\|}$ for $k \in \mathbb{N}$. Then $(u_k)_{k \in \mathbb{N}} \subset \mathbb{S}^{n-1}$
and there are a subsequence $(u_{k_i})_{i \in \mathbb{N}}$ and a vector $u \in \mathbb{S}^{n-1}$ such that $u_{k_i} \to u$ as
$i \to \infty$.

Let $\lambda \geq 0$. There is a $k_0 \in \mathbb{N}_0$ such that $\|x_k\| > \lambda$ for all $k > k_0$. Then

$$\lambda u_k = \frac{\lambda}{\|x_k\|} x_k = \frac{\lambda}{\|x_k\|} x_k + \left(1 - \frac{\lambda}{\|x_k\|}\right) 0 \in A,$$

since A is convex. Since A is closed, it follows that $\lambda u \in A$.

It remains to show that for all $a \in A$ there is a $u \in \mathbb{S}^{n-1}$ such that $a + [0, \infty)u \subset A$. For this, let $a \in A$. Then $A - a$ is closed, convex and unbounded and $0 \in A - a$. By the above, there is a $u \in \mathbb{S}^{n-1}$ such that $[0, \infty)u \subset A - a$. Hence, $a + [0, \infty)u \subset A$, which completes the proof. Note that for a set $A \subset \mathbb{R}^n$ which is convex, unbounded and closed we proved that in fact each point of A is the starting point of an infinite ray.

Finally, we assume that A is (just) convex and unbounded. Since A is nonempty, by Theorem 1.10 there is an $a \in$ relint $A \subset$ cl A. By Corollary 1.2 cl A is convex. Thus, there is a $u \in \mathbb{S}^{n-1}$ such that $a + [0, \infty)u \subset$ cl A. Let $\mu > 0$. Then $a + 2\mu u \in$ cl A and $a \in$ relint A imply $[a, a + 2\mu u) \subset$ relint $A \subset A$ by Proposition 1.2. Since $a + \mu u \in [a, a + 2\mu u)$ for all $\mu > 0$, we conclude that $a + [0, \infty)u \subset A$.

Exercise 1.2.7

Definition A set $A \subset \mathbb{R}^n$ is called connected if there are no open sets $O_1, O_2 \subset \mathbb{R}^n$ such that $A \cap O_1 \neq \emptyset \neq A \cap O_2$, $A \cap O_1 \cap O_2 = \emptyset$, $A \subset O_1 \cup O_2$.

The following basic facts are shown in textbooks on elementary topology: If $A_i \subset \mathbb{R}^n$, $i \in I \neq \emptyset$, are connected sets and $x \in A_i$ for all $i \in I$, then $\bigcup_{i \in I} A_i$ is connected. The union of all connected subsets of $A \subset \mathbb{R}^n$ containing a fixed point $x \in A$ is called the connected component of x in A. This is a connected set. The set A is the disjoint union of its connected components.

Claim Let $A \subset \mathbb{R}^n$ be a set with at most n connected components, and let $u \in$ conv(A). Then there are $a_1, \ldots, a_n \in A$ with $a \in$ conv$\{a_1, \ldots, a_n\}$.

Before we prove the claim, we show by an example that even for a path-connected set (and hence connected set) the number of points which are needed cannot be reduced in general. For this, let $x_0, \ldots, x_n$ be affinely independent points in $\mathbb{R}^n$. Let $A := \bigcup_{i=1}^n [x_0, x_i]$. Then A is path-connected. Any $n - 1$ points of A lie in the convex hull of at most n of the points $x_0, \ldots, x_n$. Then it is easy to check that $x := 1/(n+1)(x_0 + \cdots + x_n) \in$ conv(A) is not a convex combination of at most $n - 1$ points of A.

Now we turn to the proof of the claim. By Carathéodory's theorem there are points $a_1, \ldots, a_{n+1} \in A$ and numbers $\lambda_1, \ldots, \lambda_{n+1} \geq 0$ with $\lambda_1 + \cdots + \lambda_{n+1} = 1$ and $\lambda_1 a_1 + \cdots + \lambda_{n+1} a_{n+1} = a$. We can assume that $a_1, \ldots, a_{n+1}$ are affinely independent and that $\lambda_1, \ldots, \lambda_{n+1} > 0$. For $j \in \{1, \ldots, n+1\}$, let

$$C_j := a - \text{pos}\{a_1 - a, \ldots, (a_j - a)^{\vee}, \ldots, a_{n+1} - a\}.$$

As usual, $(a_j - a)^{\vee}$ means that $a_j - a$ is omitted.

Then $a_j \in \text{int}(C_j)$, for $j \in \{1, \ldots, n+1\}$, since $\lambda_1, \ldots, \lambda_{n+1} > 0$,

$$a_j - a = - \sum_{i=1, i \neq j}^{n+1} \frac{\lambda_i}{\lambda_j}(a_i - a)$$

and since the n vectors $a_1 - a, \ldots, (a_j - a)^{\vee}, \ldots, a_{n+1} - a$ are linearly independent (which can be easily checked).

Next we show that the cones $\text{int}(C_j)$, $1 \leq j \leq n+1$, are pairwise disjoint. To see this, suppose that $z \in C_1 \cap C_{d+1}$ (say). Then

$$z = a - \sum_{i=2}^{n+1} \beta_i(a_i - a) = a - \sum_{i=1}^{n} \gamma_i(a_i - a), \quad \beta_i, \gamma_i \geq 0.$$

Since $\sum_{i=1}^{n+1} \lambda_i(a_i - a) = 0$ we get

$$\sum_{i=2}^{n} \left(\beta_i - \gamma_i - \frac{\lambda_i}{\lambda_{n+1}} \beta_{n+1} \right)(a_i - a) - \left(\frac{\lambda_1}{\lambda_{n+1}} \beta_{n+1} + \gamma_1 \right)(a_1 - a) = 0.$$

This yields $\beta_{n+1} \frac{\lambda_1}{\lambda_{n+1}} + \gamma_1 = 0$, that is, $\gamma_1 = 0$ and $\beta_{n+1} = 0$. But then $z \in \text{bd}(C_1) \cap \text{bd}(C_{n+1})$.

Let A_j denote the connected component of $a_j \in \text{int}(C_j)$, $j \in \{1, \ldots, n+1\}$. If $A_j \cap \text{bd}(C_j) = \emptyset$, then $A_j \subset \text{int}(C_j)$. Since A has at most n connected components and the sets $\text{int}(C_j)$, $1 \leq j \leq n+1$, are pairwise disjoint, we get $A_j \cap \text{bd}(C_j) \neq \emptyset$ for some $j \in \{1, \ldots, n+1\}$. Hence, we have (say) $z_0 \in A_{n+1} \cap \text{bd}(C_{n+1}) \neq \emptyset$.

Since

$$\text{bd}(C_{n+1}) = a - \left\{ \sum_{i=1}^{n} \mu_i(a_i - a) : \quad \mu_i \geq 0 \text{ for } i = 1, \ldots, n \text{ and} \right.$$

$$\left. \mu_j = 0 \text{ for some } j \in \{1, \ldots, n\} \right\}$$

we can assume that (say) $z_0 = a - \sum_{i=2}^{n} \mu_i(a_i - a) \in A_{n+1} \subset A$, $\mu_2, \ldots, \mu_n \geq 0$. But then we further conclude that

$$\left(1 + \sum_{i=2}^{n} \mu_i \right) a = z_0 + \sum_{i=2}^{n} \mu_i a_i,$$

that is,

$$a = \left(1 + \sum_{i=2}^{n} \mu_i\right)^{-1} \left(z_0 + \sum_{i=2}^{n} \mu_i a_i\right) \in \mathrm{conv}\{z_0, a_2, \ldots, a_n\},$$

which proves the assertion.

Exercise 1.2.11

First, assume that c is as in Exercise 1.2.10. Let $p \in -(K-c)$, that is, $p = -(a-c)$ for some $a \in K$. If $a = c$, then $p = 0 \in n(K - c)$. Hence, suppose that $a \neq c$. Choose $b \in \mathrm{bd}(K) \cap (a + [0, \infty)(c - a))$. Then

$$\|a - c\| \leq \frac{n}{n+1}\|a - b\| = \frac{n}{n+1}\left(\|a - c\| + \|b - c\|\right),$$

thus $\|a - c\| \leq n\|b - c\|$. This shows that $c - a = \lambda n(b - c)$ for some $\lambda \in [0, 1]$. It follows that

$$p = \lambda n(b - c) \in \lambda n(K - c) \subset n(K - c),$$

which proves the asserted inclusion.

For the reverse direction, suppose that $-(K - c) \subset n(K - c)$ for some $c \in K$. Let $a \in \mathrm{int}\, K$. Then $c - a = -(a - c) \in n(K - c)$. Hence there is some $k \in K$ with $c - a = n(k - c)$. This shows that $(n + 1)c = a + nk$, that is,

$$c = \frac{1}{n+1}a + \frac{n}{n+1}k \in (a, k) \subset \mathrm{int}\, K.$$

Therefore, we get $c \in \mathrm{int}\, K$. Let $a \in K$, $b \in \mathrm{bd}\, K$ and $c \in [a, b]$. Then we get $-(a - c) \in -(K - c) \subset n(K - c)$ and hence there is some $k \in K$ with $-(a-c) = n(k-c)$. Then there is some $\lambda \in [0, 1]$ such that $(k-c) = \lambda(b-c) \neq 0$, since $c \in \mathrm{int}\, K$ and $b \in \mathrm{bd}\, K$. This shows that $-(a - c) = n\lambda(b - c)$. From this we conclude that

$$\|a - c\| \leq n\|b - c\| = n(\|a - b\| - \|a - c\|),$$

which implies that $\|a - c\| \leq \frac{n}{n+1}\|a - b\|$.

Exercise 1.3.3

Let $A, B \subset \mathbb{R}^n$ be convex.

(a) "$\subset$": By Corollary 1.3 we have

$$A + B \subset \mathrm{cl}\, A + \mathrm{cl}\, B = \mathrm{cl}(\mathrm{relint}\, A) + \mathrm{cl}(\mathrm{relint}\, B) \subset \mathrm{cl}(\mathrm{relint}\, A + \mathrm{relint}\, B),$$

and hence

$$\mathrm{cl}(A + B) \subset \mathrm{cl}(\mathrm{relint}\, A + \mathrm{relint}\, B).$$

Moreover, we also have

$$\mathrm{cl}(\mathrm{relint}\, A + \mathrm{relint}\, B) \subset \mathrm{cl}(A + B).$$

Therefore,

$$\mathrm{aff}(A + B) = \mathrm{aff}(\mathrm{cl}(A + B)) = \mathrm{aff}(\mathrm{cl}(\mathrm{relint}\, A + \mathrm{relint}\, B)).$$

Then Corollary 1.3 implies that

$$\mathrm{relint}(A + B) \subset \mathrm{relint}(\mathrm{cl}(\mathrm{relint}\, A + \mathrm{relint}\, B))$$
$$= \mathrm{relint}(\mathrm{relint}\, A + \mathrm{relint}\, B)$$
$$\subset \mathrm{relint}\, A + \mathrm{relint}\, B.$$

"$\supset$": Let $a \in \mathrm{relint}\, A$ and $b \in \mathrm{relint}\, B$. We show that for all $y \in \mathrm{aff}(A + B)$ there is some $z \in (a + b, y)$ such that $[a + b, z] \subset A + B$.

For this, let $y \in \mathrm{aff}(A + B)$. Then there are $k \in \mathbb{N}$, $y_i = a_i + b_i$ with $a_i \in A$, $b_i \in B$, $\alpha_i \in [0, 1]$, for $i = 1, \ldots, k$, such that

$$y = \sum_{i=1}^{k} \alpha_i y_i \quad \text{and} \quad \sum_{i=1}^{k} \alpha_i = 1.$$

Define

$$y_a := \sum_{i=1}^{k} \alpha_i a_i \in \mathrm{aff}\, A \quad \text{and} \quad y_b := \sum_{i=1}^{k} \alpha_i b_i \in \mathrm{aff}(B).$$

Since $a \in \mathrm{relint}\, A$ and $y_a \in \mathrm{aff}(A)$, Theorem 1.11 shows that there is some $\lambda_a \in (0, 1)$ such that

$$(1 - \lambda)a + \lambda y_a \in A \quad \text{for } \lambda \in [0, \lambda_a].$$

Since $b \in \operatorname{relint} B$ and $y_b \in \operatorname{aff}(B)$, Theorem 1.11 shows that there is some $\lambda_b \in (0, 1)$ such that

$$(1 - \lambda)b + \lambda y_b \in B \quad \text{for } \lambda \in [0, \lambda_b].$$

Let $\lambda_0 := \min\{\lambda_a, \lambda_b\} \in (0, 1)$. Then

$$(1-\lambda)(a+b)+\lambda y = (1-\lambda)a+\lambda y_a+(1-\lambda)b+\lambda y_b \in A+B \quad \text{for } \lambda \in [0, \lambda_0].$$

Hence we can choose $z := (1 - \lambda_0)(a + b) + \lambda_0 y$.

(b) Let A be bounded.

"⊃": Let $a \in \operatorname{cl} A$, $b \in \operatorname{cl} B$. Thus there are sequences $(a_r)_{r\in\mathbb{N}} \subset A$, $(b_r)_{r\in\mathbb{N}} \subset B$ with $\lim_{r\to\infty} a_r = a$ and $\lim_{r\to\infty} b_r = b$. From $(a_r + b_r)_{r\in\mathbb{N}} \subset A + B$ and $\lim_{r\to\infty}(a_r + b_r) = a + b$ we conclude $a + b \in \operatorname{cl}(A + B)$.

"⊂": Let $x \in \operatorname{cl}(A + B)$. There is a sequence $(x_r)_{r\in\mathbb{N}} \subset A + B$ with $\lim_{r\to\infty} x_r = x$. Hence there are sequences $(a_r)_{r\in\mathbb{N}} \subset A$, $(b_r)_{r\in\mathbb{N}} \subset B$ with $x_r = a_r + b_r$ and $\lim_{r\to\infty} x_r = \lim_{r\to\infty}(a_r + b_r) = x$. Since A is bounded, $\operatorname{cl} A$ is compact and $(a_r) \subset \operatorname{cl} A$ has a converging subsequence $(a_{r_k})_{k\in\mathbb{N}}$ with $\lim_{k\to\infty} a_{r_k} =: a \in \operatorname{cl} A$. Then $\lim_{k\to\infty} b_{r_k} = \lim_{k\to\infty}((b_{r_k} + a_{r_k}) - a_{r_k}) = x - a =: b$ which implies that $b \in \operatorname{cl} B$. From this we finally deduce $x = a + b \in \operatorname{cl} A + \operatorname{cl} B$.

(c) Let $A = \{(x, y) \in \mathbb{R}^2 : x \geq 0, y \geq \frac{1}{x}\} = \operatorname{cl} A$ and $B = (-\infty, 0] \times \{0\} = \operatorname{cl} B$. Then $A + B = \{(x, y) \in \mathbb{R}^2 \cdot y > 0\}$ and $\operatorname{cl}(A+B) = \{(x, y) \in \mathbb{R}^2 : y \geq 0\}$, but $\operatorname{cl}(A + B) \supsetneqq \operatorname{cl} A + \operatorname{cl} B = A + B$.

Exercise 1.4.3

"⇒": Let A be compact and $u \in \mathbb{S}^{n-1}$. Consider the map $f(x) := \langle x, u \rangle$, which is continuous on $\mathbb{R}^n$. Since A is compact, it attains its maximum value α, i.e. $\langle x, u \rangle \leq \alpha$ for all $x \in A$ and $\langle x_0, u \rangle = \alpha$ for some $x_0 \in A$. Thus $A \subset \{f \leq \alpha\}$ and $E := \{f = \alpha\}$ is a supporting hyperplane of A since $x_0 \in E \cap A$.

"⇐": It remains to show that A is bounded. Let $\{e_1, \ldots, e_n\}$ be the standard basis of $\mathbb{R}^n$. By the assumption there are supporting halfspaces $\{\langle e_i, \cdot \rangle \leq \alpha_i\}$ and $\{\langle -e_i, \cdot \rangle \leq \beta_i\}$ of A for all $i \in \{1, \ldots, n\}$. Let $R := \max\{\alpha_1, \ldots, \alpha_n, \beta_1, \ldots, \beta_n\}$ and consider $x \in A$, $x = (x_1, \ldots, x_n)^\top$. Then $x_i = \langle e_i, x \rangle \leq \alpha_i \leq R$ and $-x_i = \langle -e_i, x \rangle \leq \beta_i \leq R$ imply $|x_i| \leq R$. Thus $\|x\|^2 = \sum_{i=1}^{n} |x_i|^2 \leq nR^2$ and $\|x\| \leq \sqrt{n}R$. Therefore $A \subset B^n(0, \sqrt{n}R)$ and hence A is bounded.

Exercise 1.4.8

Let $A \cap H_i$, $i \in I$, be support sets of A. Suppose that $x_0 \in \bigcap_{i \in I}(K \cap H_i) \neq \emptyset$. Then there are $u_i \neq 0$ with $H_i = H(u_i, \langle x_0, u_i \rangle)$ and $K \subset H^-(u_i, \langle x_0, u_i \rangle)$. Let $u_1, \ldots, u_m$ be a maximal set of linearly independent vectors from $\{u_i : i \in I\}$. We define $u := u_1 + \cdots + u_m$. Then $u \neq 0$, $H(u, \langle x_0, u \rangle)$ is a supporting hyperplane of K and $\bigcap_{i \in I}(K \cap H_i) = K \cap H(u, \langle x_0, u \rangle)$.

To see this, let $x \in K$. Then $\langle x, u_i \rangle \leq \langle x_0, u_i \rangle$, $i \in I$, in particular, we get by addition of the corresponding inequalities for $u_1, \ldots, u_m$ that $\langle x, u \rangle \leq \langle x_0, u \rangle$. Therefore, we have $K \subset H^-(u, \langle x_0, u \rangle)$. Next assume that $x \in \bigcap_{i \in I}(K \cap H_i)$. Then $x \in K$ and $\langle x, u_i \rangle = \langle x_0, u_i \rangle$ for $i \in I$, hence also $\langle x, u \rangle = \langle x_0, u \rangle$, that is, $x \in H(u, \langle x_0, u_i \rangle)$. Now, let $x \in H(u, \langle x_0, u_i \rangle)$. Then we have $\langle x, u \rangle = \langle x_0, u \rangle$ and $\langle x, u_i \rangle \leq \langle x_0, u_i \rangle$ for all $i \in I$, since $x \in K$. Thus we get

$$\langle x, u \rangle = \sum_{i=1}^{m} \langle x, u_i \rangle \leq \langle x_0, u_i \rangle = \langle x_0, u \rangle,$$

hence we deduce that $\langle x, u_i \rangle = \langle x_0, u_i \rangle$ for $i = 1, \ldots, m$. Since this is a linear relation in u_i and any other $u' \in \{u_i : i \in I\}$ is a linear combination of $u_1, \ldots, u_m$, this holds for any of the vectors u_i, $i \in I$. This proves the reverse inclusion.

Exercise 1.4.11

(a) Let $u, v \in N(A, a)$, $\lambda \geq 0$ and $x \in A$. Then

$$\langle u + \lambda v, x - a \rangle = \langle u, x - a \rangle + \lambda \langle v, x - a \rangle \leq 0.$$

This shows that $N(A, a)$ is a convex cone. If $u_i \in N(A, a)$, $i \in \mathbb{N}$, and $u_i \to u$ as $i \to \infty$, then $\langle u_i, x - a \rangle \leq 0$ for $x \in A$ and $i \in \mathbb{N}$. Then also $\langle u, x - a \rangle \leq 0$ for $x \in A$. Hence $N(A, a)$ is closed. A similar argument shows that A can be enlarged to its closure without changing the normal cone.

(b) Clearly, $N(A, a) \subset N(A \cap B^n(a, \varepsilon), a)$. For the reverse inclusion, let $x \in A$. For $t \in (0, 1)$, let $x_t = a + t(x - a) = (1 - t)a + tx \in A$ and $u \in N(A \cap B^n(a, \varepsilon), a)$. Then

$$t \langle u, x - a \rangle = \langle u, x_t - a \rangle \leq 0$$

if $t > 0$ is sufficiently small (since then $x_t \in A \cap B^n(a, \varepsilon)$). Thus $\langle u, x - a \rangle \leq 0$, that is, $u \in N(A, a)$.

(c) By (a) we can assume that A is closed. Then the first assertion follows from the existence of a supporting hyperplane of A passing through a and the second assertion is clear.

Exercise 1.4.12

Let $u \in N(A + B, a + b)$. Hence $\langle u, x + y - (a + b) \rangle \leq 0$ for $x \in A$ and $y \in B$. Choosing $y = b$, this yields $u \in N(A, a)$. Choosing $x = a$, this yields $u \in N(B, b)$. Hence $u \in N(A, a) \cap N(B, b)$. The reverse inclusion follows by adding the individual inequalities.

Exercise 1.4.13

We can assume that A, B are closed.

"$\supset$":   Let $u = u_1 + u_2, u_1 \in N(A, c), u_2 \in N(B, c)$. Then

$$\langle u_1, a - c \rangle \leq 0, \qquad \langle u_2, b - c \rangle \leq 0,$$

for $a \in A, b \in B$. Then, for $x \in A \cap B$,

$$\langle u, x - c \rangle = \langle u_1, x - c \rangle + \langle u_2, x - c \rangle \leq 0 + 0 = 0.$$

Hence $u \in N(A \cap B, c)$.

"$\subset$":   Let $u \in N(A \cap B, c)$. Define closed convex sets

$$A^* := A \times [0, \infty),$$
$$B^* := \{(b, \lambda) \in B \times \mathbb{R} : \lambda \leq \langle u, b - c \rangle\}.$$

Then

$$\mathrm{relint}\, A^* := \mathrm{relint}\, A \times (0, \infty),$$
$$\mathrm{relint}\, B^* := \{(b, \lambda) \in \mathrm{relint}\, B \times \mathbb{R} : \lambda < \langle u, b - c \rangle\}.$$

Claim $\mathrm{relint}\, A^* \cap \mathrm{relint}\, B^* = \emptyset$. In fact, if not then there is some $(x, \lambda) \in \mathrm{relint}\, A^* \cap$ relint B^*, in particular, $x \in A \cap B, \lambda > 0$ and $\lambda < \langle u, b - c \rangle \leq 0$, a contradiction.

But then A^* and B^* can be properly separated. Hence there is some $(v, \gamma) \in \mathbb{R}^{n+1} \setminus \{0\}$ such that

$$\langle (a, \lambda_1), (v, \gamma) \rangle \leq \langle (b, \lambda_2), (v, \gamma) \rangle \quad \text{for } (a, \lambda_1) \in A^*, (b, \lambda_2) \in B^*$$

and there is some $(a^0, \lambda_1^0) \in A^*$ and some $(b^0, \lambda_2^0) \in B^*$ such that

$$\langle (a^0, \lambda_1^0), (v, \gamma) \rangle < \langle (b^0, \lambda_2^0), (v, \gamma) \rangle.$$

Thus we have

$$\langle a, v \rangle + \lambda_1 \gamma \le \langle b, v \rangle + \lambda_2 \gamma,$$

for all $a \in A, \lambda_1 \ge 0, b \in B, \lambda_2 \le \langle u, b - c \rangle$, and

$$\langle a^0, v \rangle + \lambda_1^0 \gamma \le \langle b^0, v \rangle + \lambda_2^0 \gamma,$$

for some $a^0 \in A, \lambda_1^0 \ge 0, b^0 \in B, \lambda_2^0 \le \langle u, b^0 - c \rangle$.

Since $(c, 1) \in A^*$ and $(c, 0) \in B^*$, we have

$$\langle c, v \rangle + \gamma \le \langle c, v \rangle + 0 \Rightarrow \gamma \le 0.$$

If $\gamma = 0$, then $\langle a, v \rangle \le \langle b, v \rangle$ for $a \in A, b \in B$ and $\langle a^0, v \rangle \le \langle b^0, v \rangle$ for some $a^0 \in A, b^0 \in B$. But then A, B can be properly separated, that is relint $A \cap$ relint $B = \emptyset$, a contradiction. This implies that $\gamma < 0$.

Since $(a, 0) \in A^*$ for $a \in A$ and $(c, 0) \in B^*$, we get $\langle a, v \rangle \le \langle c, v \rangle$ for $a \in A$, hence $\langle v, a - c \rangle \le 0$ for $a \in A$, that is, $v \in N(A, c)$. Thus we get $-v/\gamma \in N(A, c)$.

Since $(c, 0) \in A^*$ and $(b, \langle u, b - c \rangle) \in B^*$ for $b \in B$, we get

$$\langle c, v \rangle \le \langle b, v \rangle + \langle u, b - c \rangle \gamma.$$

Hence $\langle b - c, v + \gamma u \rangle \ge 0$ or $\langle b - c, v/\gamma + u \rangle \le 0$ for $b \in B$. Thus we have $v/\gamma + u \in N(B, c)$ and therefore $u \in -v/\gamma + N(B, c)$. Hence we get $u \in N(A, c) + N(B, c)$.

Exercise 1.5.3

(a) Let $A \subset \mathbb{R}^n$ be closed and convex. Let $\emptyset \ne M \subset A$ be extreme and $x \in \mathrm{cl}\, M$. Since M is convex, relint M is not empty by Theorem 1.10. Let $y \in$ relint M. There is a $z \in M$ such that $y \in (z, x)$ and thus $(z, x) \cap M \ne \emptyset$. Since M is extreme, we obtain that $[z, x] \subset M$. Hence $x \in M$.

(b) Consider $M = A \cap H(u, \alpha)$, where $\alpha \in \mathbb{R}, u \in \mathbb{S}^{n-1}$ and $H(u, \alpha)$ is a supporting hyperplane with $A \subset H^-(u, \alpha)$. Let $x, y \in A$ be such that $(x, y) \cap M \ne \emptyset$, i.e. $\lambda x + (1 - \lambda) y \in M$ for some $\lambda \in (0, 1)$. Then

$$\langle x, u \rangle \le \alpha, \quad \langle y, u \rangle \le \alpha \quad \text{and} \quad \langle \lambda x + (1 - \lambda) y, u \rangle = \alpha.$$

This implies that $\langle x, u \rangle = \alpha = \langle y, u \rangle$ and thus (by the convexity of A) $[x, y] \subset M$.

(c) Let $x, y \in A$ be such that $(x, y) \cap (M \cap N) \ne \emptyset$. Then $(x, y) \cap M \ne \emptyset$ implies that $[x, y] \subset M$ and $(x, y) \cap N \ne \emptyset$ implies that $[x, y] \subset N$. Hence $[x, y] \subset M \cap N$. The convexity of $M \cap N$ follows since the intersection of two convex sets is convex. The same argument works for the intersection of arbitrary families of faces of A.

(d) Let $x, y \in A$, $(x, y) \cap N \neq \emptyset$. Then $(x, y) \cap M \neq \emptyset$. Since M is extreme in A, we obtain $[x, y] \subset M$ and in particular $x, y \in M$. From this it follows that $[x, y] \subset N$, since N is extreme in M.

(e) We first show that if $M \cap \mathrm{relint}(N) \neq \emptyset$, then $N \subset M$. So let $x \in M \cap \mathrm{relint}(N)$ and let $z \in N \setminus \{x\}$. Then there is some $y \in N$ such that $x \in (y, z) \subset N \subset M$. Since M is extreme, it follows that $[y, z] \subset M$, hence $z \in M$. To conclude the assertion in (e), simply observe that $\mathrm{relint}(M) \cap \mathrm{relint}(N) \neq \emptyset$ implies that $M \subset N$ and $N \subset M$, that is, $M = N$.

(f) It remains to prove the existence. The uniqueness assertion follows from (e). Let F be the intersection of all faces of A which contain B. (The set A itself contains B, for instance.) Then F is a face of A by (c) and $B \subset F$. We claim that $B \subset \mathrm{relint}(F)$. Assume that $x \in B \setminus \mathrm{relint}(F)$. Hence $x \in \mathrm{relbd}(F)$. Hence there is a supporting hyperplane H of F with $x \in H$, $H \cap F \neq F$ (in the affine hull of F). We have $x \in B \subset F$, B is relatively open, and hence $B \subset H$ and therefore $B \subset H \cap F$. Then $H \cap F$ is a face of F, hence also a face of A. But then $F \subset H \cap F$ by the definition of F, a contradiction.

(g) This is clear by the preceding parts of the exercise.

(h) We can assume that $\dim(A) = n$. Let F be an $(n-1)$-dimensional extreme face. Choose $x \in \mathrm{relint}(F)$. Then there is a supporting hyperplane of A through x, hence $F \subset A \cap H$. But $H \cap A$ is a face of A and the relative interiors of these faces intersect. Hence $F = A \cap H$.

6.2 Solutions of Exercises for Chap. 2

Exercise 2.1.2

(a) $\Rightarrow$ (b):   Let f be closed, i.e., epi f is closed. Let $x \in \mathbb{R}^n$ and set $z := \liminf_{y \to x} f(y)$. For $z = \infty$ there is nothing to show. Hence, let $z \neq \infty$ and let (y_k) be a sequence with $y_k \to x$ as $k \to \infty$ and $\lim_{k \to \infty} f(y_k) = z$. Since epi f is closed, we have $(x, z) = \lim_{k \to \infty} (y_k, f(y_k)) \in$ epi f. This implies that $f(x) \leq z = \liminf_{y \to x} f(y)$.

(b) $\Rightarrow$ (c):   Let f be lower semi-continuous and $\alpha \in \mathbb{R}$. For $\{f \leq \alpha\} = \emptyset$ there is nothing to show. Otherwise, let $(x_k) \subset \{f \leq \alpha\}$ be a convergent sequence with limit x. Then we have $f(x) \leq \liminf_{y \to x} f(y) \leq \liminf_{k \to \infty} f(x_k) \leq \alpha$. Thus $f(x) \leq \alpha$ and $x \in \{f \leq \alpha\}$. Hence $\{f \leq \alpha\}$ is closed.

(c) $\Rightarrow$ (a):   Let all the sublevel sets $\{f \leq \alpha\}$, $\alpha \in \mathbb{R}$, be closed. Let $(x_k, \alpha_k) \in$ epi f with $\lim_{k \to \infty} (x_k, \alpha_k) = (x, \alpha)$. Let $\epsilon > 0$. Then there is a $k_0(\epsilon) \in \mathbb{N}$ such that $f(x_k) \leq \alpha_k \leq \alpha + \epsilon$ for $k \geq k_0(\epsilon)$. Hence,

$x_k \in \{f \le \alpha + \epsilon\}$ for $k \ge k_0(\epsilon)$. Since all sublevel sets are closed, this implies that $x \in \{f \le \alpha + \epsilon\}$, that is, $f(x) \le \alpha + \epsilon$. Since $\epsilon > 0$ was arbitrary, this implies that $f(x) \le \alpha$, and we conclude $(x, \alpha) \in$ epi f.

Exercise 2.2.4

(a) For $x, y \in A$ and $\lambda \in [0, 1]$, define $g_{xy}(\lambda) := f((1 - \lambda)x + \lambda y)$. Since f is differentiable, g_{xy} is also differentiable (in fact, in an open neighbourhood of $[0, 1]$). Then

$$g'_{xy}(\lambda) = \langle \text{grad } f((1 - \lambda)x + \lambda y), y - x \rangle.$$

Clearly, f is convex if and only if g_{xy} is convex for all $x, y \in A$; the latter is equivalent to g'_{xy} being increasing for all $x, y \in A$.

"$\Rightarrow$": Let f be convex. Then g_{xy} is convex. Hence g'_{xy} is increasing. In particular, we have $g'_{xy}(0) \le g'_{xy}(1)$, which shows that

$$\langle \text{grad } f(x), y - x \rangle \le \langle \text{grad } f(y), y - x \rangle,$$

from which the assertion follows.

"$\Leftarrow$": Let f be such that the monotonicity condition for grad f is satisfied. Let $x, y \in A$ and $\lambda_0, \lambda_1 \in [0, 1]$ with $\lambda_0 < \lambda_1$. Then

$$g'_{xy}(\lambda_1) - g'_{xy}(\lambda_0)$$
$$= \langle \text{grad } f(x + \lambda_1(y - x)) - \text{grad } f(x + \lambda_0(y - x)), y - x \rangle \ge 0.$$

Thus g'_{xy} is increasing, and hence g_{xy} is convex, which shows that f is convex.

(b) For $x, y \in A$ and $\lambda \in [0, 1]$, define $g_{xy}(\lambda) := f((1-\lambda)x+\lambda y)$. Since f is twice differentiable, g_{xy} is also twice differentiable (in fact, in an open neighbourhood of $[0, 1]$). Then

$$g''_{xy}(\lambda) = d^2 f((1 - \lambda)x + \lambda y)(y - x, y - x).$$

Clearly, f is convex if and only if g_{xy} is convex, for all $x, y \in A$; the latter is equivalent to $g''_{xy} \ge 0$.

"$\Leftarrow$": Let $\partial^2 f(x)$ be positive semi-definite for all $x \in A$, that is, $d^2 f(x)(v, v) \ge 0$ for all $v \in \mathbb{R}^n$. But then $g''_{xy} \ge 0$ for all $x, y \in A$, and therefore g_{xy} is convex. The convexity of g_{xy} now yields the convexity of f.

"$\Rightarrow$": Let f be convex. Hence g_{xy} is convex and $g''_{xy} \ge 0$ for all $x, y \in A$. Let $x \in A$ and $\varepsilon > 0$ be such that $B^n(x, \varepsilon) \subset A$. Furthermore, let $u \in \mathbb{S}^{n-1}$ and define $x' := x - \varepsilon u$ and $y' := x + \varepsilon u$ which are both elements of A.

From $g''_{x'y'}(\frac{1}{2}) = d^2 f(x)(2\varepsilon u, 2\varepsilon u) \geq 0$ it follows that $d^2 f(x)(u, u) \geq 0$ for all $u \in \mathbb{S}^{n-1}$. This implies that $d^2 f(x)(v, v) \geq 0$ for all $v \in \mathbb{R}^n$, and hence $d^2 f(x)$ is positive semi-definite for all $x \in A$.

Exercise 2.2.5

Let $x \in \operatorname{int} \operatorname{dom} f$.

(a) • $\partial f(x) \neq \emptyset$: Since $x \in \operatorname{int} \operatorname{dom} f$, we have $(x, f(x)) \in \operatorname{bd}(\operatorname{cl}(\operatorname{epi} f))$. By Theorem 1.16 there is a supporting hyperplane H of epi f through $(x, f(x))$, which is not vertical since $x \in \operatorname{int} \operatorname{dom} f$. Hence, we get $(x, f(x)) \in H :=$ $\{(z, t) \in \mathbb{R}^{n+1} : \langle (z, t), (\tilde{v}, r) \rangle = \alpha\}$, where $r \in \mathbb{R}$ and $\tilde{v} \in \mathbb{R}^n$ (not both zero), and epi $f \subset H^+$. This implies that $\langle x, \tilde{v} \rangle + f(x)r = \alpha$ and $\langle y, \tilde{v} \rangle + tr \geq \alpha$ for all $t \geq f(y)$ and $y \in \mathbb{R}^n$, in particular, we conclude that $r > 0$. Hence, $\langle y - x, \tilde{v} \rangle + (f(y) - f(x))r \geq 0$ for all $y \in \mathbb{R}^n$. Then $\langle y - x, v \rangle \leq f(y) - f(x)$ with $v := \frac{\tilde{v}}{-r}$. Thus $v \in \partial f(x)$.

• $\partial f(x)$ is closed: Let $v_i \in \partial f(x)$ with $v_i \to v \in \mathbb{R}^n$ as $i \to \infty$. Let $y \in \mathbb{R}^n$ be arbitrary. Then $f(y) - f(x) \geq \langle v_i, y - x \rangle$ for all $i \in \mathbb{N}$. By the continuity of the inner product we thus have $f(y) - f(x) \geq \langle v, y - x \rangle$ and hence $v \in \partial f(x)$.

• $\partial f(x)$ is bounded: By Theorem 2.5 there is some $\alpha > 0$ and some $L = L(\alpha, x) > 0$ such for all $y \in \mathbb{R}^n$ with $\|x - y\| \leq \alpha$ we have $|f(x) - f(y)| \leq L \cdot \|x - y\|$. Now let $v \in \partial f(x) \setminus \{0\}$. Then $\|x - (x + \alpha \frac{v}{\|v\|})\| \leq \alpha$ and hence

$$\alpha L \geq \left| f\left(x + \alpha \frac{v}{\|v\|}\right) - f(x) \right| \geq f\left(x + \alpha \frac{v}{\|v\|}\right) - f(x) \geq \left\langle v, \alpha \frac{v}{\|v\|} \right\rangle$$

by the definition of $\partial f(x)$. From this it follows that $\|v\| \leq L$. Hence $\partial f(x)$ is bounded.

• $\partial f(x)$ is convex: Let $v_1, v_2 \in \partial f(x)$ and $\lambda \in (0, 1)$. Then $f(y) - f(x) \geq \langle v_i, y - x \rangle$ for $i = 1, 2$ and all $y \in \mathbb{R}^n$. Using this we get, for all $y \in \mathbb{R}^n$,

$$f(y) - f(x) = \lambda(f(y) - f(x)) + (1 - \lambda)(f(y) - f(x))$$

$$\geq \lambda \langle v_1, y - x \rangle + (1 - \lambda) \langle v_2, y - x \rangle$$

$$= \langle \lambda v_1 + (1 - \lambda) v_2, y - x \rangle.$$

Hence, $\lambda v_1 + (1 - \lambda) v_2 \in \partial f(x)$.

(b) "$\subset$": Let $v \in \partial f(x)$, $u \in \mathbb{R}^n \setminus \{0\}$ and $t > 0$. Then $f(x + tu) - f(x) \geq t \langle v, u \rangle$ yields $f'(x; u) \geq \langle v, u \rangle$.

"⊃": Let $\langle v, u \rangle \le f'(x; u)$ for $u \in \mathbb{R}^n \setminus \{0\}$. Define $h(t) := f(x+tu)$ for $t \in \mathbb{R}$. The function h is convex and hence

$$f(x+u) - f(x) = \frac{h(1) - h(0)}{1 - 0} \ge h'(0) = f'(x; u) \ge \langle v, u \rangle.$$

Thus, choosing $u = y - x$ we see that $v \in \partial f(x)$.

(c) Let f be differentiable at x. Then $f'(x; u) = \langle \operatorname{grad} f(x), u \rangle$ for all $u \in \mathbb{R}^n$ and hence $\operatorname{grad} f(x) \in \partial f(x)$ by (b). Now let $v \in \partial f(x)$ and $u \ne 0$. Then, by (b), $\langle v, u \rangle \le f'(x; u) = -f'(x; -u) \le \langle v, u \rangle$, since $\langle v, -u \rangle \le f'(x; -u)$. Hence $\langle v, u \rangle = f'(x; u) = \langle \operatorname{grad} f(x), u \rangle$ for all $u \ne 0$. From this we conclude $v = \operatorname{grad} f(x)$.

Exercise 2.2.6

Since the argument is local (with respect to the given point $x \in \operatorname{int} \operatorname{dom} f$), we can assume that f is real-valued. We put

$$\operatorname{grad} f(x) = \sum_{i=1}^n f_i(x) e_i,$$

where $f_i(x)$ denotes the ith partial derivative of f at x in direction e_i and $e_1, \ldots, e_n$ is an orthonormal basis. We consider $g(h) := f(x+h) - f(x) - \langle \operatorname{grad} f(x), h \rangle$, $h \in \mathbb{R}^n$. Then $g(0) = 0$, g has partial derivatives at 0 and $g_i(0) = 0$ for $i = 1, \ldots, n$. Moreover, g is real-valued and convex. We write $h = \sum_{i=1}^n \eta_i e_i$ and then obtain

$$g(h) = g\left(\sum_{i=1}^n \eta_i e_i\right) = g\left(\frac{1}{n} \sum_{i=1}^n n \eta_i e_i\right)$$

$$\le \frac{1}{n} \sum_{i=1}^n g(n \eta_i e_i) = \sum_* \eta_i \frac{g(n \eta_i e_i)}{n \eta_i}$$

$$\le \left(\sum_* \eta_i^2\right)^{\frac{1}{2}} \left(\sum_* \left(\frac{g(n \eta_i e_i)}{n \eta_i}\right)^2\right)^{\frac{1}{2}}$$

$$\le \|h\| \sum_* \left|\frac{g(n \eta_i e_i)}{n \eta_i}\right|,$$

where the summation extends over all $i = 1, \ldots, n$ for which $\eta_i \ne 0$. Now observe that

$$0 = g(0) = g\left(\frac{1}{2}(h + (-h))\right) \le \frac{1}{2}(g(h) + g(-h))$$

implies that $-g(-h) \leq g(h)$ and hence

$$-\|h\| \sum_* \left| \frac{g(-n\eta_i e_i)}{-n\eta_i} \right| \leq -g(-h) \leq g(h) \leq \|h\| \sum_* \left| \frac{g(n\eta_i e_i)}{n\eta_i} \right|,$$

from which we conclude that $|g(h)|/\|h\| \to 0$ as $\|h\| \to 0$. This shows that f is differentiable at x.

Exercise 2.2.7

Let $f : \mathbb{R}^n \to \mathbb{R}$ be convex. By Exercise 2.2.6, a convex function f is differentiable in x, if all partial derivatives of f in x exist. Let $D_i(f) := \{x \in \mathbb{R}^n : f_i(x) \text{ exists}\}$, $i \in \{1, \ldots, n\}$. This is a measurable set. We want to show that the complement of this set is a null set. It is sufficient to consider the case $i = 1$. For $z \in \{0\} \times \mathbb{R}^{n-1}$, let $f_z(t) := f(z + te_1)$ for $t \in \mathbb{R}$. Then $z + te_1 \in D_1(f)$ if and only if $t \in D_1(f_z)$. Hence we get

$$\int_{\mathbb{R}^n} \mathbf{1}\{x \in D_1(f)^c\}\,dx = \int_{\mathbb{R}^{n-1}} \int_{\mathbb{R}} \mathbf{1}\{z + te_1 \in D_1(f)^c\}\,dt\,dz$$

$$= \int_{\mathbb{R}^{n-1}} \int_{\mathbb{R}} \mathbf{1}\{t \in D_1(f_z)^c\}\,dt\,dz = 0.$$

Here we use that the function $f_z : \mathbb{R} \to \mathbb{R}$ is convex and hence differentiable almost everywhere. Since finite unions of null sets are null sets, all partial derivatives exist almost everywhere, hence f is differentiable almost everywhere.

Exercise 2.2.13

By definition, $(v, -1) \in N(\text{epi } f, (x, f(x)))$ if and only if

$$\langle (y, \eta) - (x, f(x)), (v, -1) \rangle \leq 0$$

for all $(y, \eta) \in \text{epi } f$. The latter is the same as requiring that

$$\langle y - x, v \rangle \leq \eta - f(x)$$

for all $y \in \mathbb{R}^n$ and all $\eta \in \mathbb{R}$ such that $\eta \geq f(y)$. But this just means that $v \in \partial f(x)$.

Exercise 2.3.1

Let $f : \mathbb{R}^n \to \mathbb{R}$ be positively homogeneous, i.e., $f(\lambda x) = \lambda f(x)$, for $x \neq 0$ and $\lambda > 0$, and twice continuously differentiable on $\mathbb{R}^n \setminus \{0\}$.

The following assertions are needed in the argument:

Claim 1 $d^2 f$ is positively homogeneous of degree -1.

Proof Let $x \neq 0$ and $\lambda > 0$. Since f is positively homogeneous, $f(\lambda x) = \lambda f(x)$. Then $\lambda f_i(\lambda x) = \lambda f_i(x)$ and $\lambda f_{ij}(\lambda x) = f_{ij}(x)$ for $i, j = 1, \ldots, n$. Hence we obtain $d^2 f(\lambda x) = \frac{1}{\lambda} d^2 f(x)$. $\qquad \square$

Claim 2 $d^2 f(x)(x) = 0$ for $x \neq 0$.

Proof Let $x \neq 0$ and $\lambda > 0$. Then $f_i(\lambda x) = f_i(x)$ for $i = 1, \ldots, n$, and taking the derivative with respect to λ implies that $\sum_{j=1}^{n} f_{ij}(\lambda x) x_j = 0$. Hence $d^2 f(x)(x) = 0$. $\qquad \square$

Claim 3 $d^2 f(x)|x^\perp : x^\perp \to x^\perp$ for $x \neq 0$.

Proof Let $y \perp x$. Then

$$\langle d^2 f(x)(y), x \rangle = \langle y, d^2 f(x)(x) \rangle = \langle y, 0 \rangle = 0.$$

This shows that $d^2 f(x)(y) \perp x$ if $y \perp x$. $\qquad \square$

Claim 4 From $h_{B^n}(x) = \|x\|$, $x \neq 0$ and by a direct calculation, we get

$$\partial^2 h_{B^n}(x) = \frac{1}{\|x\|} \left(\left(\delta_{ij} - \frac{x_i x_j}{\|x\|^2} \right) \right)_{i,j=1}^{n}.$$

For $x \in \mathbb{S}^{n-1}$ this yields $\partial^2 h_{B^n}(x) = ((\delta_{ij} - x_i x_j))_{i,j=1}^{n}$. Now let $u \perp x$, $u \neq 0$. Then

$$d^2 h_{B^n}(x) u = \sum_{i=1}^{n} \left(u_i - x_i \sum_{j=1}^{n} x_j u_j \right) e_i = \sum_{i=1}^{n} u_i e_i = u.$$

For $r > 0$ and $x \neq 0$, we have $d^2(f + r h_{B^n})(x) = d^2 f(x) + r d^2 h_{B^n}(x)$. Since f is positively homogeneous of degree 1, we get $d^2 f(x)(x) = 0$ by Claim 2. Furthermore, $d^2 f(x)$ is symmetric, since f is twice continuously differentiable, and thus the linear map $d^2 f(x)|_{x^\perp} : x^\perp \to x^\perp$ is well defined (Claim 3) and symmetric. The mapping $x \mapsto d^2 f(x)$ is continuous and $d^2 h_{B^n}(x)|_{x^\perp} = \mathrm{id}_{x^\perp}$ by Claim 4. Hence, we can choose $\tilde{r} > 0$ such that $\partial^2(f + \tilde{r} h_{B^n})(x)$ is positive semi-definite for all $x \in \mathbb{S}^{n-1}$. This also holds for all $x \in \mathbb{R}^n \setminus \{0\}$, since $d^2(f + \tilde{r} h_{B^n})$ is positively homogeneous of degree -1 (cf. Claim 1). Exercise 2.2.4 (b) then shows that $f + \tilde{r} h_{B^n}$ is convex on every convex subset of $\mathbb{R}^n \setminus \{0\}$. Since $f + \tilde{r} h_{B^n}$ is continuous, this function is in fact convex on $\mathbb{R}^n$. Moreover, since $f + \tilde{r}h_{B^n}$ is positively homogeneous of degree 1, Corollary 2.2 yields a convex body $L \in \mathcal{K}^n$ such that $f + \tilde{r}h_{B^n} = h_L$, and hence $f = h_L - h_{\tilde{r}B^n}$.

Exercise 2.3.5

(a) This is routine, just use the definitions.
(b) Define $\bar{f} : \mathbb{R}^n \to \mathbb{R}$ by

$$\bar{f}(x) := \sup\{f_k(x) : k \in \mathbb{N}\}, \qquad x \in \mathbb{R}^n.$$

Since $(f_k(x))_{k \in \mathbb{N}}$ converges, as $k \to \infty$, we have $\bar{f}(x) < \infty$ and $\bar{f}$ is a real-valued convex function. Hence $\bar{f}$ is continuous. If $K \subset \mathbb{R}^n$ is a given compact set, $\overline{K} := K + B^n \subset B^n(0, R)$, for a suitable $R > 0$, and $\overline{K}$ is also compact. Hence there is some $a_1 > 0$ such that $|\bar{f}(x)| \le a_1$ for all $x \in B^n(0, R)$. Since $(f_k(0))_{k \in \mathbb{N}}$ converges, as $k \to \infty$, we also have $f_k(0) \ge -a_2$ for all $k \in \mathbb{N}$ and some $a_2 > 0$. Let $x, -x \in B^n(0, R)$. Then $0 = \frac{1}{2}(x + (-x))$, and hence

$$-a_2 \le f_k(0) \le \frac{1}{2}f_k(x) + \frac{1}{2}f_k(-x) \le \frac{1}{2}f_k(x) + \frac{1}{2}\bar{f}(-x) \le \frac{1}{2}f_k(x) + \frac{1}{2}a_1,$$

which shows that $a_1 \ge f_k(x) \ge -2(a_1 + a_2)$, that is, $|f_k(x)| \le (M/?) := 2(a_1 + a_2)$ for $x \in B^n(0, R) \supset \overline{K} \supset K$.

(c) Let K be given. We choose $\overline{K}$ and R, M as in (b). Then the proof of Theorem 2.5 shows that

$$|f_k(x) - f_k(y)| \le M\|x - y\|, \qquad x, y \in K, k \in \mathbb{N}_0,$$

simply choose $\varrho = 1$, $A = K$ and $C = M/2$. Let $\varepsilon > 0$ be given. Then there is a finite subset $Z_\varepsilon \subset K$ such that $K \subset \bigcup_{z \in Z_\varepsilon} B^n(z, \varepsilon/(3M))$. Let $m(\varepsilon) \in \mathbb{N}$ be such that $|f_i(z) - f_j(z)| \le \varepsilon/3$ for $z \in Z_\varepsilon$ and $i, j \ge m(\varepsilon)$. Let $x \in K$ be given. Then there is some $z \in Z_\varepsilon$ with $\|x - z\| \le \varepsilon/(3M)$. If $i, j \ge m(\varepsilon)$, we then get

$$|f_i(x) - f_j(x)| \le |f_i(x) - f_i(z)| + |f_i(z) - f_j(z)| + |f_j(z) - f_j(x)|$$
$$\le 2M\|x - z\| + \varepsilon/3 \le \varepsilon.$$

Hence, for $j \ge m(\varepsilon)$, we obtain from $i \to \infty$ that $|f(x) - f_j(x)| \le \varepsilon$. Since $m(\varepsilon)$ is independent of $x \in K$, this yields the uniform convergence on compact sets.

6.3   Solutions of Exercises for Chap. 3

Exercise 3.1.8

(a) *Claim 1* $K_i \cap M_i \neq \emptyset$ for almost all $i \in \mathbb{N}$.

Proof Suppose that the claim does not hold. Then there is a sequence $(i_k)_{k \in \mathbb{N}}$ with $K_{i_k} \cap M_{i_k} = \emptyset$. Hence there are hyperplanes H_{i_k} separating K_{i_k} and M_{i_k}, that is, $H_{i_k} = H(u_{i_k}, \alpha_{i_k})$ for $u_{i_k} \in \mathbb{S}^{n-1}$ with $K_{i_k} \subset H^-(u_{i_k}, \alpha_{i_k})$ and $M_{i_k} \subset H^+(u_{i_k}, \alpha_{i_k})$. Without loss of generality, $u_{i_k} \to u \in \mathbb{S}^{n-1}$ as $k \to \infty$. Since $K_i \to K$ and $M_i \to M$ as $i \to \infty$, there is some $R > 0$ with $K_i, M_i \subset B^n(0, R)$ for all $i \in \mathbb{N}$. Thus, the sequence $(\alpha_{i_k})_{k \in \mathbb{N}}$ is bounded and without loss of generality $\alpha_{i_k} \to \alpha$ as $k \to \infty$. But then the hyperplane $H = H(u, \alpha)$ separates K and M, a contradiction. $\square$

Claim 2 For all $x \in K \cap M$ there exist $x_i \in K_i \cap M_i$ with $x_i \to x$ as $i \to \infty$.

Proof Assume there are $x_0 \in K \cap M$ and $r > 0$ such that $B^n(x_0, r) \cap (K_i \cap M_i) = \emptyset$ for infinitely many $i \in \mathbb{N}$. Then there is a sequence $(i_k)_{k \in \mathbb{N}}$ with $(K_{i_k} \cap B^n(x_0, r)) \cap (M_{i_k} \cap B^n(x_0, r)) = \emptyset$. Since $K_i \to K \ni x_0$, we can assume without loss of generality (by Exercise 3.1.5 (a)) that $B^n(x_0, r) \cap K_{i_k} \neq \emptyset \neq B^n(x_0, r) \cap M_{i_k}$ for all $k \in \mathbb{N}$. As in the proof of Claim 1 there are separating hyperplanes $H_{i_k} = H(u_{i_k}, \alpha_{i_k})$ with $K_{i_k} \cap B^n(x_0, r) \subset H^-(u_{i_k}, \alpha_{i_k})$ and $M_{i_k} \cap B^n(x_0, r) \subset H^+(u_{i_k}, \alpha_{i_k})$, as well as $u_{i_k} \to u \in \mathbb{S}^{n-1}$ and $\alpha_{i_k} \to \alpha$ for $k \to \infty$.

For $x \in K \cap \mathrm{int}\, B^n(x_0, r)$ there are (by Exercise 3.1.5 (i)) $x_{i_k} \in K_{i_k}$ with $x_{i_k} \to x$ and without loss of generality $x_{i_k} \in B^n(x_0, r)$ (since $\mathrm{int}\, B^n(x_0, r)$ is open). Hence, from $x_{i_k} \in K_{i_k} \cap B^n(x_0, r)$ it follows that $\langle u_{i_k}, x_{i_k} \rangle \leq \alpha_{i_k}$ and thus $\langle u, x \rangle \leq \alpha$. This shows that $K \cap \mathrm{int}\, B^n(x_0, r) \subset H^-(u, \alpha)$.

Similarly, we can show that $M \cap \mathrm{int}\, B^n(x_0, r) \subset \{\langle u, \cdot \rangle \geq \alpha\}$.

Since $x_0 \in K \cap M \cap \mathrm{int}\, B^n(x_0, r)$, we have $x_0 \in H := H(u, \alpha)$. Furthermore, for $x \in K$ we have $[x_0, x] \subset K$ and $(x_0, x] \cap \mathrm{int}\, B^n(x_0, r) \neq \emptyset$. Hence $(x_0, x] \cap (K \cap \mathrm{int}\, B^n(x_0, r)) \neq \emptyset$. Since $x_0 \in H$, we get $(x_0, x] \subset H^-(u, \alpha)$ and in particular $x \in H^-(u, \alpha)$. Thus, $K \subset H^-(u, \alpha)$. In the same manner $M \subset H^+(u, \alpha)$ follows. But then H separates K and M, a contradiction. $\square$

Claim 3 If $x_{i_k} \in M_{i_k} \cap K_{i_k}$ for $k \in \mathbb{N}$ and $x_{i_k} \to x$ as $k \to \infty$, then $x \in K \cap M$.

Proof Let $x_{i_k} \in M_{i_k} \cap K_{i_k}$ with $x_{i_k} \to x$. Then $x_{i_k} \in M_{i_k}$ and $x_{i_k} \in K_{i_k}$ with $x_{i_k} \to x$. From Exercise 3.1.5 it now follows that $x \in M$ and $x \in K$ and thus $x \in K \cap M$. $\square$

The assertion now follows from these three claims and Exercise 3.1.5.

(b) Since $K_i \to K$, we have $K_i \subset B^n(0, R)$ for some $R > 0$. Hence $K \subset B^n(0, R)$. Let $M := E \cap B^n(0, R) \in \mathcal{K}^n$. Then $M \cap \mathrm{int}\, K \neq \emptyset$. Let $M_i := M$ for all $i \in \mathbb{N}$. Then by (a) we get

$$K_i \cap E = K_i \cap B^n(0, R) \cap E = K_i \cap M_i \to K \cap M = K \cap B^n(0, R) \cap E = K \cap E.$$

Exercise 3.1.15

We choose a polytope Q with $Q \subset K \subset Q + (\varepsilon/2)B^n$, and for $i \in \{1, \dots, m\}$ we choose a polytope R_i with $K_i \subset K_i + (\varepsilon/2)B^n \subset R_i \subset K_i + \varepsilon B^n$. Then we define $Q_i = Q \cap R_i$ for $i = 1, \dots, m$. We have

$$Q = Q \cap K = \bigcup_{i=1}^{m}(Q \cap K_i) \subset \bigcup_{i=1}^{m}(Q \cap R_i) = \bigcup_{i=1}^{m} Q_i \subset Q,$$

hence

$$Q = \bigcup_{i=1}^{m} Q_i.$$

If $x \in K_i \subset K$, then there is some $y \in Q$ with $\|x - y\| \le \varepsilon/2$, hence $y \in B^n(x, \varepsilon/2) \subset R_i$. This shows $y \in Q_i$, and therefore $K_i \subset Q_i + (\varepsilon/2)B^n \subset Q_i + \varepsilon B^n$. On the other hand, we have $Q_i \subset R_i \subset K_i + \varepsilon B^n$.

Next we choose a polytope C with $\varepsilon B^n \subset C \subset 2\varepsilon B^n$ and define $P_i = Q_i + C$. Then

$$K_i \subset Q_i + \varepsilon B^n \subset Q_i + C = P_i \subset K_i + \varepsilon B^n + 2\varepsilon B^n = K_i + 3\varepsilon B^n.$$

Moreover, by basic properties of convex sets,

$$P = P_1 \cup \dots \cup P_m = (Q_1 + C) \cup \dots \cup (Q_m + C) = (Q_1 \cup \dots \cup Q_m) + C = Q + C,$$

thus P is convex. The assertion now follows after adjusting ε.

Exercise 3.2.4

Let $S \subset K$ be a simplex contained in K of maximal volume. Let $a_1, \dots, a_{n+1}$ denote the vertices of S. For $i = 1, \dots, n+1$ the hyperplane H_i parallel to $\{a_1, \dots, a_{n+1}\} \setminus \{a_i\}$ through a_i is a supporting hyperplane of K. To see this, denote by H_i^- the closed halfspace bounded by H_i which contains K and by H_i^+ its closed complement. If H_i is not a supporting hyperplane, then there is a point $a_i' \in \operatorname{int}(H_i^+) \cap K$ and therefore the simplex S' which is the convex hull of $a_1, \dots, a_{i-1}, a_i', a_{i+1}, \dots, a_{n+1}$ is contained in K and has larger volume than S, a contradiction.

Now we claim that $\overline{S} := \bigcap_{i=1}^{n+1} H_i^-$ is a simplex with $\overline{S} \supset K$ and $\overline{S} = c - n(S - c)$ if $c = (a_1 + \dots + a_{n+1})/(n+1)$ is the centre of S. To show this, we can assume that $c = 0$ (after a translation). We define $\overline{a}_i := -na_i$ for $i = 1, \dots, n+1$. Then $a_{n+1} =$

$(1/n) \sum_{i=1}^{n} (-na_i)$ and the affine subspace spanned by $\bar{a}_i$, $i = 1, \ldots, n$, is parallel to the affine subspace spanned by a_i, $i = 1, \ldots, n$. This shows that H_{n+1} is the affine subspace spanned by $\bar{a}_i$, $i = 1, \ldots, n$, and the corresponding statement holds for the other indices. This shows that $\tilde{S} := \text{conv}\{\bar{a}_1, \ldots, \bar{a}_{n+1}\} \subset \overline{S}$. Conversely, let $x \in \overline{S}$. Since $\bar{a}_1, \ldots, \bar{a}_{n+1}$ are affinely independent, there are $\lambda_1, \ldots, \lambda_{n+1}$ such that

$$x = \sum_{j=1}^{n+1} \lambda_j \bar{a}_j, \quad \sum_{j=1}^{n+1} \lambda_j = 1.$$

Since $x \in H_i^-$, it follows that $\lambda_i \geq 0$. But then $x \in \tilde{S}$. Thus we obtain $\overline{S} = \tilde{S} = -nS$.

Exercise 3.3.1

(a) We use that

$$V(s_1 + \cdots + s_n) = |\det(x_1, \ldots, x_n)|,$$

i.e. the absolute value of the determinant of $(x_1, \ldots, x_n)$ equals the volume of the parallelepiped spanned by $x_1, \ldots, x_n$. (This follows from the transformation formula for integrals.)

Now, using formula (3.6), we obtain

$$\sum_{i_1=1}^{n} \cdots \sum_{i_n=1}^{n} \alpha_{i_1} \cdots \alpha_{i_n} V(s_{i_1}, \ldots, s_{i_n}) = V(\alpha_1 s_1 + \cdots + \alpha_n s_n)$$

$$= V([0, \alpha_1 x_1] + \cdots + [0, \alpha_n x_n])$$

$$= |\det(\alpha_1 x_1, \ldots, \alpha_n x_n)|$$

$$= \alpha_1 \cdots \alpha_n \cdot |\det(x_1, \ldots, x_n)|$$

for all $\alpha_1, \ldots, \alpha_n \geq 0$. Comparing the coefficients, we deduce

$$\underbrace{\sum_{\{i_1, \ldots, i_n\} = \{1, \ldots, n\}}}_{n! \text{ possibilities}} V(s_{i_1}, \ldots, s_{i_n}) = |\det(x_1, \ldots, x_n)|.$$

Since the mixed volume is symmetric, this yields the assertion. An alternative argument can be given by using the inversion formula.

(b) "$\Leftarrow$": Let $s_i = [0, x_i] + y_i \subset K_i$, $i = 1, \ldots, n$, with linearly independent $x_1, \ldots, x_n \in \mathbb{R}^n$. From the monotonicity and the translation invariance of the mixed volume, we obtain by (a)

$$V(K_1, \ldots, K_n) \geq V(s_1, \ldots, s_n) = V([0, x_1], \ldots, [0, x_n])$$

$$= \frac{1}{n!} |\det(x_1, \ldots, x_n)| > 0,$$

since $x_1, \ldots, x_n$ are linearly independent.

"$\Rightarrow$": It is sufficient to prove the assertion for polytopes. In fact, there are sequences of polytopes $(P_i^{(j)})$ with $P_i^{(j)} \subset K_i$ and $\lim_{j \to \infty} P_i^{(j)} = K_i$, $i = 1, \ldots, n$. Since the mixed volume is continuous, we have

$$\lim_{j \to \infty} V(P_1^{(j)}, \ldots, P_n^{(j)}) = V(K_1, \ldots, K_n) > 0,$$

hence there exists some $j_0 \in \mathbb{N}$ with

$$V(P_1^{(j_0)}, \ldots, P_n^{(j_0)}) > 0.$$

If the polytopes $P_1^{(j)}, \ldots, P_n^{(j)}$ contain linearly independent segments, then the convex bodies $K_1, \ldots, K_n$ also contain linearly independent segments.

Thus it is sufficient to prove the assertion for polytopes. We are going to show this by induction on n. For $n = 1$ the assertion holds.

Let $P_1, \ldots, P_n$ be polytopes with $V(P_1, \ldots, P_n) > 0$ and without loss of generality $0 \in P_n$.

By the definition of the mixed volume, we have

$$0 < V(P_1, \ldots, P_n) = \frac{1}{n} \sum_{u \in \mathbb{S}^{n-1}} h_{P_n}(u) V^{(n-1)}(P_1(u)|u^\perp, \ldots, P_{n-1}(u)|u^\perp).$$

At least one of the summands has to be positive. Thus there is some u_0 with

$$0 < h_{P_n}(u_0) V^{(n-1)}(P_1(u_0)|u_0^\perp, \ldots, P_{n-1}(u_0)|u_0^\perp).$$

On the one hand, this implies that $0 < h_{P_n}(u_0)$, and thus there is some $x_n \in P_n$ such that $0 < \langle u_0, x_n \rangle$. This shows that $s_n := [0, x_n] \subset P_n$ and $x_n \notin u_0^\perp$, since $0 \in P_n$. On the other hand, we have

$$0 < V^{(n-1)}(P_1(u_0)|u_0^\perp, \ldots, P_{n-1}(u_0)|u_0^\perp)$$

and hence, by the induction hypothesis (applied in $u_0^\perp$), there are segments $\tilde{s}_i \subset P_i(u_0)|u_0^\perp$ with linearly independent directions. This proves the existence of segments $s_i \subset P_i(u_0) \subset P_i$ with $\tilde{s}_i = s_i | u_0^\perp$ and the directions of s_i and $\tilde{s}_i$ are the same.

Since $s_1, \ldots, s_{n-1}$ have linearly independent directions, which all lie in $u_0^\perp$, but $s_n = [0, x_n]$ has a direction, which does not lie in $u_0^\perp$, the directions of $s_1, \ldots, s_n$ are linearly independent, which proves the assertion.

Exercise 3.3.9

(a) If $x \in \text{int } P$, then $N(P, x) = \{0\}$ is a convex cone and $\{y - x : y \in \mathbb{R}^n, p(P, y) = x\} = \{x - x\} = \{0\} = N(P, x)$, since $p(P, y) = x \iff y = x$.

If $x \in \text{bd } P$, then we have $\langle x, u \rangle \geq \langle y, u \rangle$ and $\langle x, v \rangle \geq \langle y, v \rangle$ for $u, v \in N(P, x)$ and $y \in P$. Thus $\langle x, u + v \rangle \geq \langle y, u + v \rangle$ for all $y \in P$ with equality if $y = x$. From $\langle x, u + v \rangle = h_P(u + v)$ it follows that $u + v \in N(P, x)$. Furthermore, $N(P, x)$ is homogeneous of degree 1. Hence $N(P, x)$ is a convex cone.

We have

$$0 \neq u \in N(P, x)$$

$$\iff \langle x, u \rangle = h_P(u)$$

$$\iff \langle z - x, u \rangle \leq 0 \text{ for all } z \in P$$

$$\iff \|z - (x + u)\|^2 \geq \|u\|^2 = \|(x + u) - x\|^2 \text{ for all } z \in P$$

$$\iff p(P, x + u) = x$$

$$\iff u \in \{y - x : y \in \mathbb{R}^n, p(P, y) = x\}.$$

(b) Let $F \in \mathcal{F}_k(P)$, $x, y \in \text{relint } F$ and $u \in N(P, x)$. There is some $\lambda > 1$ such that $y + \lambda(x - y) \in F$, since $x, y \in \text{relint } F$. Since $u \in N(P, x)$, i.e. $\langle x, u \rangle = h_P(u)$, we have $0 \geq \langle y + \lambda(x - y) - x, u \rangle = (1 - \lambda)\langle y - x, u \rangle$. This implies that $\langle y - x, u \rangle = 0$, from which it in turn follows that $\langle y, u \rangle = \langle x, u \rangle = h_P(u)$. Hence $u \in N(P, y)$ and thus $N(P, x) = N(P, y)$. Furthermore, $\langle y - x, u \rangle = 0$ implies that $\text{aff}(N(P, F)) \subset F^\perp$ and thence $N(P, F) \subset F^\perp$.

(c) This follows from Exercises 1.5.3 and 1.5.5. In particular, for all $x \in P$ there is exactly one $F \in \mathcal{F}_k(P)$ such that $x \in \text{relint } F$.

(d) We have

$$P + B^n(\epsilon) = \text{int}(P) \cup [(P + B^n(\epsilon)) \setminus \text{int}(P)]$$

and by (c), (a) and (b)

$$(P + B^n(\epsilon)) \setminus \text{int}(P) = \bigcup_{k=0}^{n-1} \bigcup_{F \in \mathcal{F}_k(P)} (p(P, \cdot)^{-1}(\text{relint } F)) \cap (P + B^n(\epsilon))$$

$$= \bigcup_{k=0}^{n-1} \bigcup_{F \in \mathcal{F}_k(P)} (\text{relint } F + (N(P, F) \cap B^n(\epsilon))).$$

Since $N(P, z) = \{0\}$ for $z \in \text{int } P$, we conclude

$$P + B^n(\epsilon) = \bigcup_{k=0}^{n} \bigcup_{F \in \mathcal{F}_k(P)} (\text{relint } F + (N(P, F) \cap B(\epsilon))).$$

By (c) this partitioning is disjoint.

(e) Let $F \in \mathcal{F}_k(P)$, $k \in \{0, \ldots, n\}$. Then, using Fubini's theorem, we obtain

$$V(\text{relint } F + (N(P, F) \cap B^n(\epsilon))) = \lambda_F(F) \, \lambda_{F^\perp}(N(P, F) \cap B^n(\epsilon))$$

$$= \lambda_F(F) \, \epsilon^{n-k} \lambda_{F^\perp}(N(P, F) \cap B^n),$$

since $N(P, F) \cap B^n(\epsilon) \subset F^\perp$. This yields

$$V(P + B^n(\epsilon)) = V(P) + \sum_{k=0}^{n-1} \epsilon^{n-k} \sum_{F \in \mathcal{F}_k(P)} \lambda_F(F) \lambda_{F^\perp}(N(P, F) \cap B^n)$$

$$= V(P) + \sum_{k=0}^{n-1} \epsilon^{n-k} \kappa_{n-k} \sum_{F \in \mathcal{F}_k(P)} \gamma(P, F) \lambda_F(F).$$

Comparing the coefficients of this polynomial with the ones in the Steiner Formula, we get

$$V_k(P) = \sum_{F \in \mathcal{F}_k(P)} \gamma(P, F) \lambda_F(F) \qquad \text{for } k = 0, \ldots, n - 1.$$

Exercise 3.4.2

The function $f(t) := V(K + tL)^{\frac{1}{n}}$ is concave on $[0, 1]$ by the Brunn–Minkowski inequality. Since

$$h(t) := V(K + tL) = \sum_{i=0}^{n} \binom{n}{i} t^i \, V(\underbrace{K, \ldots, K}_{n-i}, \underbrace{L, \ldots, L}_{i})$$

is of class C^2 on $[0, 1]$, f is of class C^2 as well. Hence, $f'' \leq 0$ on $[0, 1]$. Since $f(t) = h(t)^{\frac{1}{n}}$, we get that

$$f'(t) = \frac{1}{n}h(t)^{\frac{1}{n}-1}h'(t)$$

and

$$f''(t) = \frac{1}{n}h(t)^{\frac{1}{n}-2}\left[\frac{1-n}{n}h'(t)^2 + h(t)h''(t)\right].$$

Hence $f''(0) \leq 0$ yields $h(0)h''(0) \leq \frac{n-1}{n}h'(0)^2$. Furthermore,

$$h'(t) = \sum_{i=1}^{n}\binom{n}{i}it^{i-1}V(\underbrace{K,\ldots,K}_{n-i},\underbrace{L,\ldots,L}_{i})$$

and

$$h''(t) = \sum_{i=2}^{n}\binom{n}{i}i(i-1)t^{i-2}V(\underbrace{K,\ldots,K}_{n-i},\underbrace{L,\ldots,L}_{i}).$$

Thus, from $h(0) = V(K)$, $h'(0) = nV(K,\ldots,K,L)$, and from the second derivative $h''(0) = n(n-1)V(K,\ldots,K,L,L)$, together with the above inequality, it follows that

$$V(K)n(n-1)V(K,\ldots,K,L,L) \leq \frac{n-1}{n}n^2V(K,\ldots,K,L)^2$$

and hence

$$V(K)V(K,\ldots,K,L,L) \leq V(K,\ldots,K,L)^2.$$

Exercise 3.4.11

(a) After a rotation and translations of A and B, we can assume that

$$A = [0, a_1] \times \cdots \times [0, a_n], \quad B = [0, b_1] \times \cdots \times [0, b_n],$$

where $a_i, b_j > 0$. Hence

$$A + B = [0, a_1 + b_1] \times \cdots \times [0, a_n + b_n].$$

Then the inequality of arithmetic mean and geometric mean shows that

$$\left(\prod_{i=1}^{n} \frac{a_i}{a_i + b_i}\right)^{\frac{1}{n}} + \left(\prod_{i=1}^{n} \frac{b_i}{a_i + b_i}\right)^{\frac{1}{n}} \leq \frac{1}{n} \sum_{i=1}^{n} \frac{a_i}{a_i + b_i} + \frac{1}{n} \sum_{i=1}^{n} \frac{b_i}{a_i + b_i} = 1.$$

Rearranging terms yields $V(A + B)^{1/n} \geq V(A)^{1/n} + V(B)^{1/n}$. From this, the version for convex combinations of A and B follows easily.

(b) For the induction step, we provide a lemma on 'corresponding dissections' of A and B.

Definition Let C^n denote the system of nonempty compact subset of $\mathbb{R}^n$. For $X, Y \in C^n$, the (Brunn–Minkowski) *deficit* is defined by

$$D(X, Y) = V(X + Y) - \left(V(X)^{\frac{1}{n}} + V(Y)^{\frac{1}{n}}\right)^n.$$

The deficit $D(\cdot, \cdot)$ is symmetric and translation invariant (independently in both arguments). The following lemma describes parallel dissection by preserving volume ratios.

Lemma 6.1 *Let $X, Y \in C^n$, $V(X), V(Y) > 0$, $e \in \mathbb{S}^{n-1}$, and $\alpha, \beta \in \mathbb{R}$. Define $X^{\pm} := X \cap H^{\pm}(e, \alpha)$, $Y^{\pm} := Y \cap H^{\pm}(e, \beta)$. Suppose that $V(X^+)/V(X) = V(Y^+)/V(Y)$. Then*

$$D(X, Y) \geq D(X^-, Y^-) + D(X^+, Y^+).$$

Proof We have the relations

$$V(X) = V(X^+) + V(X^-), \quad V(Y) = V(Y^+) + V(Y^-), \tag{6.1}$$

$$(X^+ + Y^+) \cup (X^- + Y^-) \subset X + Y, \tag{6.2}$$

$$(X^+ + Y^+) \cap (X^- + Y^-) \subset H(e, \alpha + \beta). \tag{6.3}$$

From (6.2) and (6.3) we deduce that

$$V(X + Y) \geq V(X^- + Y^-) + V(X^+ + Y^+).$$

We distinguish three cases.

(i) $V(X^+)/V(X) \in (0, 1)$. From (6.1) we get

$$V(X)/V(Y) = V(X^+)/V(Y^+) = V(X^-)/V(Y^-). \tag{6.4}$$

Hence,

$$D(X, Y) = V(X + Y) - \left(V(X)^{\frac{1}{n}} + V(Y)^{\frac{1}{n}}\right)^n$$

$$\geq V(X^- + Y^-) + V(X^+ + Y^+) - V(Y)\left(\left(\frac{V(X)}{V(Y)}\right)^{\frac{1}{n}} + 1\right)^n$$

$$= V(X^- + Y^-) - V(Y^-)\left(\left(\frac{V(X^-)}{V(Y^-)}\right)^{\frac{1}{n}} + 1\right)^n$$

$$+ V(X^+ + Y^+) - V(Y^+)\left(\left(\frac{V(X^+)}{V(Y^+)}\right)^{\frac{1}{n}} + 1\right)^n$$

$$= D(X^-, Y^-) + D(X^+, Y^+).$$

(ii) $V(X^+)/V(X) = 0$. Then we have $V(X^+) = V(Y^+) = 0$ and $V(X) = V(X^-)$, $V(Y) = V(Y^-)$. Hence,

$$D(X, Y) = V(X + Y) - \left(V(X)^{\frac{1}{n}} + V(Y)^{\frac{1}{n}}\right)^n$$

$$\geq V(X^+ + Y^+) + V(X^- + Y^-) - \left(V(X^-)^{\frac{1}{n}} + V(Y^-)^{\frac{1}{n}}\right)^n$$

$$- \left(V(X^+)^{\frac{1}{n}} + V(Y^+)^{\frac{1}{n}}\right)^n$$

$$= D(X^+, Y^+) + D(X^-, Y^-).$$

(iii) $V(X^+)/V(X) = 1$. Then $V(X^+) = V(X)$ and $V(Y^+) = V(Y)$ as well as $V(X^-) = V(Y^-) = 0$. Now we proceed as in case (b).   □

(b) Suppose that X, Y are finite unions of boxes with axes parallel to the coordinate axes (say) and mutually disjoint interiors (for X and for Y, respectively). Let F, G be such families of boxes and

$$X = \bigcup_{P \in F} P, \qquad Y = \bigcup_{Q \in G} Q.$$

By (a) the asserted inequality holds for $\sharp F = \sharp G = 1$. We prove (b) by induction on $p = \sharp F + \sharp G$. Suppose the inequality is true for at most $p - 1 \geq 2$ such boxes. We may assume that $\sharp F > 1$. Choose $e \in \mathbb{S}^{n-1}$ as one of the unit vectors of the coordinate directions and $\alpha \in \mathbb{R}$ such that $H(e, \alpha)$ separates two of the boxes of F, i.e., X^+ and X^- each contain at least one element of F, where $X^{\pm} := X \cap H^{\pm}(e, \alpha)$. Further, choose $\beta \in \mathbb{R}$ such that

$$V(X)^+)/V(X) = V(Y^+)/V(Y).$$

Finally, we put

$$F^{\pm} := \{P \cap X^{\pm} : P \in F, \text{ int } P \cap H^{\pm}(e, \alpha) \neq \emptyset\},$$
$$G^{\pm} := \{Q \cap Y^{\pm} : Q \in G, \text{ int } Q \cap H^{\pm}(e, \beta) \neq \emptyset\}.$$

Thus we get

$$\sharp F^{\pm} < \sharp F, \quad \sharp G^{\pm} \leq \sharp G, \quad \sharp F^{\pm} + \sharp G^{\pm} < p.$$

Defining

$$\tilde{X}^{\pm} := \bigcup_{\tilde{P} \in F^{\pm}} \tilde{P}, \qquad \tilde{Y}^{\pm} := \bigcup_{\tilde{Q} \in G^{\pm}} \tilde{Q}$$

we obtain from Lemma 6.1 and by the induction hypothesis that

$$D(X, Y) \geq D(X^{+}, Y^{+}) + D(X^{-}, Y^{-})$$
$$\geq D(\tilde{X}^{+}, \tilde{Y}^{+}) + D(\tilde{X}^{-}, \tilde{Y}^{-})$$
$$\geq 0 + 0 = 0.$$

(c) Let $X, Y \in C^n$, $m \in \mathbb{N}$, be arbitrary. Define

$$\mathcal{W}_m := \left\{[0, 2^{-m}]^n + 2^{-m} z : z \in \mathbb{Z}^n\right\}.$$

With

$$X_m := \bigcup \{W \in \mathcal{W}_m : W \cap X \neq \emptyset\}, \quad Y_m := \bigcup \{W \in \mathcal{W}_m : W \cap Y \neq \emptyset\},$$

we get for $m \to \infty$ that

$$X_m \downarrow X, \quad Y_m \downarrow Y, \quad X_m + Y_m \downarrow X + Y,$$

and hence

$$V(X + Y)^{\frac{1}{n}} = \lim_{m \to \infty} V(X_m + Y_m)^{\frac{1}{n}} \geq \lim_{m \to \infty} \left\{V(X_m)^{\frac{1}{n}} + V(Y_m)^{\frac{1}{n}}\right\}$$
$$= V(X)^{\frac{1}{n}} + V(Y)^{\frac{1}{n}}.$$

Exercise 3.4.13

The assertion is proved by induction over $n \in \mathbb{N}$. We start with $n = 1$.

We can assume that $f \not\equiv 0$ and $g \not\equiv 0$ are bounded. Otherwise, we first consider $\min\{f, m\}$ instead of f and $\min\{g, m\}$ instead of g for $m \in \mathbb{N}$. The general case the follows by means of the increasing convergence theorem. By homogeneity, we can then assume that $\sup f = \sup g = 1$.

In the following, for a measurable function $\gamma : \mathbb{R} \to [0, \infty)$ and $t \in \mathbb{R}$ we briefly write

$$[\gamma \geq t] := \{x \in \mathbb{R} : \gamma(x) \geq t\}.$$

Fubini's theorem shows that

$$\int_{\mathbb{R}} \gamma(x)\,dx = \int_0^\infty \lambda^1([\gamma \geq t])\,dt.$$

From $f(x) \geq t$, $g(y) \geq t$ it follows that $h((1 - \lambda)x + \lambda y) \geq t$, hence

$$[h \geq t] \supset (1 - \lambda)[f \geq t] + \lambda[g \geq t].$$

For $t \in (0, 1)$, the sets on the right-hand side are nonempty, hence by the trivial one-dimensional special case of the Brunn–Minkowski inequality, we have (approximate the sets $[f \geq t]$ and $[g \geq t]$ from inside by compact sets)

$$\lambda^1([h \geq t]) \geq (1 - \lambda) \cdot \lambda^1([f \geq t]) + \lambda \cdot \lambda^1([g \geq t]).$$

Integration with respect to t over $[0, 1)$ yields

$$\int_{\mathbb{R}} h(x)\,dx = \int_0^\infty \lambda^1([h \geq t])\,dt$$

$$\geq \int_0^1 \lambda^1([h \geq t])\,dt$$

$$\geq (1 - \lambda) \int_0^1 \lambda^1([f \geq t])\,dt + \lambda \int_0^1 \lambda^1([g \geq t])\,dt$$

$$= (1 - \lambda) \int_{\mathbb{R}} f(x)\,dx + \lambda \int_{\mathbb{R}} g(x)\,dx$$

$$\geq \left(\int_{\mathbb{R}} f(x)\,dx \right)^{1-\lambda} \left(\int_{\mathbb{R}} g(x)\,dx \right)^{\lambda},$$

where we used the arithmetic-geometric-mean inequality in the last step.

We continue the induction argument. Let $n > 1$ and suppose that the assertion has already been proved in lower dimensions. We use the identification $\mathbb{R}^n = \mathbb{R}^{n-1} \times \mathbb{R}$ and define, for $s \in \mathbb{R}$ and $z \in \mathbb{R}^{n-1}$,

$$h_s(z) := h(z, s), \quad f_s(z) := f(z, s), \quad g_s(z) := g(z, s).$$

Let $z_1, z_2 \in \mathbb{R}^{n-1}$, $a, b \in \mathbb{R}$ and $c := (1 - \lambda)a + \lambda b$. By assumption, we have

$$
\begin{aligned}
h_c((1 - \lambda)z_1 + \lambda z_2) &= h((1 - \lambda)z_1 + \lambda z_2, (1 - \lambda)a + \lambda b) \\
&= h((1 - \lambda)(z_1, a) + \lambda(z_2, b)) \\
&\geq f(z_1, a)^{1-\lambda} g(z_2, b)^{\lambda} \\
&= f_a(z_1)^{1-\lambda} g_b(z_2)^{\lambda}.
\end{aligned}
$$

The induction hypothesis yields

$$\underbrace{\int_{\mathbb{R}^{n-1}} h_c(z)\, dz}_{=:H(c)} \geq \left(\underbrace{\int_{\mathbb{R}^{n-1}} f_a(z)\, dz}_{=:F(a)} \right)^{1-\lambda} \left(\underbrace{\int_{\mathbb{R}^{n-1}} g_b(z)\, dz}_{=:G(b)} \right)^{\lambda},$$

that is,

$$H((1 - \lambda)a + \lambda b) \geq F(a)^{1-\lambda} G(b)^{\lambda}$$

for $a, b \in \mathbb{R}$. The case $n = 1$ has already been proved, so that we obtain by Fubini's theorem that

$$
\begin{aligned}
\int_{\mathbb{R}^n} h(x)\, dx &= \int_{\mathbb{R}} \int_{\mathbb{R}^{n-1}} h_s(z)\, dz\, ds = \int_{\mathbb{R}} H(s)\, ds \\
&\geq \left(\int_{\mathbb{R}} F(a)\, da \right)^{1-\lambda} \left(\int_{\mathbb{R}} G(b)\, db \right)^{\lambda} \\
&= \left(\int_{\mathbb{R}^n} f(x)\, dx \right)^{1-\lambda} \left(\int_{\mathbb{R}^n} g(x)\, dx \right)^{\lambda}.
\end{aligned}
$$

This completes the induction argument and thus the proof of the asserted inequality.

6.4   Solutions of Exercises for Chap. 4

Exercise 4.2.3

(a) This has been shown in the solution of Exercise 1.1.13.
(b) Let $\mathbb{R}_+^k$ denote the set of all vectors $y = (y^{(1)}, \ldots, y^{(k)})$ with $y^{(i)} \geq 0$, $i = 1, \ldots, k$. For $y \in \mathbb{R}_+^k$ and $u_1, \ldots, u_k \in \mathbb{S}^{n-1}$, let $P_{[y]} = \bigcap_{i=1}^k H^-(u_i, y^{(i)})$. Let $y \in \mathbb{R}_+^k$ and $(y_j)_{j \in \mathbb{N}} \subset \mathbb{R}_+^k$ be such that $y_j \to y$ for $j \to \infty$. We show that $P_{[y_j]} \to P_{[y]}$ by means of Exercise 3.1.5. Hence, we have to show that

  (i) each $x \in P_{[y]}$ is the limit of a sequence of points $(x_j)_{j \in \mathbb{N}}$ with $x_j \in P_{[y_j]}$ for all $j \in \mathbb{N}$;
  (ii) every accumulation point of a sequence $(x_j)_{j \in \mathbb{N}}$ with $x_j \in P_{[y_j]}$, for $j \in \mathbb{N}$, lies in $P_{[y]}$.

  (i) Let $x \in P_{[y]}$. If $x = 0$, then $x \in P_{[y_j]}$ for all $j \in \mathbb{N}$ and we put $x_j := x$. Now let $x \neq 0$. Since $u_1, \ldots, u_k$ span $\mathbb{R}^n$ and are centred, there is some $i \in \{1, \ldots, k\}$ such $\langle x, u_i \rangle > 0$, hence

$$I(x) := \{i \in \{1, \ldots, k\} : \langle x, u_i \rangle > 0\} \neq \emptyset.$$

Then we put

$$\lambda_j := 1 \wedge \min\{\langle x, u_i \rangle^{-1} y_j^{(i)} : i \in I(x)\},$$

where $r \wedge s := \min\{r, s\}$ for $r, s \in \mathbb{R}$. Then we have

$$\langle \lambda_j x, u_i \rangle = \lambda_j \langle x, u_i \rangle \leq y_j^{(i)} \quad \text{for } i = 1, \ldots, k,$$

so that $\lambda_j x \in P_{[y_j]}$. We have $\lambda_j \leq 1$ and, for $i \in I(x)$,

$$\langle x, u_i \rangle^{-1} y_j^{(i)} \to \langle x, u_i \rangle^{-1} y^{(i)} \geq 1.$$

This shows that $\lambda_j \to 1$ for $j \to \infty$, hence $\lambda_j x \to x$ for $j \to \infty$.
  (ii) Let $x \in \mathbb{R}^n$ and $x_j \in P_{[y_j]}$ with $x_j \to x$ for $j \to \infty$. Then $\langle x_j, u_i \rangle \leq y_j^{(i)}$ for all $j \in \mathbb{N}$ and $i \in \{1, \ldots, k\}$. This implies that $\langle x, u_i \rangle \leq y^{(i)}$ for $i \in \{1, \ldots, k\}$, and therefore $x \in P_{[y]}$.

Exercise 4.4.3

The assertion is proved in [81, Theorem 3.5.2], where the argument is based on [81, Theorem 3.2.11] and this in turn uses [81, Lemma 3.2.9].

Here we provide an alternative approach.

Lemma 6.2 *Let $P \in \mathcal{P}^n$ be an n-dimensional polytope, let $P_1, \ldots, P_m \in \mathcal{P}^n$ be n-dimensional, centrally symmetric polytopes such that $P = P_1 \cup \ldots \cup P_m$, $P_i \cap P_j$ is the empty set or a face of both, P_i and P_j, for $i, j = 1, \ldots, m$. Further, we assume that each facet of P is a facet of precisely one of the polytopes $P_1, \ldots, P_m$. Then P is centrally symmetric.*

Proof Let F be a facet of P with exterior unit normal u. Then there is a polytope, say P_1, which has F as a facet. By symmetry of P_1, $-F$ is also a translate of a facet of P_1 with exterior unit normal $-u$. If this is not already a facet of P, there is a polytope $P_2 \neq P_1$ (say), which has $-F$ as a facet with exterior unit normal u. We continue this reasoning. After finitely many steps, we obtain a translate of F or of $-F$, which is a facet of P_k with exterior unit normal vector $-u$, for which there is no polytope P_j having this translate as a facet with exterior unit normal u. Hence, it must be a facet of P.

This shows that for each facet of P with exterior unit normal u, there is a facet of P with exterior unit normal $-u$ which is a translate of F or of $-F$. But then $-P$ has the same facet normals with the same facet volumes as P. Therefore $S_{n-1}(P, \cdot) = S_{n-1}(-P, \cdot)$, which proves the assertion.   □

Lemma 6.3 *If $P \in \mathcal{P}^n$ is an n-polytope, $n \geq 3$, with centrally symmetric facets. Then P is centrally symmetric.*

Proof Let $S_0 = \{u_1, \ldots, u_m\}$ be the finite set of exterior unit facet normals of P with corresponding facets $F_1, \ldots, F_m$. We can assume that $\langle u_i, u \rangle < 0$ exactly for $i = 1, \ldots, k$. Let $u \in \mathbb{S}^{n-1}$ be fixed. Then the centrally symmetric $(n-1)$ polytopes $F_1 | u^\perp, \ldots, F_k | u^\perp$ provide a decomposition of $P | u^\perp$ as assumed in Lemma 6.2. But then it follows that $P | u^\perp$ is centrally symmetric, for all $u \in \mathbb{S}^{n-1}$. Hence, there is a vector $x_u \in u^\perp$ such that $h(P, v) = h(P, -v) + \langle x_u, v \rangle$ for $v \in u^\perp$.

For $v \in \mathbb{R}^n$ we choose $u \in \mathbb{S}^{n-1} \cap v^\perp$ and define $f(v) := \langle x_u, v \rangle$. Since $f(v) = h(P, v) - h(P, -v)$, this is independent of the choice of u (note that $f(0) = 0$). For $v_1, v_2 \in \mathbb{R}^n$, there is some $u \in \mathbb{S}^{n-1} \cap v_1^\perp \cap v_2^\perp$. Hence $f(v_i) = \langle x_u, v_i \rangle$ and $f(v_1 + v_2) = \langle x_u, v_1 + v_2 \rangle$, from which the additivity of f follows. Since f is also homogeneous, f is a linear functional. But then there is some $x_0 \in \mathbb{R}^n$ such that $f(v) = \langle x_0, v \rangle$ for $v \in \mathbb{R}^n$. Therefore, we conclude that $h(P, v) = h(P, -v) + \langle x_0, v \rangle$ for $v \in \mathbb{R}^n$, that is, $P = -P + x_0$, which shows that P is centrally symmetric.
                                                                                    □

By the argument in the proof of Lemma 6.3 it also follows that if $K, L \in \mathcal{K}^n, n \geq 3$, are such that for all linear subspaces H of codimension 1 the projections $K | H$ and $L | H$ are translates of each other, the K and L are translates of each other.

Corollary 6.1 *Let $P \in \mathcal{P}^n$ be a polytope, $n \geq 3$, with $\dim P \geq k$ for some $k \geq 2$. If all k-faces of P are centrally symmetric, then also all l-faces of P are centrally symmetric, for $l \geq k$. In particular, P is centrally symmetric.*

Proof Let Q be a $(k+1)$-face of P. All facets of Q are k-faces of P, hence centrally symmetric. By Lemma 6.3, we see that Q is centrally symmetric. Now we can iterate this argument.   □

Exercise 4.4.6

(a) Let $a, b, c \in \mathbb{R}$. If $a = b = c = 0$, then Hlawka's inequality holds. Now suppose that $a \neq 0, b \neq 0$ or $c \neq 0$, and hence $|a| + |b| + |c| + |a+b+c| > 0$.
   Expanding and summarizing again, one can confirm that

$$(|a| + |b| + |c| - |a + b| - |a + c| - |b + c| + |a + b + c|)$$

$$\times (|a| + |b| + |c| + |a + b + c|)$$

$$= \underbrace{(|a| + |b| - |a + b|)}_{\geq 0} \cdot \underbrace{(|c| - |a + b| + |a + b + c|)}_{\geq 0} \qquad (6.5)$$

$$+ \underbrace{(|a| + |c| - |a + c|)}_{\geq 0} \cdot \underbrace{(|b| - |a + c| + |a + b + c|)}_{\geq 0}$$

$$+ \underbrace{(|b| + |c| - |b + c|)}_{\geq 0} \cdot \underbrace{(|a| - |b + c| + |a + b + c|)}_{\geq 0}$$

$$\geq 0.$$

This proves the assertion.

The crucial point of the preceding derivation, which is sometimes called Hlawka's identity (6.5), is not so obvious. Therefore we provide another argument, which is perhaps more straightforward. The asserted inequality is equivalent to the following inequality obtained by squaring both sides. Hence, we have to show that

$$(a + b)^2 + (a + c)^2 + (b + c)^2$$

$$+ 2|a + b|\,|a + c| + 2|a + b|\,|b + c| + 2|a + c|\,|b + c|$$

$$\leq a^2 + b^2 + c^2 + (a + b + c)^2 + 2|a|\,|b| + 2|a|\,|c| + 2|b|\,|c|$$

$$+ 2|a|\,|a + b + c| + 2|b|\,|a + b + c| + 2|c|\,|a + b + c|.$$

Note that

$$(a + b)^2 + (a + c)^2 + (b + c)^2 = a^2 + b^2 + c^2 + (a + b + c)^2,$$

and

$$|b|\,|c|+|a|\,|a+b+c| \ge |bc+a(a+b+c)| = |(a+b)(a+c)| = |a+b|\,|a+c|.$$

By cyclic permutation, we also have

$$|a|\,|b| + |c|\,|a+b+c| \ge |a+c|\,|b+c|,$$

$$|a|\,|c| + |b|\,|a+b+c| \ge |a+b|\,|b+c|.$$

Addition of these three inequalities and the equation yields the assertion.

(b) Let $x, y, z \in \mathbb{R}^n$. Since $f : \mathbb{R} \to [0, \infty)$, $x \mapsto |x|$, satisfies Hlawka's inequality, we get

$$h(x + y) + h(x + z) + h(y + z)$$

$$= \sum_{i=1}^{k} \alpha_i |\langle x + y, x_i \rangle| + \sum_{i=1}^{k} \alpha_i |\langle x + z, x_i \rangle| + \sum_{i=1}^{k} \alpha_i |\langle y + z, x_i \rangle|$$

$$= \sum_{i=1}^{k} \alpha_i (|\langle x + y, x_i \rangle| + |\langle x + z, x_i \rangle| + |\langle y + z, x_i \rangle|)$$

$$= \sum_{i=1}^{k} \alpha_i (|\langle x, x_i \rangle + \langle y, x_i \rangle| + |\langle x, x_i \rangle + \langle z, x_i \rangle| + |\langle y, x_i \rangle + \langle z, x_i \rangle|)$$

$$= \sum_{i=1}^{k} \alpha_i (f(\langle x, x_i \rangle + \langle y, x_i \rangle) + f(\langle x, x_i \rangle + \langle z, x_i \rangle) + f(\langle y, x_i \rangle + \langle z, x_i \rangle))$$

$$\le \sum_{i=1}^{k} \alpha_i (|\langle x, x_i \rangle| + |\langle y, x_i \rangle| + |\langle z, x_i \rangle| + |\langle x + y + z, x_i \rangle|)$$

$$= h(x) + h(y) + h(z) + h(x + y + z).$$

(c) Let $P \subset \mathbb{R}^n$ be a zonotope, hence $P = \sum_{i=1}^{k} s_i$ with $k \in \mathbb{N}$ and segments $s_i = [-\alpha_i x_i, \alpha_i x_i]$ for $x_i \in \mathbb{R}^n$ and $\alpha_i \ge 0$ $(i = 1, \ldots, k)$. Then the support function h_P of P satisfies

$$h_P = h\left(\sum_{i=1}^{k} s_i, \cdot\right) = \sum_{i=1}^{k} h_{s_i} = \sum_{i=1}^{k} \alpha_i h_{[-x_i, x_i]} = \sum_{i=1}^{k} \alpha_i |\langle \cdot, x_i \rangle|.$$

Now it follows from (b) that h_P satisfies Hlawka's inequality. The same argument works if P is a zonoid (simply replace summation by integration).

(d) Let $P \subset \mathbb{R}^n$ be a polytope such that h_P satisfies Hlawka's inequality. Choosing $x, y \in \mathbb{R}^n$ and $z := -(x + y)$, we obtain

$$h_P(x + y) + h_P(-y) + h_P(-x) \leq h_P(x) + h_P(y) + h_P(-(x + y)) + h_P(0)$$

which holds if and only if

$$h_P(x + y) - h_P(-(x + y))$$
$$\leq h_P(x) + h_P(y) - (h_P(-y) + h_P(-x)). \tag{6.6}$$

Replacing $x, y \in \mathbb{R}^n$ by $-x$ and $-y$, we get

$$h_P(-(x + y)) - h_P(x + y) \leq h_P(-x) + h_P(-y) - (h_P(y) + h_P(x))$$

which holds if and only if

$$h_P(x + y) - h_P(-(x + y))$$
$$\geq h_P(x) + h_P(y) - (h_P(-y) + h_P(-x)). \tag{6.7}$$

From (6.6) and (6.7) we obtain the identity

$$h_P(x + y) - h_P(-(x + y)) = (h_P(x) - h_P(-x)) + (h_P(y) - h_P(-y)).$$

This shows that the function

$$H : \mathbb{R}^n \to \mathbb{R}, \ x \mapsto h_P(x) - h_P(-x)$$

is additive, that is, $H(x + y) = H(x) + H(y)$ for $x, y \in \mathbb{R}^n$.

Since support functions are positively homogeneous, it follows that H is positively homogeneous, since for $x \in \mathbb{R}^n$ and $\lambda > 0$ we have

$$H(\lambda x) = h_P(\lambda x) - h_P(-\lambda x) = \lambda h_P(x) - \lambda h_P(-x) = \lambda H(x)$$

and

$$H(0) = h_P(0) - h_P(0) = 0.$$

Furthermore, H is homogeneous, since for $x \in \mathbb{R}^n$ and $\lambda < 0$ we also have by the positive homogeneity of support functions that

$$H(\lambda x) = h_P(\lambda x) - h_P(-\lambda x) = -\lambda h_P(-x) - (-\lambda) h_P(x) = \lambda H(x).$$

But then H is linear, hence there is some $c \in \mathbb{R}^n$ such that $H = \langle c, \cdot \rangle$. This yields

$$h_P(x) - h_P(-x) = \langle c, x \rangle$$

for $x \in \mathbb{R}^n$. It follows that $P - \frac{c}{2} = -(P - \frac{c}{2})$, hence P is centrally symmetric.

Now we show that the support function of an arbitrary face of P also satisfies Hlawka's inequality, and hence is centrally symmetric. For this, let $x, y, z \in \mathbb{R}^n$, $u \in \mathbb{S}^{n-1}$ and $\alpha > 0$. We plug $u + \alpha x, u + \alpha y, u + \alpha z$ into Hlawka's inequality. This yields

$$\begin{aligned}
0 \leq &\, h_P(u + \alpha x) + h_P(u + \alpha y) + h_P(u + \alpha z) + h_P(3u + \alpha(x + y + z)) \\
&- h_P(2u + \alpha(x + y)) - h_P(2u + \alpha(y + z)) - h_P(2u + \alpha(x + z)) \\
= &\, (h_P(u + \alpha x) - h_P(u)) + (h_P(u + \alpha y) - h_P(u)) + (h_P(u + \alpha z) - h_P(u)) \\
&+ 3(h_P(u + \tfrac{\alpha}{3}(x + y + z)) - h_P(u)) - 2(h_P(u + \tfrac{\alpha}{2}(x + y)) - h_P(u)) \\
&- 2(h_P(u + \tfrac{\alpha}{2}(y + z)) - h_P(u)) - 2(h_P(u + \tfrac{\alpha}{2}(x + z)) - h_P(u)).
\end{aligned}$$

Dividing by α and passing to the limit $\alpha \downarrow 0$, we obtain for the directional derivative

$$h'_P(u; v) = \lim_{\alpha \downarrow 0} \tfrac{1}{\alpha}(h_P(u + \alpha v) - h_P(u))$$

of h_P at u in direction v the inequality

$$\begin{aligned}
0 \leq &\, h'_P(u; x) + h'_P(u; y) + h'_P(u; z) + h'_P(u; x + y + z) \\
&- h'_P(u; x + y) - h'_P(u; y + z) - h'_P(u; x + z).
\end{aligned}$$

Since

$$h_{P(w)}(v) = h'_P(w; v)$$

for $w \in \mathbb{S}^{n-1}$ and $v \in \mathbb{R}^n$, we deduce that

$$\begin{aligned}
h_{P(u)}(x + y) &+ h_{P(u)}(y + z) + h_{P(u)}(x + z) \\
&\leq h_{P(u)}(x) + h_{P(u)}(y) + h_{P(u)}(z) + h_{P(u)}(x + y + z).
\end{aligned}$$

This shows that $h_{P(u)}$ satisfies Hlawka's inequality, where $u \in \mathbb{S}^{n-1}$ was arbitrary.

This argument shows that each face of P is centrally symmetric, and hence P is a zonotope by Exercise 4.4.3.

Exercise 4.4.8

(a) This is a special case of Exercise 4.4.7 (a), since the argument works for any function of the form

$$h(x) = \int_{\mathbb{S}^{n-1}} |\langle x, u \rangle| \, \mu(du), \quad x \in \mathbb{R}^n,$$

where μ is a finite signed measure on $\mathbb{S}^{n-1}$. If μ has an L^1-density f with respect to σ, then

$$h'(e; x) = 2 \int_{\Omega_e^{n-1}} \langle u, x \rangle f(u) \, \mathcal{H}^{n-1}(du)$$

$$= \int_{\mathbb{S}^{n-1}} \text{sgn}(\langle e, u \rangle) \langle u, x \rangle f(u) \, \mathcal{H}^{n-1}(du),$$

by the symmetry of f and the reflection invariance of $\mathcal{H}^{n-1}$. This proves the differentiability of h and the form of its differential. The continuity assertion follows from the dominated convergence theorem.

(b) By the homogeneity properties of support functions, it is essentially sufficient to show the following. Let $e = e_1, e_2, \ldots, e_n$ be an orthonormal basis of $\mathbb{R}^n$. Let $x = e_i$, $y = e_j$ for $i, j \in \{1, \ldots, n\}$. Then we have to prove that

$$h''(e; e_i, e_j) = 2 \int_{\mathbb{S}_e^{n-1}} \langle u, e_i \rangle \langle u, e_j \rangle f(u) \, \mathcal{H}^{n-2}(du).$$

Starting with the formula stated in (a), we obtain

$$h''(e; e_i, e_j)$$

$$= \lim_{t \downarrow 0} \frac{1}{t} \int_{\mathbb{S}^{n-1}} [\text{sgn}(\langle u, e + te_i \rangle) - \text{sgn}(\langle u, e \rangle)] \langle u, e_j \rangle f(u) \, \mathcal{H}^{n-1}(du).$$

Put $\mathbb{S}^{n-1}(t) := \{u \in \mathbb{S}^{n-1} : \langle u, e \rangle \langle u, e + te_i \rangle < 0\}$ for $t > 0$. For $u \in \mathbb{S}^{n-1}(t)$, the expression in brackets in the preceding integral equals $[\cdot] = 2 \, \text{sgn}(\langle u, e_i \rangle)$, and $[\cdot] = 0$ if $u \notin \mathbb{S}^{n-1}(t)$. Hence

$$h''(e; e_i, e_j) = 2 \lim_{t \downarrow 0} \frac{1}{t} \int_{\mathbb{S}^{n-1}(t)} \text{sgn}(\langle u, e_i \rangle) \langle u, e_j \rangle f(u) \, \mathcal{H}^{n-1}(du).$$

We define $\mathbb{S}_{e,e_i}^{n-1} := \mathbb{S}^{n-1} \cap e^\perp \cap e_i^\perp$, and then we consider the transformation

$$T : \mathbb{S}_{e,e_i}^{n-1} \times (0, \pi) \times [0, \arctan t] \to \mathbb{S}^{n-1}(t),$$

$$(v, \alpha, \beta) \mapsto \sin(\alpha)v + \cos(\alpha) \cdot [\cos(\beta)e_i - \sin(\beta)e],$$

which yields an injective parametrization of $\mathbb{S}^{n-1}(t)$ (up to a set of measure zero) and has the Jacobian $JT(v, \alpha, \beta) = (\sin(\alpha))^{n-3} |\cos(\alpha)|$.

We use the short notation $F(u) = \mathrm{sgn}(\langle u, e_i \rangle) \langle u, e_j \rangle f(u)$ and then obtain

$$\frac{1}{t} \int_{\mathbb{S}^{n-1}(t)} F(u) \, \mathcal{H}^{n-1}(du)$$

$$= \frac{1}{t} \int_{\mathbb{S}^{n-1}_{e,e_i}} \int_0^\pi \int_0^{\arctan(t)} F(T(v, \alpha, \beta))(\sin(\alpha))^{n-3} |\cos(\alpha)| \, d\beta \, d\alpha \, \mathcal{H}^{n-3}(dv).$$

Since $\mathrm{sgn}(\langle e_i, T(v, \alpha, \beta) \rangle) = \mathrm{sgn}(\cos(\alpha))$ if $\beta \in (0, \pi/2)$, we have

$$|F(T(v, \alpha, \beta)) - F(T(v, \alpha, 0))|$$

$$\leq |\langle e_j, T(v, \alpha, \beta) \rangle f(T(v, \alpha, \beta)) - \langle e_j, T(v, \alpha, 0) \rangle f(T(v, \alpha, 0))|,$$

which is as small as we wish if $t > 0$ is small enough. Hence, as $t \downarrow 0$ we obtain

$$\frac{1}{t} \int_{\mathbb{S}^{n-1}(t)} F(u) \, \mathcal{H}^{n-1}(du)$$

$$= o(1) + \frac{1}{t} \int_{\mathbb{S}^{n-1}_{e,e_i}} \int_0^\pi \int_0^{\arctan(t)} F(T(v, \alpha, 0))$$

$$\times (\sin(\alpha))^{n-3} |\cos(\alpha)| \, d\beta \, d\alpha \, \mathcal{H}^{n-3}(dv)$$

$$\to \int_{\mathbb{S}^{n-1}_{e,e_i}} \int_0^\pi F(T(v, \alpha, 0))(\sin(\alpha))^{n-3} |\cos(\alpha)| \, d\beta \, d\alpha \, \mathcal{H}^{n-3}(dv)$$

$$= \int_{\mathbb{S}^{n-1}_{e,e_i}} \int_0^\pi \mathrm{sgn}(\cos(\alpha)) \langle e_j, T(v, \alpha, 0) \rangle f(T(v, \alpha, 0))$$

$$\times (\sin(\alpha))^{n-3} |\cos(\alpha)| \, d\beta \, d\alpha \, \mathcal{H}^{n-3}(dv)$$

$$= \int_{\mathbb{S}^{n-1}_e} \langle e_i, u \rangle \langle e_j, u \rangle f(u) \, d\beta \, d\alpha \, \mathcal{H}^{n-3}(dv),$$

where we used that $\mathrm{sgn}(\cos(\alpha)) |\cos(\alpha)| = \cos(\alpha) = \langle e_i, T(v, \alpha, 0) \rangle$ (and a decomposition of spherical Lebesgue measure for $\mathbb{S}^{n-1}_e$).

Exercise 4.5.4

In the proof, we only need that f is weakly additive. Parts (a)–(c) of the argument can be skipped if Corollary 4.5 is used.

We say that n-polytopes $P_1, \ldots, P_k \in \mathcal{P}^n$ are almost disjoint if $\text{int}(P_i) \cap \text{int}(P_j) = \emptyset$ for $i \neq j$.

(a) Let $P_1, \ldots, P_k, P \in \mathcal{P}^n$ be n-polytopes such that $P = P_1 \cup \cdots \cup P_k$. Suppose that $P_1, \ldots, P_k$ are almost disjoint. Then $f(P) = f(P_1) + \cdots + f(P_k)$.

Proof This is shown by induction over k. For $k = 1$ there is nothing to show. Let $k = 2$. Assume that P_1, P_2, P satisfy the assumptions. Since P_1, P_2 have disjoint interiors, they can be separated by a hyperplane H, hence $P_1 \subset H^-$, $P_2 \subset H^+$, and $P_1 \cap P_2 \subset H$. Since f is simple, we get $f(P_1 \cap P_2) = 0$, and since f is additive, we conclude that $f(P) = f(P_1) + f(P_2)$.

For the induction step we assume that the assertion has been proved for unions of less than k full-dimensional polytopes. Let $P_1, \ldots, P_k, P \in \mathcal{P}^n$ be n-polytopes such that $P = P_1 \cup \cdots \cup P_k$. Suppose that $P_1, \ldots, P_k$ are almost disjoint. Again there is a hyperplane H such that $P_1 \subset H^-$, $P_2 \subset H^+$, and $P_1 \cap P_2 \subset H$. Then

$$P \cap H^+ = \bigcup_{i=2}^{k} (P_i \cap H^+),$$

since each point of $P_1 \cap H$ is also contained in some P_i with $i \geq 2$ (as P is convex). By induction hypothesis, we deduce that

$$f(P \cap H^+) = \sum_{i=2}^{k} f(P_i \cap H^+) = \sum_{i=1}^{k} f(P_i \cap H^+).$$

In the same way, we obtain

$$f(P \cap H^-) = \sum_{i=1, i \neq 2}^{k} f(P_i \cap H^-) = \sum_{i=1}^{k} f(P_i \cap H^-).$$

Adding these relations, we arrive at

$$f(P) = f((P \cap H^+) \cup (P \cap H^-)) = f(P \cap H^+) + f(P \cap H^-)$$

$$= \sum_{i=1}^{k} \left[f(P_i \cap H^+) + f(P_i \cap H^-) \right]$$

$$= \sum_{i=1}^{k} f(P_i).$$

This completes the induction argument. $\square$

(b) Let $Q_1, \ldots, Q_r \subset$ be finite unions of n-polytopes. Then there are almost disjoint n-polytopes $P_1, \ldots, P_m$ such that $Q_i = \bigcup_{P_j \subset Q_i} P_j$ for $i = 1, \ldots, r$.

Proof Each polytope involved in one of Q_i is an intersection of finitely many halfspaces. In addition to these closed halfspaces we consider the closure of the complement of each of these. Consider all polytopes arising as a nonempty intersection of these closed halfspaces. These polytopes can be used as $P_1, \ldots, P_m$. $\qquad\square$

(c) The functional f has an additive extension to finite unions of n-polytopes.

Proof Let Q be a finite union of n-polytopes. Suppose there are two representations, $Q = P_1 \cup \cdots \cup P_m$ and $Q = P'_1 \cup \cdots \cup P'_k$, in terms of almost disjoint n-polytopes in both cases. Defining $f(\emptyset) = 0$, as usual, we obtain

$$\sum_{i=1}^{m} f(P_i) = \sum_{i=1}^{m} f\left(\bigcup_{j=1}^{k} (P_i \cap P'_j)\right) = \sum_{i=1}^{m} \sum_{j=1}^{k} f(P_i \cap P'_j).$$

Here we used (a) and the assumption that f is simple. Moreover, in the union $\bigcup_{j=1}^{k} (P_i \cap P'_j)$ lower-dimensional intersections can be omitted, since they are contained in the union of the full-dimensional intersections. By symmetry of the double sum, we see that $\sum_{i=1}^{m} f(P_i) = \sum_{j=1}^{k} f(P'_j)$. Hence, we can define

$$f(Q) - \sum_{i-1}^{m} f(P_i).$$

Moreover, we see from (b) that f is additive on finite unions of n-polytopes. $\qquad\square$

What we have proved directly up to this point follows almost immediately from the fact that any weakly additive functional on $\mathcal{P}^n$ has an additive extension to $U(\mathcal{P}^n)$ (see Corollary 4.5).

Let $n \geq 1$. Then $f(\{x\}) = 0$ and therefore f is nonnegative on nonempty convex bodies. Here we use that f is increasing. Moreover, if Q_1, Q_2 are finite unions of n-polytopes with $Q_1 \subset Q_2$, it follows that $f(Q_1) \leq f(Q_2)$. (One can use special representations of Q_1, Q_2 as finite unions of almost disjoint n-polytopes such that each n-polytope contained Q_1 is also employed in Q_2. This can be done by the same type of dissection as in (b).)

Let $W = [0, 1]^n$ denote the unit cube. Since W is the almost disjoint union of m^n translates of $(1/m)W$, we have

$$f(W) = m^n f((1/m)W), \quad f((1/m)W) = c \cdot V_n((1/m)W), \quad c := f(W) \geq 0.$$

Let P be an n-polytope. Let $W_1, \ldots, W_k$ denote the translates of $(1/m)W$ which are contained in P. Then, using (c) we get

$$f(P) \geq f\left(\bigcup_{i=1}^{k} W_i\right) = kf((1/m)W) = k\,c\,V_n((1/m)W) = c\,V_n\left(\bigcup_{i=1}^{k} W_i\right).$$

Let $\varepsilon > 0$ be given. By basic properties of cubical dissections of $\mathbb{R}^n$ and Lebesgue measure, there is some $m \in \mathbb{N}$ such that

$$V_n\left(\bigcup_{i=1}^{k} W_i\right) \geq V_n(P) - \varepsilon,$$

and hence

$$f(P) \geq c \cdot (V_n(P) - \varepsilon).$$

Covering P by almost disjoint translates of cubes $(1/m)W$, for sufficiently large $m \in \mathbb{N}$, we also get

$$f(P) \leq c \cdot (V_n(P) + \varepsilon).$$

This shows that $f(P) = c \cdot V_n(P)$ for each n-polytope P.

Let $K \in \mathcal{K}^n$ be n-dimensional. Then there are n-polytopes $P_1, P_2 \in \mathcal{P}^n$ such that $P_1 \subset K \subset P_2$ and $V_n(P_2) - \varepsilon \leq V_n(K) \leq V_n(P_1) + \varepsilon$. But then

$$c[V_n(K) - \varepsilon] \leq cV_n(P_1) = f(P_1) \leq f(K) \leq f(P_2) = cV_n(P_2) \leq c[V_n(K) + \varepsilon],$$

where we used that f is increasing. Thus we conclude that $f = c \cdot V_n$, first for n-dimensional convex bodies, but then generally, since f and V_n are simple.

6.5 Solutions of Exercises for Chap. 5

Exercise 5.1.6

We start by showing that β_k is an open map, that is, images of open sets are open. Let $A \subset \mathrm{SO}(n)$ be open. For this we have to show that $\beta_k^{-1}(\beta_k(A))$ is open. We have

$$\beta_k^{-1}(\beta_k(A)) = \{\vartheta \in \mathrm{SO}(n) : \vartheta U_0 \in \beta_k(A)\}$$

$$= \{\vartheta \in \mathrm{SO}(n) : \text{there is some } \vartheta' \in A \text{ such that } \vartheta U_0 = \vartheta' U_0\}.$$

Define $H := \{\vartheta \in SO(n) : \vartheta U_0 = U_0\}$. Then the condition $\vartheta U_0 = \vartheta' U_0$ is equivalent to $\vartheta^{-1}\vartheta' \in H$. Writing $\sigma := \vartheta^{-1}\vartheta'$, we have $\vartheta' = \vartheta\sigma$. Hence we get

$$\beta_k^{-1}(\beta_k(A)) = \{\vartheta \in SO(n) : \text{there is some } \sigma \in H \text{ such that } \vartheta\sigma \in A\}$$

$$= \bigcup_{\sigma \in H} \{\vartheta : \vartheta\sigma \in A\}$$

$$= \bigcup_{\sigma \in H} \{\vartheta'\sigma^{-1} : \vartheta' \in A\}$$

$$= \bigcup_{\sigma \in H} \psi_\sigma(A),$$

where $\psi_\sigma : SO(n) \to SO(n), \vartheta \mapsto \vartheta\sigma^{-1}$. Since ψ_σ is a homeomorphism, the image set $\psi_\sigma(A)$ is open, hence so is $\beta_k^{-1}(\beta_k(A))$. This proves that β_k is an open map.

Since $SO(n)$ is compact with countable base and β_k is continuous, open, and surjective, the image set is compact and has a countable base (which can be chosen as the image of a base of $SO(n)$ under β_k). In order to see that also the Hausdorff separation property is obtained, we show that the set

$$R := \{(\vartheta, \vartheta') \in SO(n)^2 : \beta_k(\vartheta) = \beta_k(\vartheta')\}$$

is closed. For this, we first observe that as in the part (a), we have

$$R = \{(\vartheta, \vartheta') : \vartheta^{-1}\vartheta' \in H\}.$$

Since the map $(\vartheta, \vartheta') \mapsto \vartheta^{-1}\vartheta'$ is continuous and H is closed, it follows that R is closed.

For $\vartheta, \rho \in SO(n)$ and $U = \beta_k(\vartheta) \in G(n, k)$, we have

$$\varphi(\rho, U) = \rho U = \rho\vartheta U_0 = \beta_k(\rho\vartheta).$$

Using the map $\alpha : SO(n) \times SO(n) \to SO(n), (\rho, \vartheta) \mapsto \vartheta\rho$, we obtain

$$\varphi \circ (\text{id} \times \beta_k) = \beta_k \circ \alpha,$$

where id is the identity and $\text{id} \times \beta_k$ is the product map. Since α is continuous and β_k is continuous, open and surjective (and hence the same is true for $\text{id} \times \beta_k$), it follows that φ is continuous.

For $L, L' \in G(n, k)$ there is some $\rho \in SO(n)$ such that $L' = \rho L$. This can be seen by choosing orthonormal bases adjusted to L and to L' respectively, and by thus defining an appropriate rotation.

The independence of the topology on the choice of U_0 can be seen by using that the continuous operation of $SO(n)$ on $G(n, k)$ is transitive.

References

1. Adiprasito, K., Bárány, I., Mustafa, N.H.: Theorems of Carathéodory, Helly, and Tverberg without dimension. In: Proceedings of the Thirtieth Annual ACM-SIAM Symposium on Discrete Algorithms, pp. 2350–2360. SIAM, Philadelphia (2019). arXiv:1806.08725
2. Aleksandrov, A.D.: Konvexe Polyeder. Akademie-Verlag, Berlin (1958)
3. Alesker, S., Fu, J.H.G.: In: Gallego, E., Solanes, G. (eds.) Integral Geometry and Valuations. Advanced Courses in Mathematics. CRM Barcelona, viii+112 pp. Birkhäuser/Springer, Basel (2014). Lectures from the Advanced Course on Integral Geometry and Valuation Theory held at the Centre de Recerca Matemàtica (CRM), Barcelona, 6–10 September 2010
4. Artstein-Avidan, S., Giannopoulos, A., Milman, V.D.: Asymptotic Geometric Analysis. Part I. Mathematical Surveys and Monographs, vol. 202, xx+451 pp. American Mathematical Society, Providence (2015)
5. Bakelman, I.J.: Convex Analysis and Nonlinear Geometric Elliptic Equations. Springer, Berlin (1994)
6. Bárány, I.: Helge Tverberg is eighty: a personal tribute. Eur. J. Combin. **66**, 24–27 (2017)
7. Bárány, I., Karasev, R.: Notes about the Carathéodory number. Discrete Comput. Geom. **48**, 783–792 (2012)
8. Bárány, I., Onn, S.: Carathéodory's theorem, colourful and applicable. In: Intuitive Geometry (Budapest, 1995). Bolyai Society Mathematical Studies, vol. 6, pp. 11–21. János Bolyai Mathematical Society, Budapest (1997)
9. Bárány, I., Blagojević, P.V.M., Ziegler, G.M.: Tverberg's theorem at 50: extensions and counterexamples. Not. Am. Math. Soc. **63**, 732–739 (2016)
10. Barvinok, A.: A Course in Convexity. AMS, Providence (2002)
11. Benson, R.V.: Euclidean Geometry and Convexity. McGraw-Hill, New York (1966)
12. Bernig, A.: Algebraic integral geometry. In: Global Differential Geometry. Springer Proceedings in Mathematics, vol. 17, pp. 107–145. Springer, Berlin (2012)
13. Betke, U., Weil, W.: Isoperimetric inequalities for the mixed area of plane convex sets. Arch. Math. **57**, 501–507 (1991)
14. Billingsley, P.: Convergence of Probability Measures. Wiley, New York (1968)
15. Blaschke, W.: Kreis und Kugel. 2. Aufl. Walter der Gruyter, Berlin (1956)
16. Boissonnat, J.-D., Yvinec, M.: Algorithmic Geometry, English edition. Cambridge University Press, Cambridge (1998)
17. Boltyanski, V., Martini, H., Soltan, P.S.: Excursions into Combinatorial Geometry. Springer, Berlin (1997)
18. Bonnesen, T., Fenchel, W.: Theorie der konvexen Körper. Springer, Berlin (1934)
19. Böröczky, K. Jr.: Finite Packing and Covering. Cambridge University Press, Cambridge (2004)

© Springer Nature Switzerland AG 2020
D. Hug, W. Weil, *Lectures on Convex Geometry*, Graduate Texts in Mathematics 286, https://doi.org/10.1007/978-3-030-50180-8

20. Böröczky, K., Hug, D.: A reverse Minkowski-type inequality. Proc. Am. Math. Soc. (2020). https://doi.org/10.1090/proc/15133. Article electronically published on July 29, 2020
21. Brass, P., Moser, W., Pach, J.: Research Problems in Discrete Geometry. Springer, New York (2005)
22. Brazitikos, S., Giannopoulos, A., Valettas, P., Vritsiou, B.-H.: Geometry of Isotropic Convex Bodies. Mathematical Surveys and Monographs, vol. 196, xx+594 pp. American Mathematical Society, Providence (2014)
23. Brøndsted, A.: An Introduction to Convex Polytopes. Springer, Berlin (1983)
24. Burago, Y.D., Zalgaller, V.A.: Geometric Inequalities. Springer, Berlin (1988)
25. Busemann, H.: Convex Surfaces. Interscience Publishers, New York (1958). Reprint of the 1958 original. Dover Publications, Inc., Mineola (2008), 196 pp.
26. Cohn, D.L.: Measure Theory. Birkhäuser, Boston (1997)
27. Cordero-Erausquin, D., Klartag, B., Merigot, Q., Santambrogio, F.: One more proof of the Alexandrov-Fenchel inequality. C. R. Math. Acad. Sci. Paris **357**(8), 676–680 (2019). arXiv:1902.10064
28. de Berg, M., Cheong, O., van Kreveld, M., Overmars, M.: Computational Geometry. Algorithms and Applications, 3rd edn. Springer, Berlin (2008)
29. Edelsbrunner, H.: Algorithms in Combinatorial Geometry. Springer, Berlin (1987)
30. Eggleston, H.G.: Convexity. Cambridge University Press, London (1958)
31. Ewald, G.: Combinatorial Convexity and Algebraic Geometry. Springer, New York (1996)
32. Fejes Tóth, L.: Lagerungen in der Ebene, auf der Kugel und im Raum. 2. verb. u. erw. Aufl. Springer, Berlin (1972)
33. Gänssler, P., Stute, W.: Wahrscheinlichkeitstheorie. Springer, Berlin (1977)
34. Gantmacher, F.R.: The Theory of Matrices, vol. 2. AMS Chelsea Publishing, Providence (reprinted, 2000)
35. Gardner, R.J.: Geometric Tomography. Cambridge University Press, Cambridge (1995) Revised 2nd edn. (2006)
36. Goodman, J.E., O'Rourke, J.: Handbook of Discrete and Computational Geometry. CRC Press, Boca Raton (1997). Third edition, edited by Jacob E. Goodman, Joseph O'Rourke and Csaba D. Tóth, 2018, 1927 pp.
37. Groemer, H.: Geometric Applications of Fourier Series and Spherical Harmonics. Cambridge University Press, Cambridge (1996)
38. Gruber, P.M.: Convex and Discrete Geometry. Grundlehren der mathematischen Wissenschaften, Bd. 336. Springer, Berlin (2007)
39. Grünbaum, B.: Convex Polytopes. Interscience Publishers, London (1967). 2nd edn. (prepared by Volker Kaibel). Springer, New York (2003)
40. Grünbaum, B., Shephard, G.C.: Tilings and Patterns: An Introduction. Freeman, New York (1989)
41. Guédon, O., Nayar, P., Tkocz, T., Ryabogin, D., Zvavitch, A.: Analytical and Probabilistic Methods in the Geometry of Convex Bodies. IMPAN Lecture Notes, vol. 2, 183 pp. Polish Academy of Sciences, Institute of Mathematics, Warsaw (2014)
42. Hadwiger, H.: Altes und Neues über konvexe Körper. Birkhäuser, Basel (1955)
43. Hadwiger, H.: Vorlesungen über Inhalt, Oberfläche und Isoperimetrie. Springer, Berlin (1957)
44. Hadwiger, H., Debrunner, H., Klee, V.: Combinatorial Geometry in the Plane. Holt, Rinehart and Winston, New York (1964)
45. Hörmander, L.: Notions of Convexity. Birkhäuser, Basel (1994)
46. Horn, R.A., Johnson, C.R.: Matrix Analysis, 2nd edn. Cambridge University Press, Cambridge (2013)
47. Huppert, B., Willems, W.: Lineare Algebra, 2nd edn. Vieweg+Teubner Verlag, Springer Fachmedien Wiesbaden GmbH, Berlin (2010)
48. Jensen, E.B.V., Kiderlen, M.: Tensor Valuations and Their Applications in Stochastic Geometry and Imaging. Lecture Notes in Mathematics, vol. 2177. Springer, Cham (2017)
49. Joswig, M., Theobald, Th.: Algorithmische Geometrie: polyedrische und algebraische Methoden. Vieweg, Wiesbaden (2008)

50. Kelly, L., Weiss, M.L.: Geometry and Convexity. Wiley/Interscience Publishers, New York (1979)
51. Klain, D.A.: A short proof of Hadwiger's characterization theorem. Mathematika **42**, 329–339 (1995)
52. Klain, D.A.: Containment and inscribed simplices. Indiana Univ. Math. J. **59**, 1231–1244 (2010)
53. Klain, D.A., Rota, G.-C.: Introduction to Geometric Probability. Cambridge University Press, Cambridge (1997)
54. Klein, R.: Algorithmische Geometrie. Addison-Wesley-Longman, Bonn (1997)
55. Koldobsky, A.: Fourier Analysis in Convex Geometry. Mathematical Surveys and Monographs. American Mathematical Society, Providence (2005)
56. Koldobsky, A., Yaskin, V.: The Interface Between Convex Geometry and Harmonic Analysis. CBMS Regional Conference Series in Mathematics. American Mathematical Society, Providence (2008)
57. Leichtweiß, K.: Konvexe Mengen. Springer, Berlin (1980)
58. Leichtweiß, K.: Affine Geometry of Convex Bodies. J.A. Barth, Heidelberg (1998)
59. Leonard, I.E., Lewis, J.E.: Geometry of Convex Sets, x+321 pp. Wiley, Hoboken (2016)
60. Leppmeier, M.: Kugelpackungen von Kepler bis heute. Eine Einführung für Schüler, Studenten und Lehrer. Vieweg, Braunschweig (1997)
61. Lindquist, N.F.: Support functions of central convex bodies. Portugaliae Math. **34**, 241–252 (1975)
62. Linhart, J.: Kantenlängensumme, mittlere Breite und Umkugelradius konvexer Körper. Arch. Math. **29**, 558–560 (1977)
63. Lutwak, E.: Containment and circumscribing simplices. Discrete Comput. Geom. **19**, 229–235 (1998)
64. Lyusternik, L.A.: Convex Figures and Polyhedra. Dover Publications, New York (1963)
65. Marti, J.T.: Konvexe Analysis. Birkhäuser, Basel (1977)
66. Martini, H.: Convex polytopes whose projection bodies and difference sets are polars. Discrete Comput. Geom. **6**(1), 83–91 (1991)
67. Martini, H., Weissbach, B.: On quermasses of simplices. Stud. Sci. Math. Hung. **27**, 213–221 (1992)
68. Matoušek, J.: Lectures on Discrete Geometry. Graduate Texts in Mathematics, vol. 212. Springer, New York (2002)
69. McMullen, P.: Non-linear angle-sum relations for polyhedral cones and polytopes. Math. Proc. Camb. Philos. Soc. **78**, 247–261 (1975)
70. McMullen, P.: Valuations and Euler type relations on certain classes of convex polytopes. Proc. Lond. Math. Soc. **35**, 113–135 (1977)
71. McMullen, P., Shephard, G.C.: Convex Polytopes and the Upper Bound Conjecture. Cambridge University Press, Cambridge (1971)
72. Meyer, C.: Matrix Analysis and Applied Linear Algebra. SIAM. Philadelphia (2000)
73. Nachbin, L.: The Haar Integral. Van Nostrand, Princeton (1965)
74. O'Rourke, J.: Computational Geometry in C. Cambridge University Press, Cambridge (1994)
75. Pach, J., Agarval, P.K.: Combinatorial Geometry. Wiley-Interscience Series. Wiley, New York (1995)
76. Rockafellar, R.T.: Convex Analysis. Princeton University Press, Princeton (1970)
77. Rogers, C.A.: Packing and Covering. Cambridge University Press, Cambridge (1964)
78. Rubin, B.: Introduction to Radon Transforms. Cambridge University Press, New York (2015)
79. Rudin, W.: Real and Complex Analysis, 3rd edn., xiv+416 pp. McGraw-Hill Book Co., New York (1987)
80. Sah, C.-H.: Hilbert's Third Problem: Scissors Congruence. Pitman, San Francisco (1979)
81. Schneider, R.: Convex Bodies: The Brunn-Minkowski Theory, 2nd expanded edn. Cambridge University Press, Cambridge (2014)
82. Schneider, R., Weil, W.: Integralgeometrie. Teubner, Stuttgart (1992)
83. Schneider, R., Weil, W.: Stochastische Geometrie. Teubner, Stuttgart (2000)

84. Schneider, R., Weil, W.: Stochastic and Integral Geometry. Springer, Berlin (2008)
85. Shenfeld, Y., van Handel, R.: Mixed volumes and the Bochner method. Proc. Am. Math. Soc. **147**(12), 5385–5402 (2019). arXiv:1811.08710
86. Shenfeld, Y., van Handel, R.: Extremals in Minkowski's quadratic inequality (2019). Preprint available from arXiv:1902.10029v1
87. Simon, B.: Convexity: An Analytic Viewpoint. Cambridge Tracts in Mathematics, vol. 187. Cambridge University Press, Cambridge (2011)
88. Soltan, V.: Lectures on Convex Sets, x+405 pp. World Scientific Publishing, Hackensack (2015)
89. Stoer, J., Witzgall, Ch.: Convexity and Optimization in Finite Dimensions I. Springer, Berlin (1970)
90. Thomas, R.R.: Lectures on Geometric Combinatorics. Lecture Notes. University of Washington, Seattle (2004)
91. Thompson, A.C.: Minkowski Geometry. Cambridge University Press, Cambridge (1996)
92. Valentine, F.A.: Convex Sets. McGraw-Hill, New York (1964). Deutsche Fassung: Konvexe Mengen. BI, Mannheim (1968)
93. Webster, R.: Convexity. Oxford University Press, New York (1994)
94. Ziegler, G.M.: Lectures on Polytopes. Springer, Berlin (1995). Revised 6th printing 2006
95. Zong, C.: Strange Phenomena in Convex and Discrete Geometry. Springer, New York (1996)
96. Zong, C.: Sphere Packings. Springer, New York (1999)
97. Zong, C.: The Cube: A Window to Convex and Discrete Geometry. Cambridge University Press, Cambridge (2006)

Index

© Springer Nature Switzerland AG 2020
D. Hug, W. Weil, *Lectures on Convex Geometry*, Graduate Texts
in Mathematics 286, https://doi.org/10.1007/978-3-030-50180-8